DATE DUE

NO 14 '97			
AP 24 '98			
MY 30 '00			
NO 12 '08			

Fundamentals of Air Conditioning Systems

12

Fundamentals of Air Conditioning Systems

by Billy C. Langley, Ed.D., CM

Published by
THE FAIRMONT PRESS, INC.
700 Indian Trail
Lilburn, GA 30247

...blication Data

Langley, Billy C., 1931-
Fundamentals of air conditioning systems / Billy C. Langley.
p. cm.
Includes index.
ISBN 0-88173-176-5
1. Air conditioning--Equipment and supplies. I. Title.
TH7687.7.L373 1995 697.9'3--dc20 94-23108
CIP

Fundamentals Of Air Conditioning Systems by Billy C. Langley.

Published by The Fairmont Press, Inc.
700 Indian Trail
Lilburn, GA 30247

Printed in the United States of America

10 9 8 7 6 5 4 3 2 1

ISBN 0-88173-176-5 FP

ISBN 0-13-147422-7 PH

Distributed by PTR Prentice Hall
Prentice-Hall, Inc.
A Paramount Communications Company
Englewood Cliffs, NJ 07632

Prentice-Hall International (UK) Limited, London
Prentice-Hall of Australia Pty. Limited, Sydney
Prentice-Hall Canada Inc., Toronto
Prentice-Hall Hispanoamericana, S.A., Mexico
Prentice-Hall of India Private Limited, New Delhi
Prentice-Hall of Japan, Inc., Tokyo
Simon & Schuster Asia Pte. Ltd., Singapore
Editora Prentice-Hall do Brasil, Ltda., Rio de Janeiro

Table of Contents

Chapter

1. Air Conditioning and Psychrometrics 1
2. Heat Load Calculation Factors 37
3. Residential Heat Load Calculations 83
4. Residential Equipment Sizing and Selection 111
5. Residential Equipment Location 143
6. Commercial Heat Load Calculation 165
7. Commercial Equipment Sizing and Selection 189
8. Commercial Equipment Location 213
9. Refrigerant Lines 233
10. Duct Systems and Designs 261
11. Duct Pipe, Fittings, and Insulation 313
12. Indoor Air Quality 333
13. System Cost Estimating 367

Index 387

Preface

Fundamentals of Air Conditioning Systems has been written to provide a thorough and practical study of the procedures used when calculating the heat loss/gain of a building. Both residential and commercial applications are presented. The text begins with a study of psychrometrics as it is applied to air conditioning and heating. The book then presents the procedures and steps taken to estimate the size, determine the location, and estimate the cost of making the complete installation. The book is written so that it may be used as a study guide, a textbook, a curriculum guide, or a course for independent study. *Fundamentals of Air Conditioning Systems* can serve as a comprehensive textbook for anyone wanting, or needing, knowledge pertaining to system sizing, design, installation and cost estimating. It can also serve as a reference manual for the more experienced personnel.

Fundamentals of Air Conditioning Systems is an accumulation of more than thirty-five years of field experience in the areas of service, installation, business, teaching, and research. It has a dual viewpoint of field experience and education using practical and everyday terminology and language. The reader can use this method to prepare for actual work in the field.

In writing the text the author covers theory, and practical applications using examples to help reinforce the theory presented. When needed, specific manufacturer information is incorporated.

CHAPTER ORGANIZATION

Chapter One presents psychrometrics and how they relate to air conditioning and heating systems and what part they play in the overall operation of the system. Heat load calculation factors are discussed in Chapter Two, along with the purpose of an air conditioning system and what is needed to maintain indoor conditions at the desired level.

Chapter Three presents a thorough discussion of the residential heat load calculation procedure. The heat load for an actual residential building is used as an example. The procedures used for residential equipment sizing is discussed in Chapter Four. There is discussion concerning what is important when sizing this type of equipment. The procedure used to determine the best residential equipment location is discussed in Chapter

Five. Presented are the steps and considerations required to properly locate the equipment. Also discussed are the most popular air distribution systems for residential installations.

Chapter Six thoroughly discusses commercial heat load calculation procedures. An actual commercial building is used so that the reader can "get the feel" of working in the field. Commercial equipment sizing is covered in Chapter Seven. The procedures used to actually size a unit that will properly heat and cool the building are discussed. Chapter Eight discusses the proper procedure to use when determining the proper location of commercial equipment. Also discussed are the most popular types of commercial air distribution systems.

Refrigerant lines are presented in Chapter Nine. The need to properly size refrigerant lines and some of the pitfalls of line sizing are discussed.

Chapter Ten presents the methods used in designing duct systems. In this chapter the duct system for an actual building is properly designed. The different methods and procedures are covered.

Duct pipe, fittings, and insulation are discussed in Chapter Eleven. Both suitable and unsuitable fittings and duct systems for various applications are presented. Also, the proper procedure that should be used to insulate the air distribution system is presented.

Chapter Twelve presents indoor air quality and variables and how to correct some of the problems encountered. The different types of pollutants and some methods of controlling them are also presented.

System cost estimating is discussed in Chapter Thirteen. In this chapter the reader learns one method used for estimating the cost of installing a complete and actual heating and cooling system. The cost of installing an actual system is discussed so the reader can see how it is done in the field.

Chapter 1

Air Conditioning and Psychrometrics

The basic principles or air conditioning are misunderstood by most people, even by many in the field. Air conditioning is not simply just the process of cooling the air. It covers six different functions, defined below. All of these functions are required before the term "air conditioning" can be properly applied to a system.

Air conditioning was not used until about 1920, and then it was used mostly in public places such as theaters, trains, and some department stores. This is when the general public became aware of the benefits of air conditioning.

INTRODUCTION

Air conditioning is usually applied to the cooling effect of a system. However, air conditioning, by definition, requires that six functions take place. These six functions are cooling, heating, air circulation, air cleaning, humidification, and dehumidification. The person who is to estimate an air conditioning job must have, at least, a working knowledge of the functions and how they affect the overall total operation of the system. Some of these functions are required during all modes of operation.

When the system is operating in the cooling mode, the processes involved are air circulation, air cleaning, cooling, and dehumidifying the air. When the system is in the heating mode, the processes are heating, circulation, cleaning, and humidifying the air. It should be obvious that air circulation is a very important aspect of the total system operation.

DEFINITION

Air conditioning is defined as the simultaneous mechanical control of temperature, humidity, and air motion. Unless all of these functions are accomplished the system is not an air conditioning system. When this definition is considered, a building that requires an indoor condition of 135°F dry bulb temperature and a relative humidity of 75% is an air conditioning system just the same as a system which is designed to provide an indoor temperature of 75°F dry bulb and a relative humidity of 40%. Thus, an air conditioning system can provide any desired combination of the conditions indoors regardless of what the outdoor conditions are.

HUMAN COMFORT

There are basically two reasons for the use of an air conditioning system. The first is to improve some industrial or manufacturing process, and the second is to provide human comfort. The indoor conditions required are determined by the need and the desire of the occupants or the process. Thus, for human comfort applications, an understanding of the required body functions is fundamental to a proper understanding of air conditioning.

Human comfort depends on the rate at which the body loses heat. The human body is a heat machine that uses food as its fuel. The food that we eat contains the components carbon and hydrogen. The energy in these components is released by oxidation of the food. When we breathe in air, the oxygen that is taken from the air is necessary for the oxidation of the food. When we consider the products of combustion that occur in a gas flame, we know that the products of combustion (oxidation) are carbon dioxide (CO_2) and water vapor (H_2O). The medical profession terms this function as metabolism.

A natural function of the human body is to maintain itself at a constant temperature. The temperature that the mechanism attempts to maintain is 98.6°F. This temperature regulating mechanism is a very delicate part of the human body. The body always produces more heat than it needs; therefore, heat rejection is an ongoing process with the human body. The purpose of an air conditioning system is to help the body's system to regulate this heat loss. This theory applies to both the heating and cooling functions of the system. During the summertime the purpose of the air conditioning system is to increase the cooling rate of the body. During the wintertime its job is to decrease the rate of cooling.

There are three ways that the human body rejects this extra heat: (1) convection, (2) radiation, and (3) evaporation. All three of these methods are used simultaneously.

CONVECTION

Convection is the process by which the air next to the body reaches a higher temperature than the air at some distance from it and moves to the cooler area. The warmer air is lighter and rises, removing some of the heat from the skin as it moves upward. When the warm air rises it is replaced with cooler air which absorbs heat from the skin. This is a continuous process which maintains the interior of the human body at a constant temperature of 98.6°F even though the skin temperature may vary from about 40°F to 105°F. The skin temperature will vary in a direct relationship to the temperature, velocity, and humidity of the surrounding air.

RADIATION

Heat radiates from the human body in the same manner that it does from the sun to the surface of the earth. The heat travels in rays from the warmer body to the surface of a cooler body. Radiation and convection are different heat transfer methods but they work together to help maintain the temperature of the human body. During the transfer of heat with this method the temperature of the air between the two objects has little or no bearing on the heat transfer process. A good example of this is when a person stands close to a radiant heater and feels the heat on the side closest to the heater. That side will be warm while the other side will remain cool or cold. However, the temperature of the air between the heater and the person will be the same as that on the side away from the heater.

EVAPORATION

Evaporation is the means used by the body to regulate its temperature when it is not in an air conditioned space. When the human body experiences excessive heat, moisture will be released to the skin and then evaporated. The heat that causes this moisture to evaporate is taken from the skin. This removal of heat is what cools the body. This evaporation process turns the moisture into a low-pressure steam or water vapor in a continuous process. The appearance of these drops of moisture on the skin is an indication that the body is generating more heat than it can reject at the normal rate to maintain the desired body temperature.

There is no set rule for the best conditions for everyone. This can be observed at a large meeting when some of the people are hot, some are sweating, and some others are cold, all inside a building with a constant temperature throughout.

From the above discussion it should be obvious that there are three conditions that affect the heat dissipation from the human body. These three conditions are: (1) temperature, (2) relative humidity, and (3) air motion. When a change in either one or more of these conditions occurs the heat dissipation process will also be changed accordingly. If one of these factors is increased the other two must take on more of the work of dissipating the heat.

Air Temperature

When the air temperature is lower than the skin temperature, the heat dissipation will be increased because of the convection process. As the temperature of the air drops more heat will be lost through the skin because of convection. Heat always flows from a warmer body to a cooler body. When this temperature is increased the amount of heat transferred will increase accordingly. When the temperature difference is very great the body will lose heat faster than it can generate it and the person will begin to feel cool and eventually cold.

When the surrounding air temperature is greater than the temperature of the skin, the skin will absorb heat from the air. The convection process is therefore reversed and the temperature of the body will increase. Thus, the temperature of the air has a great effect on the comfort of the human body. Studies have shown that the temperature that is comfortable for most people is between 72°F and 80°F.

The temperature of the area surrounding the person will also have an effect on comfort. This is because heat is radiated either to or from the body in a relationship to the temperature of the surroundings. When the surroundings are at a temperature greater than the body the heat will be radiated into the body causing it to feel warm. When the temperature of the surrounding surfaces is lower than the body the heat will be radiated from the body causing it to feel cool. As the temperature difference between the two bodies increases the amount of heat transferred will also increase. When the temperature difference between the body and the surrounding objects is decreased the rate of heat transfer will also decrease.

Relative Humidity

The relative humidity in the surrounding air has a tremendous effect on the amount of heat rejected by the body through the radiation process.

Relative humidity is a measurement of the moisture in the air and is an indication of the ability of the air to absorb more moisture. The relative humidity inside a building is an indication of the amount of air conditioning needed for the building.

Example

There are two cubic feet of air with a 70°F temperature, containing 8 grains of water vapor. What is the relative humidity of the air (Figure 1-1)? (A grain is a very small amount of moisture. It requires 7000 grains of moisture to weigh one pound.)

Figure 1-1. Cubic foot of saturated air.

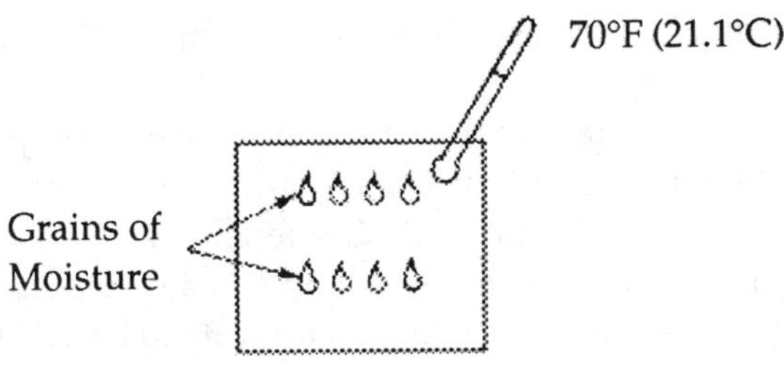

The relative humidity can be found by the following procedure. There are only 8 grains of water vapor in two cubic feet of air. If those two feet of air held all the water vapor they could hold at that temperature, they would actually contain 16 grains of water vapor and would be considered saturated (Figure 1-2).

Figure 1-2. Cubic feet of saturated air.

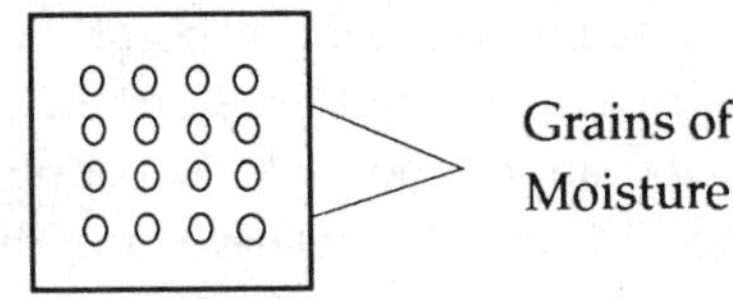

The relative humidity is determined by dividing the amount of water vapor actually present in the air by the amount that the air could hold when saturated at that same temperature (Figure 1-3).

Figure 1-3. Determining relative humidity.

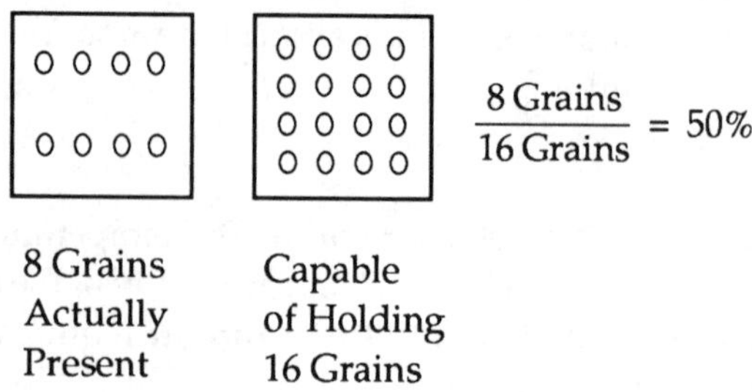

This indicates that the relative humidity of our sample is 50%. By definition, relative humidity is an indication of the amount of moisture in the air compared to the amount that could be held at that temperature. The relative humidity in a sample of air will change with a change in temperature.

This can be shown by raising the temperature of our sample of air to 92°F with no additional water vapor. The psychrometric chart shows that two cubic feet of air at 92°F will hold 32 grains of water vapor when saturated. The relative humidity in this example is determined by dividing 8 grains by 32 grains to find the relative humidity to be 25% (Figure 1-4).

Figure 1-4. Relative humidity at a higher temperature.

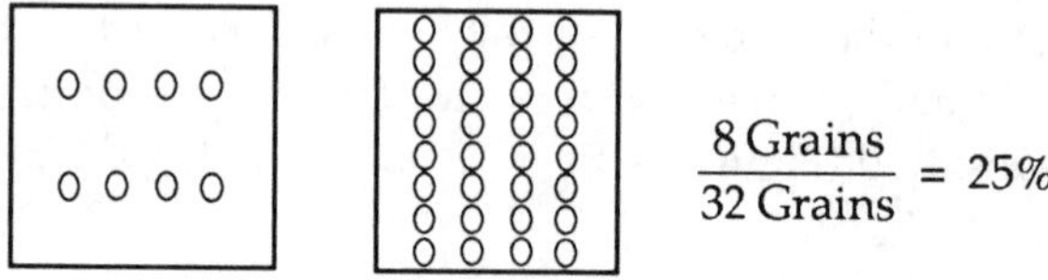

When air having a low relative humidity surrounds the body the body will reject more heat through the evaporation process. Likewise, when the surrounding air has a high relative humidity the body will reject less heat through the evaporation process. Most people are comfortable in air that is conditioned to 80°F and 50% relative humidity.

Air Movement

When the rate of air flow over a body is increased, the amount of heat lost by evaporation will be increased. This is because the evaporation process depends on the ability of the air to absorb moisture. When the air passes over a body it absorbs moisture from the skin and the moist air is replaced with air having a lower moisture content which will then absorb more moisture. When the air flow rate is increased the amount of moisture evaporated also is increased.

However, when the air is static, or remains still, the air closest to the skin will absorb moisture only until it is saturated. When the saturation point is reached the heat lost by evaporation will be reduced and will eventually stop completely. Under these conditions most people will feel uncomfortable.

Air movement also affects the heat lost through convection. This is because the warm air next to the skin is constantly being replaced with cooler air that will absorb more heat.

The movement of air also removes heat from the surroundings, which tends to increase the amount of heat lost by radiation. Air motion is one of the conditions that affect the comfort of human beings.

AIR CONDITIONING TERMINOLOGY

It is necessary to learn the terms used in the air conditioning industry before a full understanding of air conditioning can be reached. The following are the most common definitions used.

Dry Air

Dry air contains no moisture. Air naturally contains some moisture.

Absolute Humidity

The actual quantity or weight of the moisture in a quantity of air is termed absolute humidity. Absolute humidity is expressed in terms of weight of moisture in grains per pound of dry air. It takes 7000 grains of moisture to weigh one pound. When there is either a rise or a fall in the temperature of the air, the quantity of moisture will not be affected as long as the dew point temperature is not reached.

Relative Humidity

The ratio between the moisture present in a quantity of air and the amount that it could hold under the same pressure and temperature is known as the relative humidity. Relative humidity is expressed as a percentage. Note the difference between absolute humidity and relative humidity. The relative humidity of a quantity of air can be calculated by dividing the amount of moisture actually present by the maximum amount that the air could hold under the same conditions.

Example

If two cubic feet of air hold 8 grains of moisture but are capable of holding 16 grains under the same conditions, what would be the relative humidity?

$$RH = \frac{\textit{absolute humidity}}{\textit{maximum amount air can hold}} = \frac{8}{16} = 50\%$$

An increase or decrease in the temperature of the air would increase or decrease the relative humidity of the air. There is one exception and that is if the dew point temperature is reached. If the air temperature is reduced below this temperature, moisture will be condensed out of the air. Air is saturated when it contains all the moisture it can hold at that same temperature and pressure. Saturated air has a relative humidity of 100%.

Humidification

Humidification is the process of adding moisture to the air. The moisture is usually added with a humidifier.

Dehumidification

Dehumidification is the process used to remove moisture from air. This is a major purpose of the refrigeration system in a cooling system. Moisture is removed by placing the evaporator coil in the air stream. When the air passes through the coil its temperature is reduced to below the dew point temperature and moisture is condensed out of the air.

Saturated Air

Air that is holding all of the moisture that it can hold under a set of conditions is said to be saturated. At this temperature, the dry bulb temperature, the wet bulb temperature, and the dew point temperature will be the same.

Dry Bulb Temperature

The dry bulb temperature is indicated by an ordinary thermometer. It is a measure of the amount of sensible heat in the air.

Wet Bulb Temperature

The wet bulb temperature of a quantity of air is determined by an ordinary dry bulb thermometer with its bulb covered with a cloth sock (Figure 1-5).

Figure 1-5. Wet bulb thermometer.

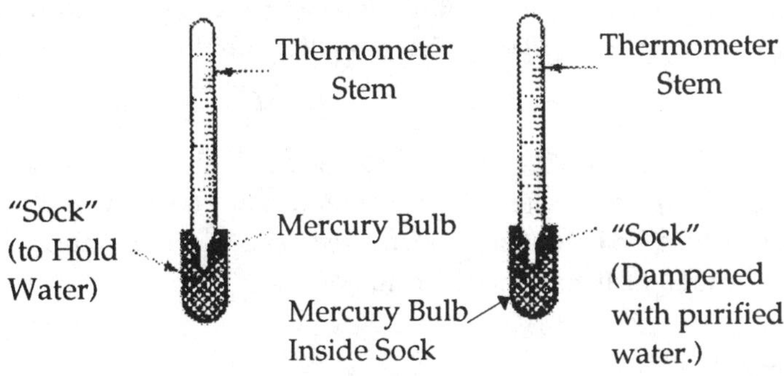

The cloth sock is wetted with distilled water and then exposed to a rapidly moving air stream. The temperature indicated by a wet bulb thermometer will usually be lower than the dry bulb temperature of the same quantity of air. There is one exception; when the saturation temperature is reached, then they both will read the same temperature.

To determine when the correct wet bulb temperature has been reached, note when three consecutive readings are found to be the same. It can then be assumed that the correct temperature has been reached.

Wet Bulb Depression

The wet bulb depression is the difference between the dry bulb temperature and the wet bulb temperature of the air, except when the saturation point is reached. Then the temperatures will be the same.

Dew Point Temperature

The temperature at which the moisture will start to condense out of the air is known as the dew point temperature. The amount of moisture in a given quantity or air will always be the same at any dew point temperature. The dew point temperature can, then, be used to determine the amount of moisture in a quantity of air. At the dew point temperature the air is holding all the moisture that it can hold at that temperature. The dew point temperature will always remain the same when there is no addition or removal of moisture from the air. When there is no change in the moisture content of the air, there will be no change in the latent heat content of the air.

Saturation Temperature

The dew point temperature and the saturation temperature are the same. When a quantity of air is at its dew point it is said to be saturated. At the saturation temperature the air has a relative humidity of 100%.

Total Heat (Enthalpy)

The total heat in a given quantity of air is the sum of both the sensible and latent heat. Zero degrees F is usually taken as a reference point from which heat content is measured. The wet bulb thermometer indicates the total heat content of the air.

Saturated Gas (Vapor)

When the temperature of a gas or water vapor is at the boiling point that corresponds to the pressure over it, the gas is said to be saturated. This is when both gas and liquid exist in the same container. There is no superheat present.

Pound of Dry Air

This term means one pound of dry air. The following examples are some of the terms used to indicate this condition: pound of air, total heat per pound of dry air, moisture per pound of dry air, and latent heat per pound of dry air. Each of these terms means a pound of dry air and does not mean a pound of the mixture.

Cubic Foot of Air

Cubic foot of air means one cubic foot of air and is used to express the quantity of air flowing past a given point.

Ventilation

Ventilation is the process used to add or remove air from an enclosed space. The air does not necessarily need to be conditioned. It can be either used air or fresh air.

Effective Temperature

Effective temperature is a measure of personal comfort. It is usually determined by trial and error by the occupants of the building. This occurs when the correct combination of dry bulb temperature, relative humidity, and air motion has been provided.

Comfort Zone

The comfort zone is that range of temperature and humidity in which most people are comfortable. The outer limits of the effective temperature are not clearly defined because they depend, to some extent, on the conditions out of doors.

Overall Coefficient of Heat Transmission

The overall coefficient of heat transmission is the quantity of Btu transmitted each hour through one square foot of any material, or combination of materials, for each degree of temperature difference between the two sides of the material. The temperature difference means the air temperature difference and not the surface temperature on each side of the material.

Static Pressure

Gas or air flowing through an enclosed space, such as a duct system, exerts a pressure at right angles against the walls of the duct. This pressure is known as static pressure. It is usually measured in inches of water column (WC). The measuring device may be a manometer or a Pitot tube inserted into the duct. The loss due to the friction of the air flowing through the duct is known as pressure drop or loss of static head.

Velocity Pressure

Velocity pressure is the result of air flowing through a duct system. This pressure is caused by the energy motion or kinetic energy in the air movement. It is measured in inches of water column (WC).

Total Pressure

This is the sum of both the static pressure and the velocity pressure inside a duct system. It is a measure of the total energy of the air flowing through it.

Stack Effect

Heated air has a tendency to rise in a vertical direction, especially when captured in a duct. This upward flow is known as stack effect. The stack effect is especially important when designing the heating system for tall buildings, in which case the entire building acts as a stack, or chimney because of the warm air inside it. The warm air rises to the upper floors where it leaks from the building through cracks and other openings. At the same time cooler air is drawn in at the lower floors through cracks and other openings. Because of the stack effect, extra heating must be supplied to the lower floors to heat the incoming cooler air.

Plenum Chamber

The plenum chamber is essentially an equalizing chamber that can be located on either or both the supply and return air openings of the equipment. The discharge air plenum is always under a constant positive pressure. The return plenum is almost always under a negative pressure.

Psychrometrics

Psychrometrics is the science that involves the different relationships of air and water vapor that is in the atmosphere around us. This phenomenon involves measuring and determining the quantities of the different properties of air. It can also be used to determine the proper conditions of air that will provide the most comfort or the best environment for an industrial process.

Psychrometric Chart

The psychrometric chart is simply a diagram representing the different relationships possible between the heat, moisture content, and the water vapor in a given quantity of air. There are many different versions of this chart. Some of the various properties of air are located in different areas on each individual chart. However, the procedure for locating the various properties on the chart is the same. The properties that are shown on a psychrometric chart are: dry bulb temperature, wet bulb temperature, dew point temperature, relative humidity, total heat of the air, vapor pressure, and the actual moisture content of the air.

Identification of Lines and Scales on a Psychrometric Chart

The following illustrations should help in locating the different property lines and scales on the psychrometric chart. The psychrometric chart is generally considered to be shaped like a shoe with the toe pointing to the left and the heel located on the right-hand side (Figure 1-6). The bottom of the chart is called the sole, the rear of the chart is called the back, the flat line at the top is called the top, and the line between the toe and the top is called the instep.

Dry Bulb Temperature Lines

Along the sole of the chart is located the dry bulb temperature line (Figure 1-7). The dry bulb lines extend vertically upward from the sole. Usually there is a single line for each degree of dry bulb temperature.

Wet Bulb Temperature Lines

The wet bulb temperature scale is located along the instep of the chart (Figure 1-8). These lines extend diagonally downward to the right

Figure 1-6. Psychrometric chart scales.

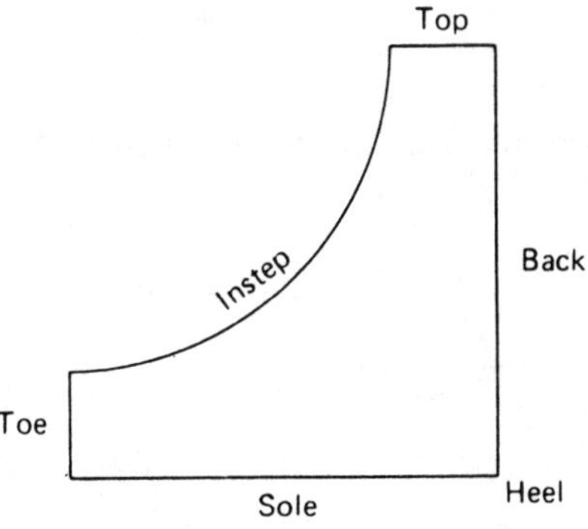

Figure 1-7. Dry bulb lines. (Courtesy of Research Products Corp.)

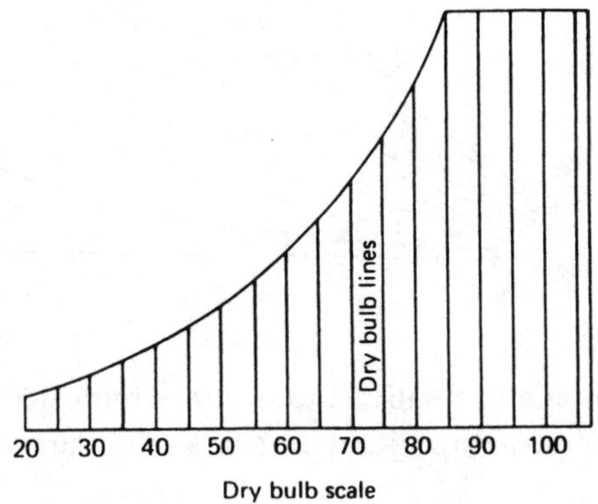

Figure 1-8. Wet bulb lines. (Courtesy of Research Products Corp.)

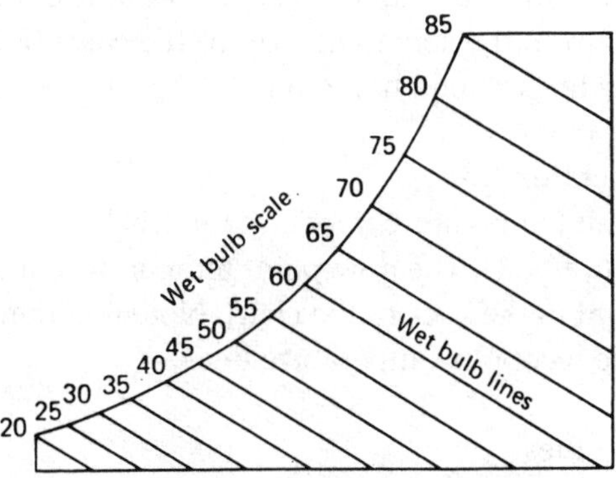

from the instep. There is an individual line for each degree of wet bulb temperature.

Relative Humidity Lines

The relative humidity lines are the only curved lines on a psychrometric chart (Figure 1-9).

Figure 1-9. Relative humidity lines. (Courtesy of Research Products Corp.)

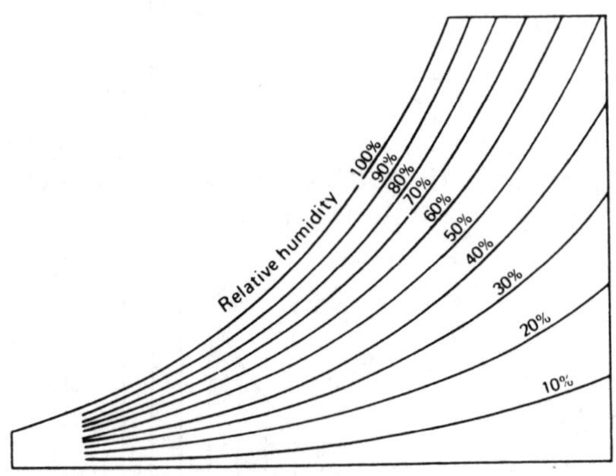

The various percentage of relative humidity is indicated directly on the lines. There is no coordinate scale such as this with the other properties of air.

Absolute Humidity Lines

The absolute humidity line is located on the heel of the chart and extends vertically from the sole to the top of the chart (Figure 1-10). The absolute humidity lines extend horizontally from the heel to the instep.

Dew Point Temperature Lines

The dew point temperature lines and the wet bulb temperature lines are the same (Figure 1-11). The dew point temperature lines extend horizontally to the right to the back of the chart. Note that they do not extend diagonally like the wet bulb temperature lines.

Specific Volume Lines

The specific volume scale is located along the sole and the back of the chart. The range is normally from 12.5 to 14.5 cubic feet (Figure 1-12).

Figure 1-10. Absolute humidity lines. (Courtesy of Research Products Corp.)

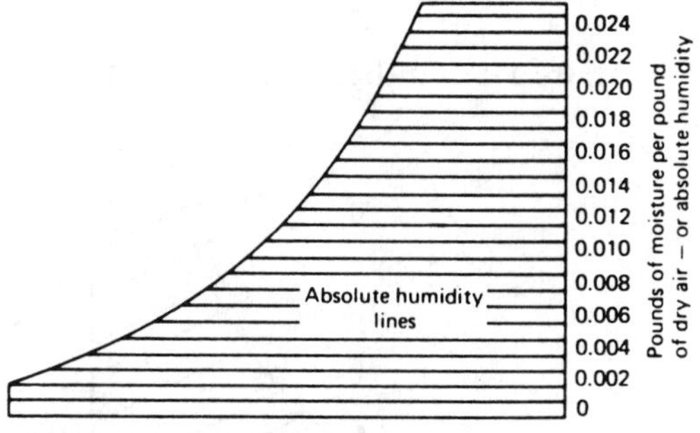

Figure 1-11. Dew point lines. (Courtesy of Research Products Corp.)

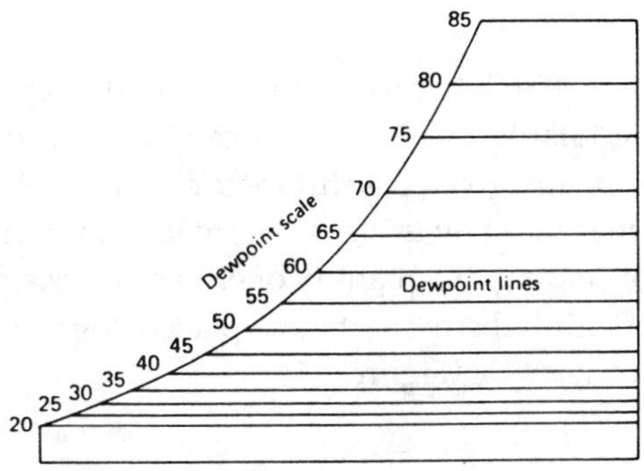

Figure 1-12. Cubic feet of air lines.

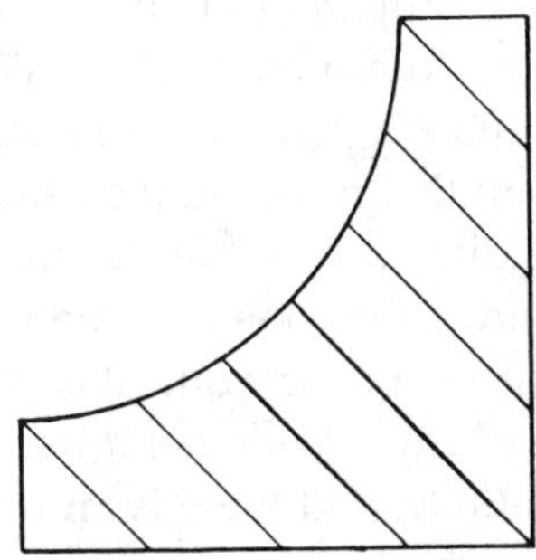

The cubic feet lines extend diagonally upward to the left from the sole and heel to the instep of the chart. These lines indicate the cubic feet per pound of dry air.

Enthalpy Lines

The enthalpy scale is located along the instep of the chart (See Figure 1-13).

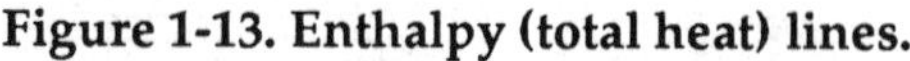

Figure 1-13. Enthalpy (total heat) lines.

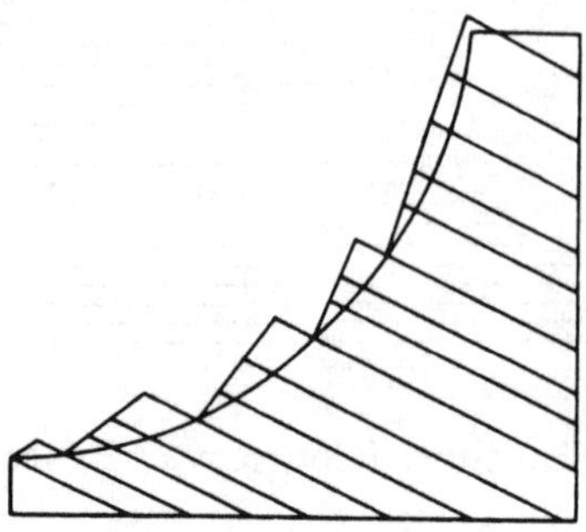

These lines and the wet bulb temperature lines are the same. The enthalpy is a measure of the total heat of the air. It can be used to determine any changes that occur in any given psychrometric process. Both sensible and latent heat can be measured by using the enthalpy of the air. The enthalpy is a quick means of finding the quantity of either of these properties.

When all seven of these charts are placed together the complete psychrometric chart is made (Figure 1-14).

USE OF THE PSYCHROMETRIC CHART

When any two of the properties of air are known the others can be determined by use of the psychrometric chart. When any dry bulb temperature and any wet bulb temperature are chosen, the intersection of these two lines is the point that indicates the conditions that exist in that quantity of air. The conditions that exist at this point will always be exactly the same. Also, any other point on the psychrometric chart will indicate the exact same properties any time it is determined. The possible number of combinations of any two temperatures that can be located on a psychrometric chart is infinite, and there is an infinite number of points that may be located on the chart.

Example 1

Given a dry bulb temperature of 90°F and a 55°F dew point temperature, find the wet bulb temperature of the mixture (Figure 1-15).

Figure 1-14. Psychrometric chart. (Courtesy of The Trane Company.)

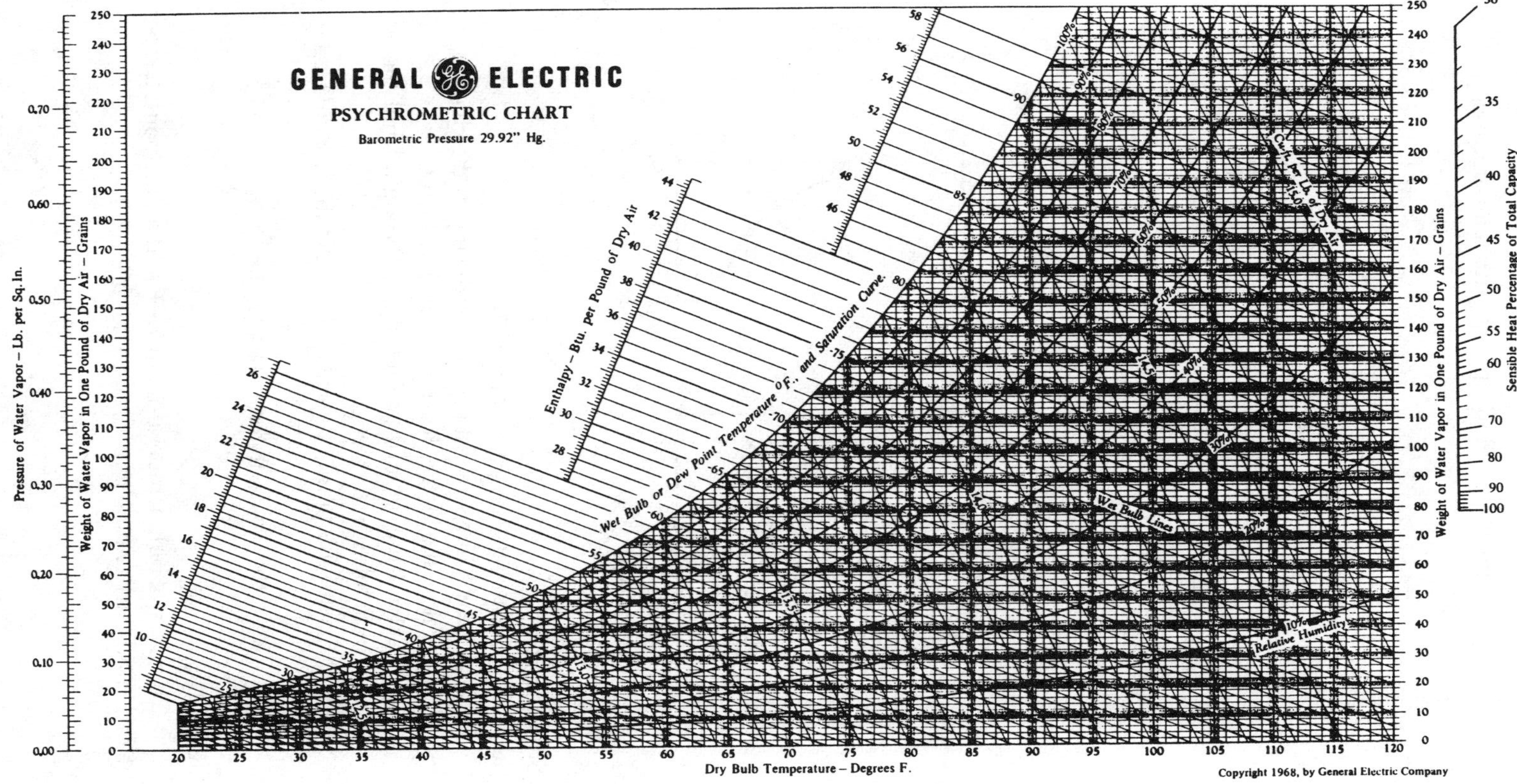

Figure 1-15. Finding wet bulb temperature.

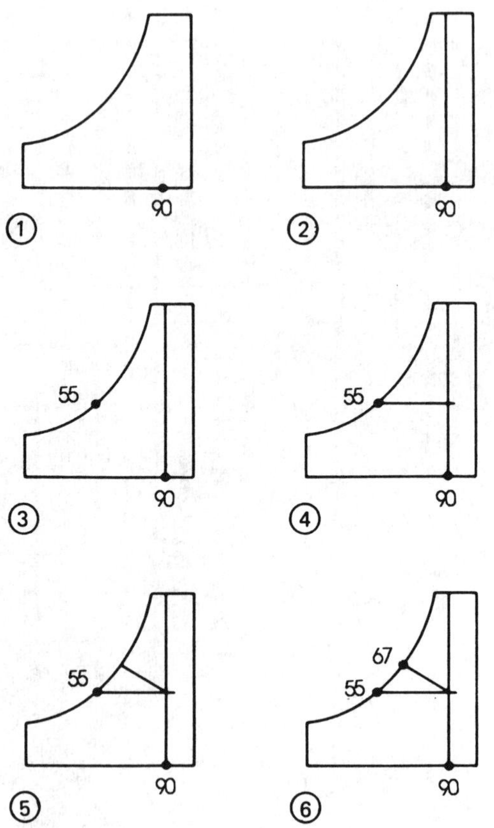

Solution

Step 1. Locate 90°F on the dry bulb scale.
2. Draw a line straight upward to the top of the chart.
3. Follow the instep down until 55°F is found.
4. Extend this point horizontally to the right until the 90°F dry bulb line is intersected.
5. Extend this point upward to the left to the wet bulb scale on the instep.
6. Read the wet bulb temperature of 67°F.

Example 2

Given a wet bulb temperature of 75°F and a dew point temperature of 75°F, find the dry bulb temperature (Figure 1-16).

Figure 1-16. Finding dry bulb temperature.

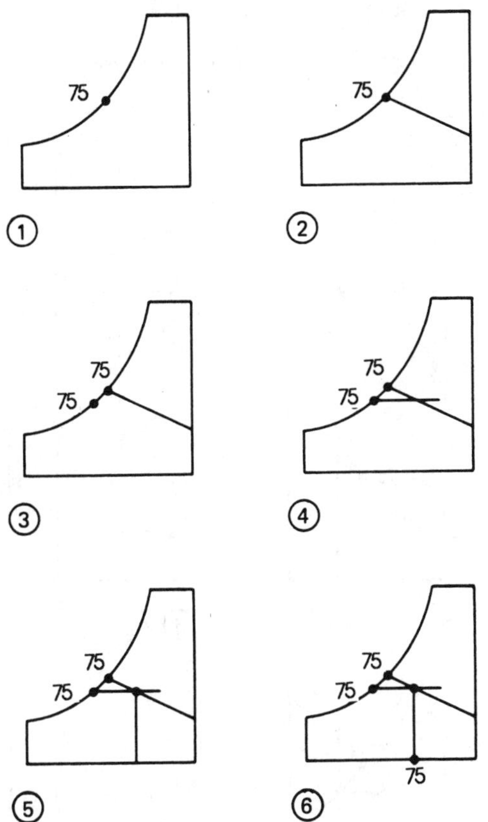

Solution

Step 1. Locate 75°F on the wet bulb scale

2. Draw a line diagonally down to the right to the back of the chart.
3. Locate 75°F on the dew point scale.
4. Extend this point horizontally to the right until the wet bulb line is crossed.
5. Extend this point vertically downward until the dry bulb line is crossed.
6. Read the dry bulb temperature of 75°F.

This is an example of saturated air.

Example 3
Given a wet bulb temperature of 75°F and a dry bulb temperature of 80°F, find the dew point temperature (Figure 1-17).

Figure 1-17. Finding dew point temperature.

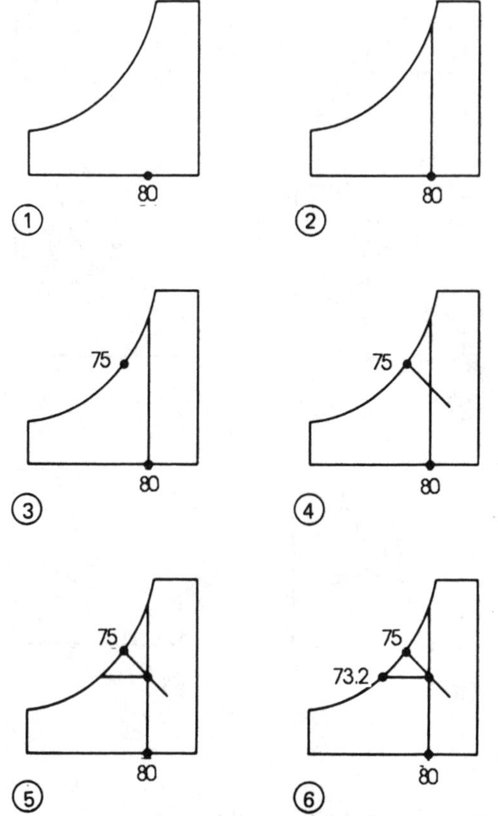

Solution

Step 1. Locate 80°F on the dry bulb scale.
2. Draw a line vertically upward to the instep of the chart.
3. Locate 75°F on the wet bulb scale.
4. Extend this point diagonally downward until the dry bulb line is crossed.
5. Extend this point horizontally to the left to the instep.
6. Read a dew point temperature of 73.2°F.

Example 4
Given a dry bulb temperature of 90°F and a dew point temperature of 55°F, find the relative humidity (Figure 1-18).

Figure 1-18. Finding relative humidity.

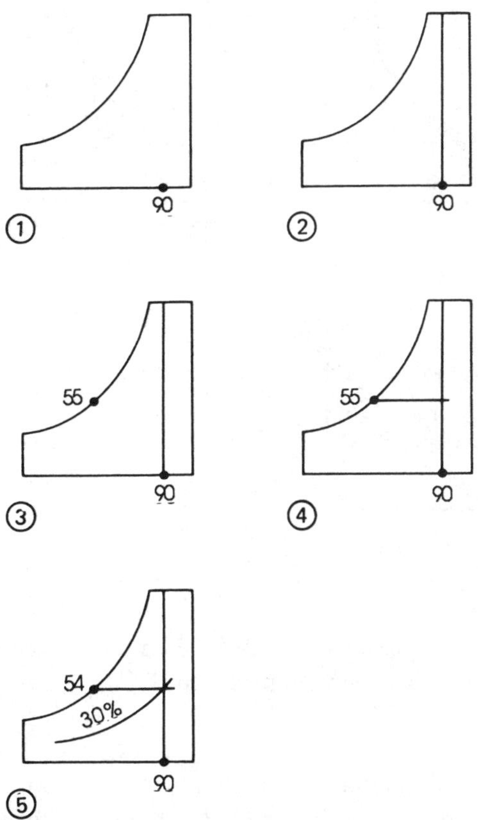

Solution

Step 1. Locate 90°F on the dry bulb scale.

2. Draw a line vertically upward to the instep of the chart.
3. Locate 55°F on the dew point scale.
4. Extend this point horizontally to the right until the dry bulb line is crossed.
5. Read the relative humidity at this point to be 30%.

Example 5

Given a wet bulb temperature of 75°F and a dew point temperature of 75°F, find the total heat content of the air (Figure 1-19).

Solution

Step 1. Locate 75°F on the wet bulb scale.

2. Draw a line diagonally upward to the left to the total heat scale.
3. Read the total heat content to be 38.6 Btu/lb of dry air.

Figure 1-19. Finding total heat content using wet bulb temperature.

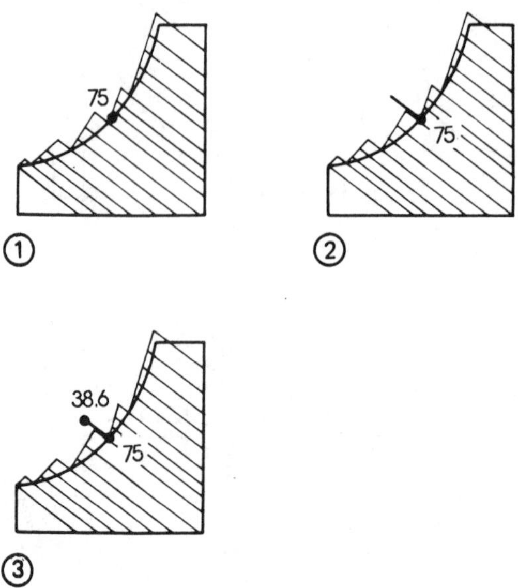

Example 6

Given a dry bulb temperature of 90°F and a dew point temperature of 55°F, find the total heat content of the air (Figure 1-20).

Solution

Step 1. Locate 90°F on the dry bulb scale.
2. Draw a line vertically upward to the top of the chart.
3. Locate 55°F on the dew point scale.
4. Extend this point horizontally to the right until the dry bulb line is intersected
5. Extend this point diagonally upward to the left to the total heat scale.
6. Find the total heat content to be 33.2 Btu/lb of dry air. Change in Total Heat: When the system is to operate in the cooling mode, our main interest is in how much heat must be removed from the air to cool it to match the inside design requirements for the building. During the heating season, heat is added to the air to match the inside design requirements. If we were in the cooling mode and the outside design condition is 70°F wet bulb and the inside design requirements calls for a 60°F wet bulb, the amount of total heat that must be removed per pound of dry air is determined by the following: (Figure 1-21).

Figure 1-20. Finding total heat content using dry bulb temperature.

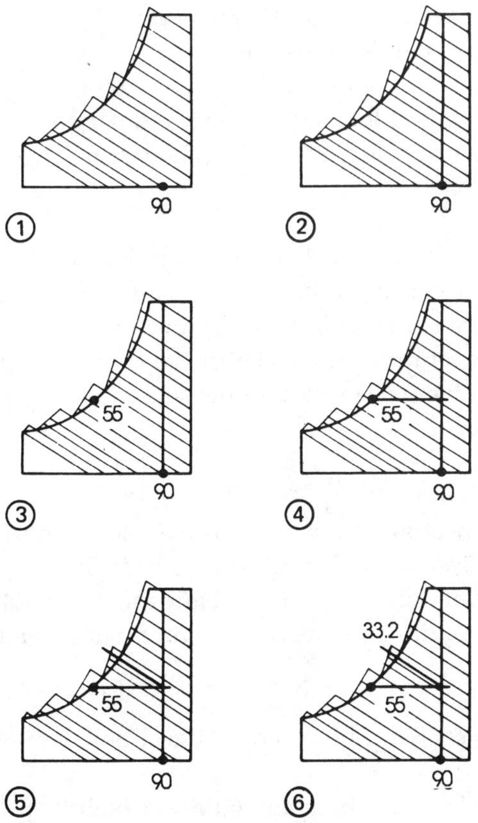

Figure 1-21. Change in total heat.

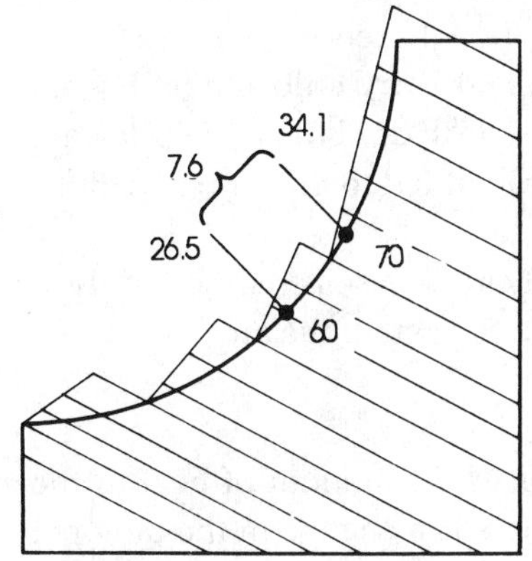

Total heat at 70°F wet bulb = 34.1 Btu/lb of air.
Total heat at 60°F wet bulb = 26.5 Btu/lb of air.
The difference is 7.6 Btu/lb of dry air.

Thus, the total heat that must be removed per pound of air is 7.6 Btu when cooling the air down from 70°F wet bulb to 60°F wet bulb.

Sensible Heat

Sensible heat is the heat that can be added to or removed from a substance without causing a change of state of the substance. If that substance is air, changing the sensible heat will result in only a change in the dry bulb temperature. A change in the dry bulb temperature will cause a change in the sensible heat of the substance only because there is no change of state.

Example 7

During the heating season air is to be heated from 60°F dry bulb temperature and 50°F wet bulb temperature to an 85°F dry bulb temperature and a 60°F wet bulb temperature. How much sensible heat must be added per pound of dry air to reach these conditions? (Figure 1-22)

Solution

Step 1. Locate 60°F dry bulb and 50°F wet bulb temperatures on the chart.
2. Locate 85°F dry bulb and 60°F wet bulb temperatures on the chart.
3. Extend these points diagonally upward to the left to the total heat scale.
4. Read 20.32 Btu and 26.23 Btu on the total heat scale.
5. Determine the difference.
Total heat at 60°F dry bulb and 50°F wet bulb = 20.32 Btu/lb.
Total heat at 85°F dry bulb and 60°F wet bulb = 26.23 Btu/lb.
Sensible heat added to the air = 5.91 Btu/lb of dry air.

This change in heat was sensible heat only because there was no change in the moisture content of the air.

Latent Heat

A good indicator of the amount of heat in the air is the dew point temperature. When there is a change in the dew point temperature there

Figure 1-22. Sensible heat change on heating air.

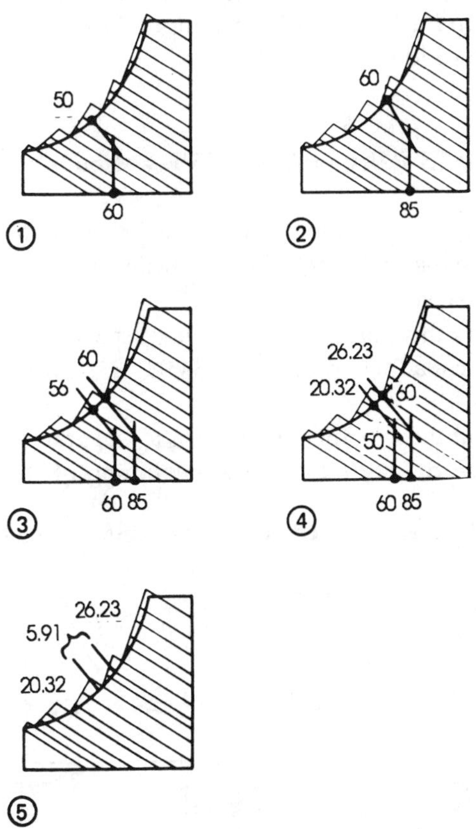

will also be a change in the moisture content of the air. Thus, the dew point temperature can only be changed by changing the amount of moisture in the air. From this it can be seen that as long as the dew point temperature is unchanged there will be no change in the moisture content of the air.

On most charts, the moisture content scale indicates the grains of moisture per pound of dry air. The dew point scale is the same scale as the wet bulb temperature scale.

The latent heat of vaporization is the amount of heat in Btu required to change a liquid to a gas at a constant temperature. If a quantity of air has a given number of grains of moisture per pound of dry air, there will be a definite amount of heat required to change this moisture into a vapor. This amount of heat is known as the latent heat of the air and moisture mixture.

Example 8

Air at 73°F dry bulb and 55°F wet bulb is to be conditioned to reach a 75°F dry bulb and a 70°F wet bulb temperature. How many grains of moisture must be added? (Figure 1-23.)

Solution

Step 1. Locate 73°F dry bulb and 55°F wet bulb temperature on the chart.

2. Extend this point to the total heat scale and read 23.25 Btu.
3. Locate the 75°F dry bulb and 70°F wet bulb temperature point on the chart.
4. Extend this point to the total heat scale and read 34.08 Btu.
5. The heat added to this air is 34.08 - 23.25 = 10.83 Btu/lb of dry air.
6. Extend these points to the right to the grains of moisture scale and read 104 grains at the beginning of the process and 36

Figure 1-23. Latent heat change and grains of moisture added.

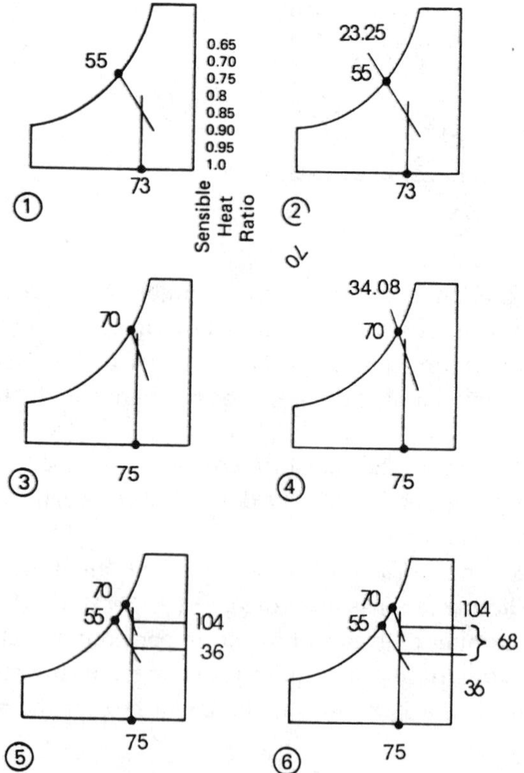

grains of moisture per pound of dry air at the end of the process.

7. The grains of moisture added per pound of dry air is 104 – 36 = 68 grains of moisture added.

The amount of heat is only latent heat because there was no change in the dry bulb temperature of the air.

Sensible Heat Ratio

The sensible heat ratio is the amount of sensible heat removed in Btu removed divided by the total amount of heat removed in Btu. The sensible heat ratio is an indication of the percentage of sensible heat removed compared to the total heat removed. In most comfort air conditioning applications, the sensible heat ratio is above 50%. The reason for this is that most comfort air conditioning systems remove more sensible heat than latent heat.

In most cases, the type of installation will determine the sensible heat ratio. A residential application may have a sensible factor of 0.7 or 0.8 representing a sensible heat factor of 70 or 80 percent. That is, 70% or 80% of the heat removed is sensible heat. A restaurant may have a sensible heat factor of 0.5 or 0.6.

On most psychrometric charts, the sensible heat ratio scale is located at the back of the chart. When two points are determined on the chart a line can be drawn through them which extends to the sensible heat ratio scale. The factor is read directly from the scale (Figure 1-24).

Figure 1-24. Sensible heat ratio line.

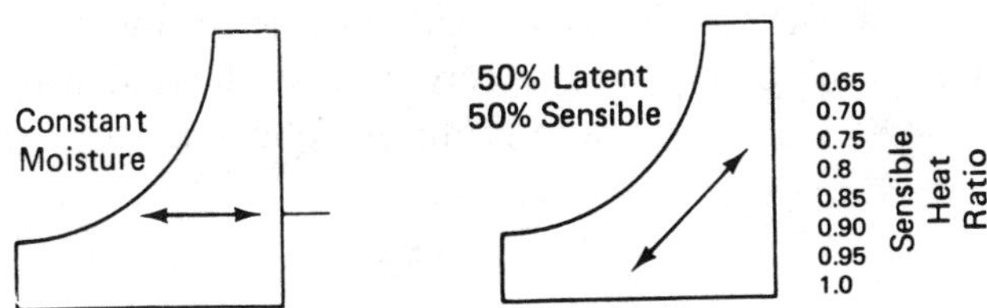

If the line arawn is at approximately a 45° angle, the sensible heat ratio is 50%, or 0.50, indicating that half of the heat removed is sensible heat and half is latent heat.

Example 9

The desired inside conditions are 82°F dry bulb temperature and 65°F wet bulb temperature. The supply air conditions are 55°F dry bulb temperature and 50°F wet bulb temperature. What is the sensible heat factor? (Figure 1-25.)

Figure 1-25. Finding sensible heat ratio (line method).

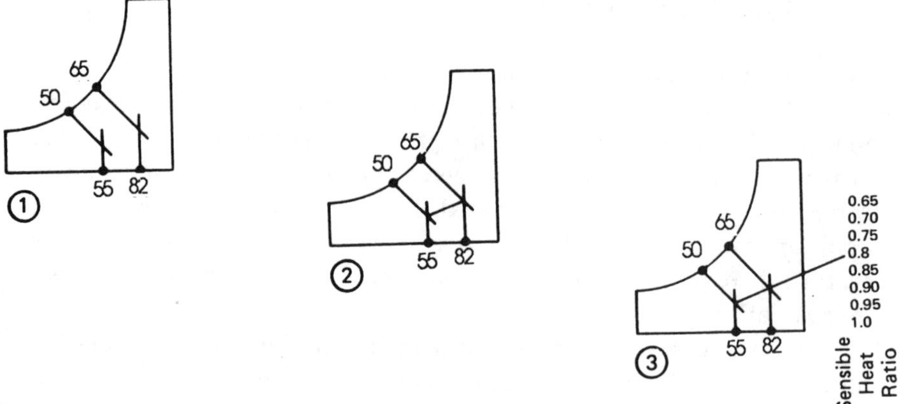

Solution

Step 1. Plot the given indoor and supply air conditions on the chart.
2. Draw a straight line between the two points.
3. Extend the line to the sensible heat ratio scale.
4. Read 0.81 on the sensible heat ratio scale.

Another method that can be used for determining the sensible heat factor is to use a ratio of the total heat removed to the sensible heat removed from the air. Using the same conditions as those in Example 9 we can calculate the sensible heat ratio using the following steps. (Figure 1-26.)

Solution

Step 1. Plot the given conditions on the chart.
2. Extend the final air conditions (55°F dry bulb and 50°F wet bulb) point horizontally to the right until the 82°F dry bulb line is crossed.
3. Extend the 82°F dry bulb line vertically downward to the sole of the chart. This procedure forms a reversed "L" figure.
4. Extend the indoor air point, the supply point, and the intersection of the two sides of the "L" to the total heat scale.

Figure 1-26. Calculated sensible heat ratio.

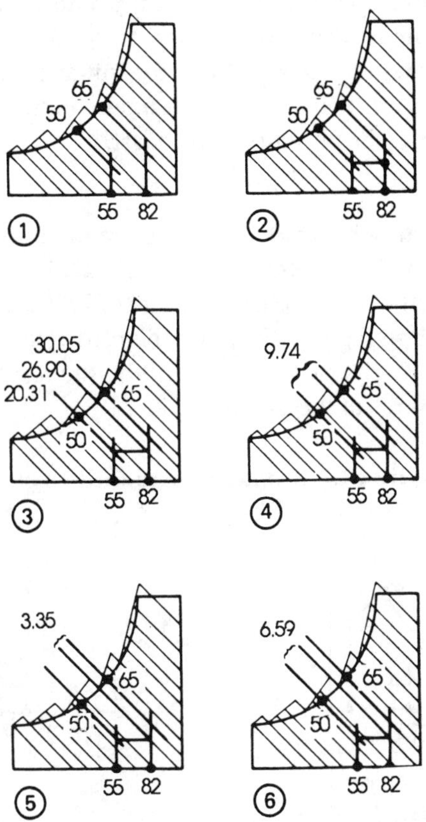

5. Read the total heat for each of the two lines extended; indoor conditions, 30.05 Btu; supply air conditions, 20.31 Btu; and intersection of the "L" 26.90 Btu.
6. Calculate the total amount of heat removed from the air: 30.05 - 20.31 = 9.74 Btu.
7. Calculate the latent heat removed 30.05 - 26.90 = 3.15 Btu latent heat.
8. Calculate the amount of sensible heat removed: 26.90 – 20.31 = 6.59 Btu.
9. Calculate the sensible heat ratio: 6.59 ÷ 9.74 = 0.67, or 67% of the heat removed is sensible heat.

Mixing Two Quantities of Air at Different Conditions

Air conditioning also involves the process of mixing two different quantities of air to obtain the desired results. The most common method of mixing two quantities of air is when the conditioned air is mixed with air in the conditioned space or when the supply air is mixed with fresh, or bypass, air. The air is also mixed when fresh air is supplied in some amount for ventilation purposes. When the initial condition and the final mixture conditions are known, then the final condition of the air can be determined by using the psychrometric chart.

Example 10

The outdoor air which has a dry bulb temperature of 92°F and a wet bulb temperature of 72°F (point A in Figure 1-27) is to be mixed with the return air which has a dry bulb temperature of 72°F and a relative humidity of 15% (point B in Figure 1-27). The mixture is to be made up of 25% outdoor air and 75% return air. What will be the resulting dry bulb and wet bulb temperatures of the mixture of air?

Figure 1-27. Mixing two quantities of air.

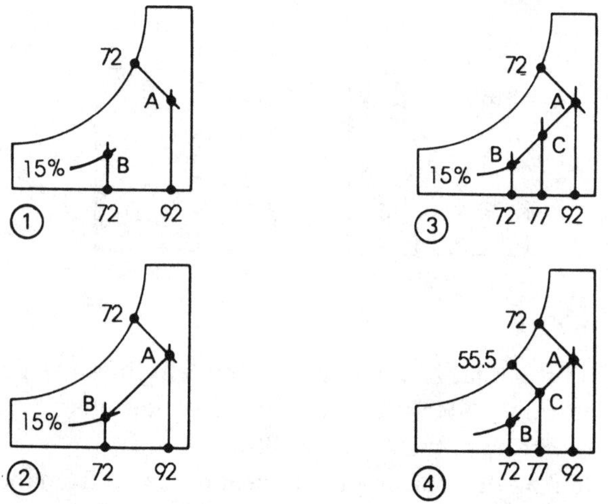

Solution

Step 1. Plot points A and B on the chart.

2. Draw a line between the two points.

3. Locate the dry bulb temperature by adding the correct percentage of each of the dry bulb temperatures; thus 25% of 92°F = 23, 75% of 72°F = 54. The resulting dry bulb temperature of the mixture is 23°F + 54°F = 77°F.

4. Locate 77°F dry bulb on the dry bulb scale and extend this point to the mixture line between the two original points, point C.

5. Extend point C to the wet bulb scale and read 55.5°F wet bulb temperature of the mixture.

It should be noted that it is not correct to attempt to determine the wet bulb temperature by using the wet bulb temperatures of the air because the wrong temperature will always be indicated. Use only the dry bulb temperatures when using this method.

Another method that can be used to calculate the wet bulb temperature of the resulting mixture is by indirect application. Using this method, it is necessary to first find the total heat for each of the conditions of the air to be mixed. Then apply the percentage method to calculate the total heat content of the mixture. From the total heat of the mixture it is then possible to determine the corresponding correct wet bulb temperature.

Example 11

The total heat at 72°F wet bulb temperature is 35.85 Btu; the total heat at 49.5°F wet bulb is 20 Btu. What is the correct wet bulb temperature?

Solution

25% of 35.85 = 8.96 Btu
75% of 20 = 15 Btu
Total heat of the mixture = 23.96 Btu.

The corresponding wet bulb temperature is 56°F.

Apparatus Dew Point and Air Quantity

The point on the psychrometric chart where the sensible heat ratio line and the saturation curve intersect is known as the *apparatus dew point* (ADP). This point is representative of the lowest temperature that the air can be supplied to the building and still gain the required amount of sensible and latent heat during the process. If the supply air has a higher ADP the amount of air could be adjusted so that it would pick up the desired amount of sensible heat. However, this air flow may not pick up the desired amount of latent heat. As a result, the indoor relative humidity would be too high. If the ADP was too low, the relative humidity would be lower than desired. However, if supply air at any condition falls on the sensible heat ratio line between point (A) and the ADP point (B), the amount of air can be adjusted to pick up the exact amount of both sensible and latent heat from the air (Figure 1-28).

In actuality, the ADP temperature will be very close to the temperature of the cooling coil surface. Thus, an extremely large coil would be

Figure 1-28. Psychrometric chart showing ADP.

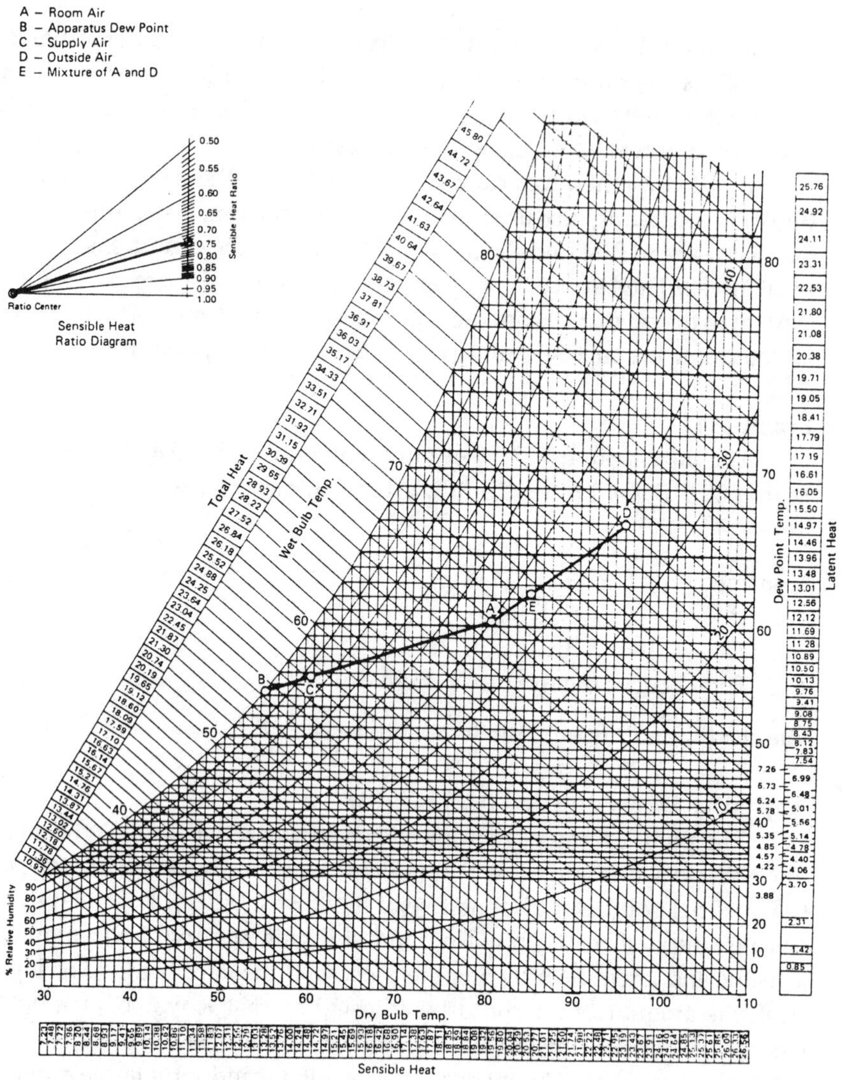

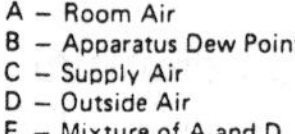

needed to supply air at the ADP temperature. As the return air enters the cooling coil it will be at its highest temperature and the maximum temperature differential between the coil surface and the air temperature will exist. As the air flows through the coil its temperature drops and the temperature differential between the air and the coil also drops. The first row of tubing in the coil will do most of the cooling. Each following row will do progressively less work toward cooling the air. It can be seen that to cool the air to the ADP temperature an extremely large number of tubes would be needed and the last few tubes would do little toward cooling the air because of the smaller temperature differential.

In most instances, the number of rows of tubing in a coil falls between a minimum of three and a maximum of eight. If the coil is to remove mostly sensible heat, fewer rows can be used. Also, when most of the heat to be removed is latent, a larger number of rows are required.

SUMMARY

The term air conditioning is defined as: The simultaneous mechanical control of temperature, humidity, air purity, and air motion. Unless all of these conditions are controlled, the term air conditioning cannot be properly applied to the system.

The two primary reasons for using air conditioning systems are: (1) to improve control of an industrial process, and (2) to maintain human comfort inside a building.

There are three ways that a body gives up its heat to an air conditioning system: (1) convection, (2) radiation, and (3) evaporation. All three of these methods are used at the same time, in most cases.

There is no set rule as to the best comfort conditions for all people. The atmospheric conditions that are comfortable for a young person may be absolutely unsuitable for an older person, or someone who is ill or in bad health.

There are three conditions that affect the ability of the body to give up its heat. They are: (1) temperature, (2) relative humidity, and (3) air motion. A change in either one of these three conditions will either speed up or slow down the cooling process of the body.

Air at a temperature lower than the temperature of the skin will speed up the convection process.

Relative humidity regulates the amount of body heat the body can give up through the radiation process.

The evaporation of perspiration from the skin is increased with an increase in the air movement. The evaporation of moisture from the skin is dependent on the ability of the air to absorb that moisture.

Air that contains no water vapor is considered to be dry air.

Absolute humidity is the actual quantity or weight of moisture that is present in the form of water vapor in the air sample.

Relative humidity is the ratio between the moisture content of a given quantity of air at any given set of conditions when compared to the maximum amount of moisture that the same quantity of air could hold under the same conditions.

Humidification is the process of adding moisture to air.

Dehumidification is the process of removing moisture from a quantity of air.

Air that contains all of the moisture that it is capable of holding at that pressure and temperature is called saturated air. The temperature read on an ordinary thermometer is known as dry bulb temperature. It is a measure of the sensible heat contained in the air.

The temperature of the air that is measured with a thermometer with the bulb covered with a wet sock is known as the wet bulb temperature.

The difference between the dry bulb temperature and the wet bulb temperature is known as the wet bulb depression.

The temperature at which the moisture will start to condense out of the air is known as the dew point temperature.

Air that is at its dew point temperature is also at its saturated temperature and is considered to be saturated with moisture.

Total heat is the sum of the latent and the sensible heat of an object or substance.

When a gas or vapor is at its boiling temperature which corresponds to its pressure, it is also at its saturation temperature.

The term pound of dry air refers to 1 lb of dry air and does not refer to a point of the air-moisture mixture.

The term cubic foot of air indicates 1 cubic foot of air and is expressing the amount of moisture present in 1 cubic foot of dry air when it is saturated.

Ventilation is the process of supplying or removing air from a space. It does not need to be conditioned air; it can be used air or fresh air.

The effective temperature is not measured on a thermometer. It is the personal comfort as felt by an individual.

The range of temperature and humidity in which most people feel comfortable is known as the comfort zone.

The amount of heat in Btu that is transmitted each hour through 1 square foot of any material, or combination of materials, for each degree of temperature difference between the two sides of the material is known as the overall coefficient of heat transmission.

The pressure that a gas or air exerts at right angles against the walls of a duct is known as static pressure.

The pressure inside a duct that is caused by the velocity of the air is known as velocity pressure.

The sum of static pressure and velocity pressure inside a duct system is known as the total static pressure on the duct system.

The upward draft or flow that is created by the tendency of heated air to rise in a vertical or inclined duct is known as stack effect.

The plenum chamber is an equalizing chamber that may be used for the supply and/or return air compartment to which all the different ducts are connected in an air conditioning or ventilation system.

Psychrometrics is the science that deals with the relationship that exists in a mixture of water vapor and air.

Heat that can be added to or removed from a substance without causing a change of state is known as the sensible heat of that substance.

An indication of the moisture content of a given quantity of air is the dew point temperature. When there is a change in the dew point temperature there will always be a change in the moisture content of the mixture.

The amount of sensible heat removed in Btu divided by the total heat removed from the air is known as the sensible heat ratio. The sensible heat ratio is an indicator of the percentage of sensible heat removed compared to the total heat removed.

The apparatus dew point is the point on the psychrometric chart where the sensible heat ratio line and the saturation curve intersect. It is representative of the lowest temperature at which the air can be supplied to the structure and still pick up the required amount of both sensible heat and latent heat.

Chapter 2

Heat Load Calculation Factors

When estimating the heat loss and heat gain of a building or a space it is necessary to have a working knowledge of the different sources of these factors. Each source of heat gain or loss must be considered separately and must be taken into account when estimating the size of a heating and/or cooling unit for the building.

INTRODUCTION

The purpose of an air conditioning system is to maintain the indoor air conditions at the desired level of temperature, circulation, cleanliness, and humidity. Such conditions are those required for (1) human comfort, (2) a product, or (3) a process that is performed in the conditioned space. To do this task, the equipment must be sized, chosen, installed, and controlled all year long. Usually the equipment is sized according to the peak load requirements. However, it is almost impossible to accurately measure the actual peak load or a partial load for any given space. Thus, these loads are said to be estimated and are termed the heat load estimate.

SURVEY

When making preparations to complete a heat load estimate, the first step is to make a comprehensive survey, including the building and the immediate surroundings to obtain accurate factors on which to make the estimation. When all the factors, including the actual instantaneous heat load through any given part of the building, are carefully weighed, the equipment can be properly selected. The proper selection of the equipment will help in providing an efficient and trouble-free installation. It is suggested that some type of survey and checklist be used so that all the factors will have been obtained. See Figure 2-1.

HEATING AND COOLING ESTIMATE
COMMERCIAL

Customer ____________
Address ____________ Town ____________
Prepared by ____________ Date ____________

DESIGN CONDITIONS		INSULATION	STRUCTURE FACES	OVERHANG OR SHADING
COOLING D.B. °F	HEATING D.B. °F	Ceiling ____	□ North ____	
Outside Temp. ____	Inside Temp. ____	Wall ____	□ East ____	
Inside Temp. ____	Outside Temp. ____	Floor ____	□ South ____	
Temp. Diff. ____	Temp. Diff. ____	Window ____	□ West ____	

CONSTRUCTION		COOLING FACTOR			HEATING FACTOR	QUANTITY	COOLING Btuh	HEATING Btuh
1. GLASS--SOLARGAIN Windows Doors Skylights (Sq.Ft.) (Note A)	SC	No Shading	Inside Shading & Overhang 0' 1' 2' 3' 4'	Outside Shades				
North Northeast		32 50	14 0 0 0 0 23 21 19 19 17	0 17				
East Southeast		50 80	23 20 19 17 14 36 26 19 8 0	23 21				
South Southwest		100 155	45 33 10 0 0 70 51 39 17 0	26 42				
West Northwest		190 128	86 82 80 67 57 58 54 51 51 49	57 37				
Horizontal		185						
DESIGN D. B. TEMP. DIFFERENTIAL		U	20 22 25	30	65 60 55 50			
2. GLASS TRANSMISSION (Sq. Ft.)			U X TD		U X TD			
Standard--Single Glazing Insulating--Double Glazing		1.13 .78	22.6 24.9 28.2 15.6 17.2 19.5	33.9 23.4	73.4 67.8 62.2 56.5 50.7 46.8 42.9 39.0			
Storm or Insulating Glass W/T Bk.		.56	11.2 12.3 14.0	16.8	36.4 33.6 30.8 28.0			
3. DOORS--(Sq.Ft.)			U X TD		U x TD			
Solid Wood, Hollow Core or Metal		.55	11.0 12.1 13.8	16.5	35.8 33.0 30.3 27.5			
4a. FLOOR SLAB (Linear Ft. Exp. Edge)			U X Zero		U X TD			
No Edge Insulation R-4 Edge Insulation		.81 .68			56.6 48.6 44.6 40.5 44.2 40.8 37.4 34.0			

Figure 2-1. Commercial Heating and Cooling Estimate Form. (Courtesy of TU Electric Co.)

4b. FLOORS: ENC. CRAWL SPACE (Sq. Ft.)(*Note B)		U X Zero				U X (TD X 20)						
No Insulation	.270					12.2	10.8	9.4	8.1			
R-7 Insulation	.093					4.2	3.7	3.2	2.8			
R-13 Insulation	.060					2.7	2.4	2.1	1.8			
5. WALLS: (Sq. Ft.) Net		U X TD				*U X TD						
No Insulation--Masonry	.389	7.8	8.6	9.7	11.7	25.3	23.3	21.4	19.5			
No Insulation---Wood Siding	.320	6.4	7.0	8.0	9.6	20.8	19.2	17.6	16.0			
No Insulation---Brick Veneer	.240	4.8	5.3	6.0	7.2	15.6	14.4	13.2	12.0			
R-4 Insulation	.138	2.8	3.0	3.4	4.1	8.9	8.3	7.6	6.9			
R-7 Insulation	.109	2.2	2.4	2.7	3.3	7.1	6.6	6.0	5.5			
R-11 Insulation	.075	1.5	1.7	1.9	2.3	4.9	4.5	4.1	3.8			
R-13 Insulation	.065	1.3	1.4	1.6	2.0	4.2	3.9	3.6	3.3			
6a. CEILINGS-W/Attic (Sq. Ft.)		U X (TD x 40)				U X TD						
No Insulation	.598	35.9	37.1	38.9	41.9	38.9	35.9	32.9	29.9			
R-4 Insulation	.176	10.6	10.9	11.4	12.3	11.4	10.6	9.7	8.8			
R-7 Insulation	.114	6.8	7.1	7.4	8.0	7.4	6.8	6.3	5.7			
R-11 Insulation	.079	4.7	4.9	5.1	5.5	5.1	4.7	4.3	4.0			
R-13 Insulation	.068	4.1	4.2	4.4	4.8	4.4	4.1	3.7	3.4			
R-19 Insulation	.048	2.9	3.0	3.1	3.4	3.1	2.9	2.6	2.4			
R-22 Insulation	.042	2.5	2.6	2.7	2.9	2.7	2.5	2.3	2.1			
R-26 Insulation	.036	2.2	2.2	2.3	2.5	2.3	2.2	2.0	1.8			
6b. Ceilings--Attic (Sq. Ft.)(*N.B)		U X (TD X 45)				U X TD						
No Insulation	.470	30.6	31.5	32.9	35.3	30.6	28.2	25.9	23.5			
R-4 Insulation	.160	10.4	10.7	11.2	12.0	10.4	9.6	8.8	8.0			
R-7 Insulation	.109	7.1	7.3	7.6	8.2	7.1	6.5	6.0	5.5			
R-11 Insulation	.076	4.9	5.1	5.3	5.7	4.9	8.6	4.2	3.8			
R-13 Insulation	.064	4.2	4.3	4.5	4.8	4.2	3.8	3.5	3.2			
R-19 Insulation	.047	3.1	3.1	3.3	3.5	3.1	2.8	2.6	2.4			
R-22 Insulation	.041	2.7	2.7	2.9	3.0	2.7	2.5	2.3	2.0			
7. VENTILATION OR INFILTRATION (*Note C) CFM		22	24	28	33	72	66	61	55			
8. SENSIBLE HEAT Btuh-Sub-total		Line 1 thru 7										

(Continued)

Figure 2-1. (*Concluded*)

CONSTRUCTION	COOLING FACTOR	HEATING FACTOR	QUANTITY	COOLING Btuh	HEATING Btuh
8. SENSIBLE HEAT Btuh-Sub-Total	Line 8 Reverse Side				
9. SENSIBLE HEAT Btu W/DUCT LOSS Sub-Total	Line 8 x Factor From Table 3				
10. VENTILATION OR INFILTRATION LATENT HEAT GAIN-EXTERIOR (*Note C)	23				
11. TOTAL HEAT Btuh W/OUT INTERNAL LOAD	LINE 9 + 10				
12. INTERIOR SENSIBLE HEAT GAIN					
A. People-Ave. Number (*Table 2)					
B. Lighting-kW	3413				
C. Motors- HP	2550				
D. Other Internal Load-kW	3413				
E. Gas Cooking (*Note D)	5880				
F. Electric Cooking (*Note D)	2400				
13. INTERIOR SENSIBLE HEAT GAIN Sub-Total	Line A thru F				
14. INTERIOR SENSIBLE HEAT GAIN W/DUCT LOSS Sub-Total	Line 13 X Factor From Table 3				
15. LATENT HEAT GAIN--INTERIOR					
A. People-Ave. (*Table 2)					
B. Other Interior Load					
C. Gas Cooking-Number of Burners	2520				
D. Electric Cooking (*Note D)	1000				
16. LATENT HEAT GAIN INTERIOR Sub-Total	Line 15 A thru D				
17. TOTAL INTERIOR HEAT GAIN	Line 14 + 16				
18. TOTAL LOAD Btuh	Cooling Line 11 + 17	Heating Line 9 thru 14			

INSTRUCTIONS FOR HEATING & COOLING ESTIMATES - FOR 219C

*Note A - 1. For Solor Gain use only the one direction producing the greatest Btuh.
2. Solor factors may be multiplied by .90 for plate glass, by .85 for double pane glass, by .50 for glass block.
3. No solor load is used if overhang exceeds 4′ per floor for nominal windows.
4. Windows shaded all day by buildings or other permanent objects should be included under window transmission only.

*Note B - For light colored walls or roofs, COOLING factor may be multiplied by .75.

*Note C - 1. Infiltration formula: Volume = CFM
2. For extra heavy people loads, calculate ventilation CFM requirements from Table 1. Use this CFM quantity in place of infiltration CFM above, if it is greater.
3. If the quantity of ventilation air is stated by owner, architect or engineer, use the amount they specify.

*Note D - Use 50% of nameplate kW for quantity and multiply by sensible and latent factors.

TABLE 1 - Ventilation CM Per Person

No Smoking	20
Some Smoking	40
Heavy Smoking	60

TABLE 2 - People - Btuh Per Person

	Sensible	Latent
Inactive	200	250
Moderate Activity	300	500
Active	400	800

TABLE 3 - Duct Loss Factors

Ductwork Location	Inches of Duct Insulation	Duct Factor Cooling	Duct Factor Heating
Attic - Vented	1	1.15	1.25
- Vented	2	1.10	1.15
- Unvented	1	1.20	1.20
- Unvented	2	1.15	1.10
Crawl Space - Unvented	1	1.05	1.10
Within Conditioned Area	1	1.00	1.00
Within Slab	1	1.20	1.25

TABLE 4 - Equipment Efficiency

Resistance Heat: 3,413 MBtu/kWh
Heat Pump: EER/SEER - Cooling
HSPF - Heating
Gas Furnace: AFUE x 10

Fuel Cost:
Electricity: Ave. kWh $________
Gas: Ave. MCF $________

1 MBtu = 1,000 Btu

HEAT SOURCES

Heat sources, generally, are considered to be of two types: (1) those resulting in an internal heat load on the conditioned space, and (2) those that result as an external heat load. Those listed in the external heat load category are a source of load on the indoor equipment but they do not affect the air after it has passed through the equipment. The listing of heat sources is as follows:

1. Heat sources resulting in an internal heat load:
   a. Heat conduction through the walls, glass, roof, etc.
   b. Excess sun heat load
   c. Duct heat gain
   d. The occupants
   e. The lighting load
   f. The equipment and other appliances
   g. The infiltration of out-of-doors air
2. Heat sources resulting in an external heat load:
   a. All ventilation air
   b. Heat from any source added to the air after it leaves a space

Any heat that enters a space could possibly come from any one or all of the heat sources listed above. This is one reason that an accurate heat load survey is so important. It is the purpose of the air conditioning equipment to remove this heat gain or add any heat that may be lost from the space, and to keep the space at the design conditions. The estimator must make certain that the total load used for the calculation is for the peak load and not an accumulation of the peak loads and another less significant heat load period. This is quite easily overlooked because the peak load in one area may not occur at the same time as the peak load in another area. Thus, the larger load can be used for conditioning both areas which occur at different times. As an example, the sun load may be at its peak at noon, and the peak customer load may not be present until later in the evening. If these two peak loads were added together, the total heat load would be too large. Another example is that the lighting heat load for some buildings may be greater than the sun heat load, and they might not occur at the same time during a twenty-four hour period. It can be seen that the heat load on any given space, or building, will vary during the day and also during the period of heavy use. Because of this variation, the heat load is based on 1 hour, which is the peak load for that hour, and not over the complete twenty-four hours. This is the load that the air conditioning system must be able to handle.

CONDUCTION LOAD

The flow of heat is always from a warmer object to a cooler object. Thus, when the indoor temperature is either lower, or higher, than the temperature of the surrounding ambient outdoor air, there will be heat conduction through the walls, windows, roof, and sometimes the floor of the building, to or from the outdoor air. There are four factors that determine the amount of heat that will flow through a part of a building having different temperatures on each side. These factors are:

1. The area of the wall in square feet.
2. The heat-conducting characteristics of the wall. This is called the overall coefficient of heat transmission, commonly referred to as the U factor.
3. The temperature difference between the two sides of the partition.
4. Wind velocity of the outside air.

The U factor is applied to all walls, doors, windows, roofs, and all parts of the building. The following basic formula is used to calculate the heat transfer through the structure.

$$Q = A \times U \times (T_1 - T_2)$$

Where:

Q = heat transmission per hour in Btu

A = area of wall in square feet

U = number of Btu that will flow through one square foot of the material in one hour, with a temperature difference of one degree Fahrenheit between the two sides

T_1 = outdoor temperature

T_2 = indoor temperature (T_1 and T_2 should be reversed when calculating heat gains.)

Many times the temperature differential ($T_1 - T_2$) is given as TD, or ΔT. When used in this manner the formula becomes:

$$Q = A \times U \times \Delta T$$

The U factors for various types of materials have been placed in tabular form and can be used to determine the U factor for many different types of construction (Table 2-1).

Table 2-1. Typical Thermal Properties of Common Building and Insulating Materials—Design Values[a]

(Copyright 1993 by the American Society of Heating, Refrigerating and Air-Conditioning Engineers, Inc., from Fundamentals Handbook. Used by permission.)

Description	Density, lb/ft^3	Conductivity[b] (*k*), Btu • in / h • ft^2 • °F	Conductance (*C*), Btu / h • ft^2 • °F	Resistance[c] (*R*) Per Inch Thickness (1/*k*), °F • ft^2 • h / Btu • in	Resistance[c] (*R*) For Thickness Listed (1/*C*), °F • ft^2 • h / Btu	Specific Heat, Btu / lb • °F
BUILDING BOARD						
Asbestos-cement board	120	4.0	—	0.25	—	0.24
Asbestos-cement board .... 0.125 in.	120	—	33.00	—	0.03	
Asbestos-cement board .... 0.25 in.	120	—	16.50	—	0.06	
Gypsum or plaster board .... 0.375 in.	50	—	3.10	—	0.32	0.26
Gypsum or plaster board .... 0.5 in.	50	—	2.22	—	0.45	
Gypsum or plaster board .... 0.625 in.	50	—	1.78	—	0.56	
Plywood (Douglas Fir)[d]	34	0.80	—	1.25	—	0.29
Plywood (Douglas Fir) .... 0.25 in.	34	—	3.20	—	0.31	
Plywood (Douglas Fir) .... 0.375 in.	34	—	2.13	—	0.47	
Plywood (Douglas Fir) .... 0.5 in.	34	—	1.60	—	0.62	
Plywood (Douglas Fir) .... 0.625 in.	34	—	1.29	—	0.77	
Plywood or wood panels .... 0.75 in.	34	—	1.07	—	0.93	0.29
Vegetable fiber board						
Sheathing, regular density[e] .... 0.5 in.	18	—	0.76	—	1.32	0.31
.... 0.78125 in.	18	—	0.49	—	2.06	
Sheathing intermediate density[c] ... 0.5 in.	22	—	0.92	—	1.09	0.31
Nail-base sheathing′ .... 0.5 in.	25	—	0.94	—	1.06	0.31
Shingle backer .... 0.375 in.	18	—	1.06	—	0.94	0.31
Shingle backer .... 0.3125 in.	18	—	1.28	—	0.78	
Sound deadening board .... 0.5 in.	15	—	0.74	—	1.35	0.30
Tile and lay-in panels, plain or acoustic .	18	0.40	—	2.50	—	0.14
.... 0.5 in.	18	—	0.80	—	1.25	
.... 0.75 in.	18	—	0.53	—	1.89	
Laminated paperboard 30	0.50	—	2.00		0.33	
Homogeneous board from repulped paper	30	0.50	—	2.00	—	0.28

Table 2-1. (Cont'd)

Description	Density, lb/ft³	Conductivity[b] (k), Btu • in / h • ft² • °F	Conductance (C), Btu / h • ft² • °F	Resistance[c] (R) Per Inch Thickness (1/k), °F • ft² • h / Btu • in	Resistance[c] (R) For Thickness Listed (1/C), °F • ft² • h / Btu	Specific Heat, Btu / lb • °F
Hardboard[e]						
Medium density 50	0.73	—	1.37	—	0 31	
High density, service-tempered grade and service grade 55	0.82	—	1.22	—	0.32	
High density, standard-tempered grade63	1.00	—	1.00	—	0.32	
Particleboard[e]						
Low density 37	0.71	—	1.41	—	0.31	
Medium density 50	0.94	—	1.06	—	0.31	
High-density 62.5	1.18	—	0.85	—	0.31	
Underlayment ....0.625 in.	40	—	1.22	—	0.82	0.29
Waferboard 37	0.63	—	1.59	—	—	—
Wood subfloor ....0.75 in.	—	—	1.06	—	0.94	0.33
BUILDING MEMBRANE						
Vapor—permeable felt	—	—	16.70	—	0.06	
Vapor—seal, 2 layers of mopped 15-lb felt ..	—	—	8.35	—	0.12	
Vapor—seal, plastic film	—	—	—	—	Negl.	
FINISH FLOORING MATERIALS						
Carpet and fibrous pad	—	—	0.48	—	2.08	0.34
Carpet and rubber pad	—	—	0.81	—	1.23	0.33
Cork tile ....0.125 in.	—	—	3.60	—	0.28	0.48
Terrazzo ....I in.	—	—	12.50	—	0.08	0.19
Tile—asphalt, linoleum, vinyl, rubber	—	—	20.00	—	0.05	0.30
vinyl asbestos						0.24
ceramic						0.19
Wood, hardwood finish ....0.75 in.	—	—	1.47	—	0.68	

Table 2-1. (*Cont'd*)

Description	Density, lb/ft³	Conductivity[b] (*k*), Btu • in / h • ft² • °F	Conductance (*C*), Btu / h • ft² • °F	Resistance[c] (*R*) Per Inch Thickness (1/*k*), °F • ft² • h / Btu • in	Resistance[c] (*R*) For Thickness Listed (1/*C*), °F • ft² • h / Btu	Specific Heat, Btu / lb • °F
INSULATING MATERIALS						
Blanket and Batt[f, g]						
Mineral fiber, fibrous form processed from rock, slag, or glass						
approx. 3-4 in.	0.4-2.0	—	0.091	—	11	
approx. 3.5 in.	0.4-2.0	—	0.077	—	13	
approx. 3.5 in.	1.2-1.6	—	0.067	—	15	
approx. 5.5-6.5 in.	0.4-2.0	—	0.053	—	19	
approx. 5.5 in.	0.6-1.0	—	0.048	—	21	
approx. 6-7.5 in.	0.4-2.0	—	0.045	—	22	
approx. 8.25-10 in.	0.4-2.0	—	0.033	—	30	
approx. 10-13 in.	0.4-2.0	—	0.026	—	38	
Board and Slabs						
Cellular glass	8.0	0.33	—	3.03	—	0.18
Glass fiber, organic bonded	4.0-9.0	0.25	—	4.00	—	0.23
Expanded perlite, organic bonded	1.0	0.36	—	2.78	—	0.30
Expanded rubber (rigid)	4.5	0.22	—	4.55	—	0.40
(smooth skin surface) (CFC-12 exp.)	1.8-3.5	0.20	—	5.00	—	0.29
Expanded polystyrene, extruded (smooth skin surface) (HCFC-142b exp.)[h]	1.8-3.5	0.20	—	5.00	—	0.29
Expanded polystyrene, molded beads	1.0	0.26	—	3.85	—	—
	1.25	0.25	—	4.00	—	—
	1.5	0.24	—	4.17	—	—
	1.75	0.24	—	4.17	—	—
	2.0	0.23	—	4.35	—	—

Table 2-1. (*Cont'd*)

Description	Density, lb/ft³	Conductivity[b] (k), Btu • in / h • ft² • °F	Conductance (C), Btu / h • ft² • °F	Resistance[c] (R): Per Inch Thickness ($1/k$), °F • ft² • h / Btu • in	Resistance[c] (R): For Thickness Listed ($1/C$), °F • ft² • h / Btu	Specific Heat, Btu / lb • °F
Cellular polyurethane/polyisocyanurate[i] (CFC-11 exp.) (unfaced)	1.5	0.16-0.18	—	6.25-5.56	—	0.38
Cellular polyisocyanurate[i] (CFC-11 exp.)(gas-permeable facers)	1.5-2.5	0.16-0.18	—	6.25-5.56	—	0.22
Cellular polyisocyanurate[j] (CFC-11 exp.)(gas-impermeable facers)	2.0	0.14	—	7.04	—	0.22
Cellularphenolic (closed cell)(CFC-11, CFC-113 exp.)	3.0	0.12	—	8.20	—	—
Cellular phenolic (open cell)	1.8-2.2	0.23	—	4.40	—	—
Mineral fiber with resin binder	15.0	0.29	—	3.45	—	0.17
Mineral fiberboard, wet felted						
Core or roof insulation	16-17	0.34	—	2.94	—	—
Acoustical tile	18.0	0.35	—	2.86	—	0.19
Acoustical tile	21.0	0.37	—	2.70	—	—
Mineral fiberboard, wet molded Acoustical tile[k]	23.0	0.42	—	2.38	—	0.14
Wood or cane fiberboard						
Acoustical tile[k] 0.5 in.	—	—	0.80	—	1.25	0.31
Acoustical tile[k] 0.75 in.	—	—	0.53	—	1.89	
Interior finish (plank, tile)	15.0	0.35	—	2.86	—	0.32
Cement fiber slabs (shredded wood with Portland cement binder)	25-27.0	0.50-0.53	—	2.0-1.89	—	—
Cement fiber slabs (shredded wood with magnesia oxysulfide binder)	22.0	0.57	—	1.75	—	0.31

Table 2-1. (*Cont'd*)

Description	Density, lb/ft³	Conductivity[b] (k), Btu • in / h • ft² • °F	Conductance (C), Btu / h • ft² • °F	Resistance[c] (R) Per Inch Thickness (1/k), °F • ft² • h / Btu • in	For Thickness Listed (1/C), °F • ft² • h / Btu	Specific Heat, Btu / lb • °F
Loose Fill						
Cellulosic insulation (milled paper or wood pulp)	2.3-3.2	0.27-0.32	—	3.70-3.13	—	0.33
Perlite, expanded	2.0-4.1	0.27-0.31	—	3.7-3.3	—	0.26
	4.1-7.4	0.31-0.36	—	3.3 2.8	—	—
	7.4-1 1.0	0.36-0.42	—	2.8-2.4	—	—
Mineral fiber (rock, slag, or glass)[g]						
approx. 3.75-5 in	0.6-2.0	—	—	—	11.0	0.17
approx. 6.5-8.75 in	0.6-2.0	—		—	19.0	
approx. 7.5-10 in	0.6-2.0	—	—	—	22.0	
approx. 10.25-13.75 in	0.6-2.0	—	—	—	30.0	
Mineral fiber (rock, slag, or glass)[g] approx. 3.5 in. (closed sidewall application)	2.0-3.5	—	—	—	12.0-14.0	—
Vermiculite, exfoliated	7.0-8.2	0.47	—	2.13	—	0.32
	4.0-6.0	0.44	—	2.27	—	—
Spray Applied						
Polyurethane foam	1.5-2.5	0.16-0.18	—	6.25-5.56	—	—
Ureaformaldehyde foam	0.7-1.6	0.22-0.28	—	4.55-3.57	—	—
Cellulosic fiber	3.5-6.0	0.29-0.34	—	3.45-2.94	—	—
Glass fiber	3.5-4.5	0.26-0.27	—	3.85-3.70	—	—
METALS (See Chapter 36, Table 3)						

Table 2-1. (*Cont'd*)

Description	Density, lb/ft³	Conductivity[b] (*k*), Btu • in / h • ft² • °F	Conductance (*C*), Btu / h • ft² • °F	Resistance[c] (*R*) Per Inch Thickness (1/*k*), °F • ft² • h / Btu • in	Resistance[c] (*R*) For Thickness Listed (1/*C*), °F • ft² • h / Btu	Specific Heat, Btu / lb • °F
ROOFING						
Asbestos-cement shingles	120	—	4.76	—	0.21	0.24
Asphalt roll roofing	70	—	6.50	—	0.15	0.36
Asphalt shingles	70	—	2.27	—	0.44	0.30
Built-up roofing ....0.375 in.	70	—	3.00	—	0.33	0.35
Slate ....0.5 in.	—	—	20.00	—	0.05	0.30
Wood shingles, plain and plastic film faced	—	—	1.06	—	0.94	0.31
PLASTERING MATERIALS						
Cement plaster, sand aggregate	116	5.0	—	0.20	—	0.20
Sand aggregate ....0.375 in.	—	—	13.3	—	0.08	0.20
Sand aggregate ....0.75 in.	—	—	6.66	—	0.15	0.20
Gypsum plaster:						
Lightweight aggregate ....0.5 in.	45	—	3.12	—	0.32	—
Lightweight aggregate ....0.625 in.	45	—	2.67	—	0.39	—
Lightweight aggregate on metal lath ....0.75 in.	—	—	2.13	—	0.47	—
Perlite aggregate	45	1.5	—	0.67	—	0.32
Sand aggregate	105	5.6	—	0.18	—	0.20
Sand aggregate ....0.5 in.	105	—	11.10	—	0.09	—
Sand aggregate ....0.625 in.	105	—	9.10	—	0.11	—
Sand aggregate on metal lath ....0.75 in.	—	—	7.70	—	0.13	—
Vermiculite aggregate	45	1 7	—	0 59	—	—
MASONRY MATERIALS						
Masonry Units						
Brick, fired clay	150	8.4-10.2	—	0.12-0.10	—	—
	140	7.4-9.0	—	0. 14-0. 11	—	—

Table 2-1. (*Cont'd*)

Description	Density, lb/ft³	Conductivity[b] (*k*), Btu • in / h • ft² • °F	Conductance (*C*), Btu / h • ft² • °F	Resistance[c] (*R*) Per Inch Thickness (1/*k*), °F • ft² • h / Btu • in	Resistance[c] (*R*) For Thickness Listed (1/*C*), °F • ft² • h / Btu	Specific Heat, Btu / lb • °F
	130	6.4-7.8	—	0.16-0.12	—	—
	120	5.6 6.8	—	0.18-0.15	—	0.19
	110	4.9-5.9	—	0.20-0.17	—	—
Brick, fired clay *continued*	100	4.2-5.1	—	0.24-0.20	—	—
....	90	3.6-4.3	—	0.28-0.24	—	—
....	80	3.0-3.7	—	0.33-0.27	—	—
....	70	2.5-3.1	—	0.40-0.33	—	—
Clay tile, hollow		0.40 0.33				
1 cell deep ....3 in.	—	—	1.25	—	0.80	0.21
1 cell deep ....4 in.	—	—	0.90	—	1.11	—
2 cells deep..6 in	—	—	0.66	—	1.52	—
2 cells deep..8 in.	—	—	0.54	—	1.85	—
2 cells deep..10 in	—	—	0.45	—	2.22	—
3 cells deep..12 in	—	—	0.40	—	2.50	—
Concrete blocks[1]						
Limestone aggregate						
8 in., 36 lb, 138 lb/ft³ concrete, 2 cores	—	—	—	—	—	—
Same with perlite filled cores	—	—	0.48	—	2.1	—
12 in., 55 lb, 138 lb/ft³ concrete, 2 cores	—	—	—	—	—	—
Same with perlite filled cores	—	—	0.27	—	3. 7	—
Normal weight aggregate (sand and gravel)						
8 in., 33-36 lb, 126-136 lb/ft³ concrete, 2 or 3 cores	—	—	0.90-1.03	—	1.11-0.97	0.22

Table 2-1. (*Cont'd*)

Description	Density, lb/ft³	Conductivity[b] (*k*), Btu • in / h • ft² • °F	Conductance (C), Btu / h • ft² • °F	Resistance[c] (*R*) Per Inch Thickness (1/*k*), °F • ft² • h / Btu • in	Resistance[c] (*R*) For Thickness Listed (1/C), °F • ft² • h / Btu	Specific Heat, Btu / lb • °F
Same with perlite filled cores	—	—	0.50	—	2.0	—
Same with verm. filled cores	—	—	0.52-0.73	—	1.92-1.37	—
12 in., 50 lb, 125 lb/ft³ concrete, 2 cores	—	—	0.81	—	1.23	0.22
Medium weight aggregate (combinations of normal weight and lightweight aggregate)						
8 in., 26-29 lb, 97-112 lb/ft³ concrete, 2 or 3 cores	—	—	0.58-0.78	—	1.71-1.28	—
Same with perlite filled cores	—	—	0.27-0.44	—	3.7-2.3	—
Same with verm. filled cores	—	—	0.30		3.3	—
Same with molded EPS (beads) filled cores	—	—	0.32	—	3.2	—
Same with molded LPS inserts in cores	—	—	0.37	—	2.7	—
Lightweight aggregate (expanded shale, clay, slate or slag, pumice)						
6 in., 16-17 lb 85-87 lb/ft³ concrete, 2 or 3 cores	—	—	0.52-0.61	—	1.93-1.65	—
Same with perlite filled cores	—	—	0.24	—	4.2	—
Same with verm filled cores	—	—	0.33	—	3 0	—
8 in., 19-22 lb, 72-86 lb/ft³ concrete,	—	—	0.32-0.54	—	3.2-1.90	0.21
Same with perlite filled cores	—	—	0.15-0.23	—	6.8-4.4	—
Same with verm. filled cores	—	—	0.19-0.26	—	5.3-3.9	—
Same with molded EPS (beads) filled cores	—	—	0.21	—	4.8	—
Same with UF foam filled cores	—	—	0.22	—	4.5 -	—
Same with molded EPS inserts in cores	—	—	0.29	—	3.5	—
12 in., 32-36 lb, 80-90 lb/ft³ concrete, 2 or 3 cores	—	—	0.38-0.44	—	2.6-2.3	—
Same with perlite filled cores	—	—	0.11-0.16	—	9.2-6.3	—

Table 2-1. ***(Cont'd)***

Description	Density, lb/ft³	Conductivity[b] (k), Btu • in / h • ft² • °F	Conductance (C), Btu / h • ft² • °F	Resistance[c] (R) Per Inch Thickness (1/k), °F • ft² • h / Btu • in	Resistance[c] (R) For Thickness Listed (1/C), °F • ft² • h / Btu	Specific Heat, Btu / lb • °F
Same with verm. filled cores	—	—	0.17	—	5.8	—
Stone, lime, or sand						
Quartzitic and sandstone	180	72	—	0.01	—	—
	160	43	—	0.02	—	—
	140	24	—	0.04	—	—
	120	13	—	0.08	—	0.19
Calcitic, dolomitic, limestone, marble, and granite	180	30	—	0.03	—	—
	160	22	—	0.05	—	—
	140	16	—	0.06	—	—
	120	11	—	0.09	—	0.19
	100	8	—	0.13	—	—
Gypsum partition tile						
3 by 12 by 30 in., solid	—	—	0.79	—	1.26	0.19
3 by 12 by 30 in., 4 cells	—	—	0.74	—	1.35	—
4 by 12 by 30 in., 3 cell.	—	—	0.60	—	1.67	—
Concretes						
Sand and gravel or stone aggregate concretes	150	10.0-20.0	—	0.10-0.05	—	0.19-0.24
(concretes with more than 50% quartz	140	9.0- 10.0	—	0.11-0.06	—	0.19-0.24
or quartzite sand have conductivities	130	7.0-13.0	—	0.14-0.08	—	—
in the higher end of the range)	140	11.1	—	0.09	—	—
Limestone concretes	120	7.9	—	0.13	—	—
	100	5.5	—	0.18	—	—
Gypsum-fiber concrete (87.5% gypsum, 12.5%	51	1.66	—	0.60	—	0.21
wood chips)	120	9.7	—	0.10	—	—
Cement/lime, mortar, and stucco	100	6.7	—	0.15	—	—
	80	4.5	—	0-.22	—	—

Table 2-1. (*Cont'd*)

Description	Density, lb/ft³	Conductivity[b] (*k*), Btu • in / h • ft² • °F	Conductance (C), Btu / h • ft² • °F	Resistance[c] (*R*) Per Inch Thickness (1/*k*), °F • ft² • h / Btu • in	Resistance[c] (*R*) For Thickness Listed (1/C), °F • ft² • h / Btu	Specific Heat, Btu / lb • °F
Lightweight aggregate concretes						
Expanded shale, clay, or slate; expanded	1206.4-9.1	—	0.16-0.11	—	—	
slags; cinders; pumice (with density up to	100	4.7 6.2	—	0.21-0.16	—	0.20
100 lb/ft³); and scoria) (sanded concretes	80	3.3-4.1	—	0.30-0.24	—	0.20
have conductivities in the higher	40	1.13	—	0.78	—	—
end of the range)						
Perlite, vermiculite, and polystyrene						
beads	50	1.8-1.9	—	0.55-0.53	—	—
	40	1.4-1.5	—	0.71-0.67	—	0.15-0.23
	30	1.1	—	0.91	—	—
	20	0.8	—	1.25	—	—
Foam concretes	120	5.4	—	0 19	—	—
	100	4.1	—	0.24	—	—
	80	30	—	033	—	—
	70	2.5	—	0.40	—	—
Foam concretes and cellular concretes	60	2.1	—	0 48	—	—
	40	1.4	—	0.71	—	—
	20	0.8	—	1.25	—	—
SIDING MATERIALS (on flat surface)						
Shingles						
Asbestos-cement.	120	—	4.75	—	0.21	
Wood, 16 in., 7.5 exposure	—	—	1.15	—	0.87	0.31
Wood, double, 16-in., 12-in. exposure.....	—	—	0.84	—	1.19	0 28
Wood, plus insul. backer board, 0.3125 in	—	—	0.71	—	1.40	0 31
Siding						
Asbestos-cement, 0.25 in., lapped.	—	—	4.76	—	0.21	0.24
Asphalt roll siding.	—	—	6.50	—	0.15	0.35

Table 2-1. *(Cont'd)*

Description	Density, lb/ft³	Conductivity[b] (*k*), Btu • in / h • ft² • °F	Conductance (*C*), Btu / h • ft² • °F	Resistance[c] (*R*) Per Inch Thickness (1/*k*), °F • ft² • h / Btu • in	Resistance[c] (*R*) For Thickness Listed (1/*C*), °F • ft² • h / Btu	Specific Heat, Btu / lb • °F
Asphalt Insulating siding (0.5 in. bed.)	—	—	0.69	—	1.46	0.35
Hardboard siding, 0.4375 in	—	—	1.49	—	0.67	0.28
Wood, drop, I by 8 in	—	—	1.27	—	0.79	0.28
Wood, bevel, 0.5 by 8 in., lapped	—	—	1.23	—	0.81	0.28
Wood, bevel, 0.75 by 10 in., lapped	—	—	0.95	—	1.05	0.28
Wood, plywood, 0.375 in., lapped	—	—	1.59	—	0.59	0.29
Aluminum or Steel, over sheathing						
Hollow-backed	—	—	1.61	—	0.61	0.29
Insulating-board backed nominal 0.375 in.	—	—	0.55	—	1.82	0.32
Insulating-board backed nominal 0.375 in., foil backed	—	—	0.34	—	2.96	
Architectural (soda-lime float) glass	158	6.9	—	—	—	0.21
WOODS (12% moisture content)[e, n]						
Hard woods						0.39°
Oak.	41.2-46.8	1.12-1.25	—	0.89-0.80	—	
Birch	42.6-45.4	1.16-1.22	—	0.87-0.82	—	
Maple	39.8-44.0	1.09-1.19	—	0.92-0.84	—	
Ash	3 8.4-41.9	1. 06-1.14	—	0.94-0.88	—	
Softwoods						0.39°
Southern Pine	35.6-41.2	1.00-1.12	—	1.00-0.89	—	
Douglas Fir-Larch	33.5-36.3	0.95-1.01	—	1.06-0.99	—	
Southern Cypress	31. 4-32.1	0. 90-0.92	—	1.11-1.09	—	
Hem-Fir, Spruce-Pine-Fir	24.5-31.4	0.74-0.90	—	1.35-1.11	—	
West Coast Woods, Cedars	21.7-31.4	0.68-0.90	—	1.48-1.11	—	
California Redwood	24.5-28.0	0.74-0.82	—	1.35-1 22		

[a]Values are for a mean temperature of 75°F Representative values for dry materials are intended as design (not specification) values for materials in normal use. Thermal values of insulating materials may differ from design values depending on their in-situ properties (e.g., density and moisture content, orientation, etc.) and variability experienced during manufacture. For properties of a particular product, use the value supplied by the manufacturer or by unbiased tests.

[b]To obtain thermal conductivities in Btu/h•ft•°F, divide the k-factor by 12 in./ft.

[c]Resistance values are the reciprocals of C before rounding off C to two decimal places.

[d]Lewis (1967).

[e]U.S. Department of Agriculture (1974).

[f]Does not include paper backing and facing, if any. Where insulation forms a boundary (reflective or otherwise) of an airspace, see Tables 2-2 and 2-3 for the insulating value of an airspace with the appropriate effective emittance and temperature conditions of the space.

[g]Conductivity varies with fiber diameter. Batt, blanket, and loose-fill mineral fiber insulations are manufactured to achieve specified R-values, the most common of which are listed in the table. Due to differences in manufacturing processes and materials, the product thicknesses, densities, and thermal conductivities vary over considerable ranges for a specified R-value.

[h]This material is relatively new and data are based on limited testing.

[i]For additional information, see Society of Plastics Engineers (SPI) *Bulletin* U108.

Values are for aged, unfaced board stock. For change in conductivity with age of expanded polyurethane/polyisocyanurate.

[j]Values are for aged products with gas-impermeable facers on the two major surfaces.

[k]Insulating values of acoustical tile vary, depending on density of the board and on type, size, and depth of perforations.

An aluminum foil facer of 0.001 in. thickness or greater is generally considered impermeable to gases.

[l]Values for fully grouted block may be approximated using values for concrete with a similar unit weight.

[m]Values for metal siding applied over flat surfaces vary widely, depending on amount of ventilation of airspace beneath the siding; whether airspace is reflective or non-reflective; and on thickness, type, and application of insulating backing-board used. Values given are averages for use as design guides, and were obtained from several guarded hot box tests (ASTM C236) or calibrated hot box (ASTM C976) on hollow-backed types and types made using backing-boards of wood fiber, foamed plastic, and glass fiber, Departures of ±50% or more from the values given may occur.

[n]See Adams (1971), MacLean (1941), and Wilkes (1979). The conductivity values listed are for heat transfer across the grain. The thermal conductivity of wood varies linearly with the density, and the density ranges listed are those normally found for the wood species given. If the density of the wood species is not known, use the mean conductivity value. For extrapolation to other moisture contents, the following empirical equation developed by Wilkes (1979) may be used:

$$k = 0.17791 + (1.874 \times 10^{-2} + 5.753 \times 10^{-4} M)\rho + 1 + 0.01M$$

where ρ is density of the moist wood in lb/ft^3, and M is the moisture content in percent.

[o]From Wilkes (1979), an empirical equation for the specific heat of moist wood at 75°F is as follows:

$$c_\rho = [(0.299 + 0.01M) + 1 + 0.01M] + \Delta c_\rho$$

where Δc_ρ accounts for the heat sorption and is denoted by

$$\Delta c_\rho = M\,(1.921 \times 10^{-3} - 3.168 \times 10^{-5}\,M)$$

where M is the moisture content in percent by mass.

A complete listing can be found in the ASHRAE Guide and Data Book which is published annually by the American Society of Heating, Refrigerating and Air-Conditioning Engineers.

When calculating the heat load on a building, the heat transferred through each type of material must be calculated separately, then added together to determine the heat load for the building. Using the principles of heat flow, calculating heat flow by use of the overall thermal resistance method is preferred.

The total resistance to the flow of heat through building materials such as a flat ceiling, floor, or wall (or a curved surface if the curve is small) is the sum of all the resistances in the structure (R-values). To calculate these resistances use the formula:

$$R = R_1 + R_2 + R_3 + \ldots$$

Where R_1, R_2, etc., are the individual resistances of the parts of the partition, and the R is the resistance of the construction materials from the inside surface to the outside surface.

The U-factor is the reciprocal of R_t. Or use the formula:

$$U = 1 \div R_t$$

When materials having a higher value of R are used, the corresponding values of U become smaller. This is the reason why it is sometimes preferable to specify resistance rather than conductance. Also, a whole number is more easily understood than are fractions or decimals.

Example 1

Calculate the U-factor of the 2 × 4 stud wall shown in Figure 2-2.

The studs are on 16 in. OC (on center). There is a 3.5 in. mineral fiber batt insulation (R-13) in the stud space. The inside finish is 0.5 in. gypsum wallboard; the outside is finished with rigid foam insulating sheathing (R-4) and 0.5 in. by 8-in. wood bevel lapped siding. The insulated cavity occupies approximately 75% of the transmission area; the studs, plates, and sills occupy 21%; and the headers occupy 4%.

Solution

Obtain the R- values of the various building components from Tables 2-1 and 2-2. Assume the R-value of the wood framing is R-1.25 per inch. Also, assume that the headers are solid wood, in this case, and group them with the studs, plates and sills (Table 2-3).

Figure 2-2. Insulated wood frame wall. (Copyright 1993 by the American Society of Heating, Refrigerating and Air Conditioning Engineers, Inc., from Fundamentals Handbook. Used by permission)

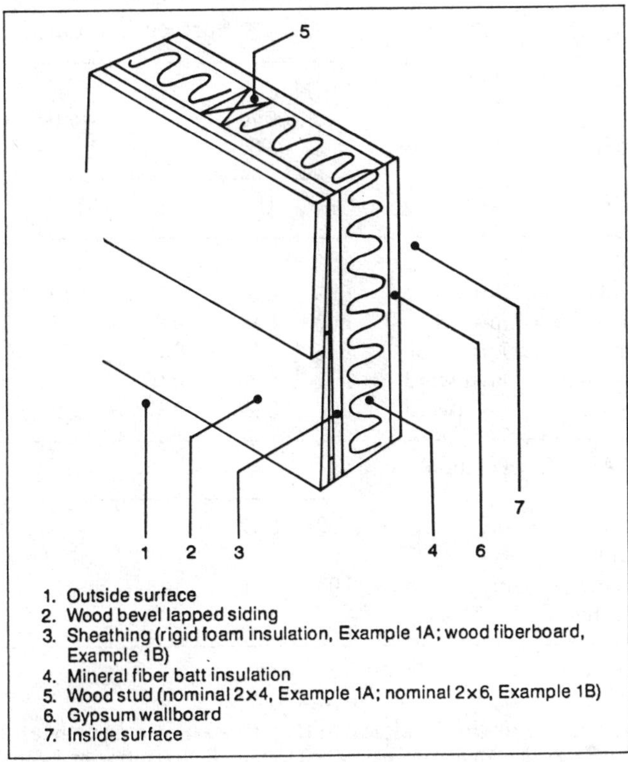

Since the U-factor is the reciprocal of the R-value, $U_1 = 0.052$ and $U_2 = 0.095$ Btu/hr/ft^2/°F.

Example 2

Calculate the U-factor of a 2 by 6 stud wall, similar to the one considered in Example 1, except that the sheathing is 0.5 in. wood fiberboard and the studs are at 24 in. OC. There is a 5.5 in. mineral fiber batt insulation (R-21) in the stud space. Assume that the headers are double 2 by 8 framing (with 0.5 in. air space), with 2.0 in. air space between the headers and the wallboard.

Solution

Obtain the R-values from Tables 2-1 and 2-2. Assume the R-value of the wood framing is 1.25 per inch. In this case, the headers must be treated separately (Table 2-4).

Table 2-2. Surface Conductances and Resistances for Air
(Copyright 1993 by the American Society of Heating, Refrigerating and Air-Conditioning, from Fundamentals Handbook. Used by permission.)

Position of Surface	Direction of Heat Flow	Surface Emittance, ε — Non-reflective ε = 0.90 h_i	R	Reflective ε = 0.20 h_i	R	Reflective ε = 0.05 h_i	R
STILL AIR							
Horizontal	Upward	1.63	0.61	0.91	1.10	0.76	1.32
Sloping—45°	Upward	1.60	0.62	0.88	1.14	0.73	1.37
Vertical	Horizontal	1.46	0.68	0.74	1.35	0.59	1.70
Sloping —45°	Downward	1.32	0.76	0.60	1.67	0.45	2.22
Horizontal	Downward	1.08	0.92	0.37	2.70	0.22	4.55
MOVING AIR (Any position)		h_o	R	h_o	R	h_o	R
15-mph Wind (for winter)	Any	6.00	0.17	—	—	—	—
7.5-mph Wind (for summer)	Any	4.00	0.25	—	—	—	—

Table 2-3.
Calculation of the Resistance (*R* value) of Various Materials.
(Copyright 1993 by the American Society of Heating, Refrigerating and Air Conditioning Engineers, Inc., from Fundamentals Handbook. Used by permission)

Element	*R* (Insulated Cavity)	*R* (Studs, Plates, and Headers)
1. Outside surface, 15 mph wind	0.17	0.17
2. Wood bevel lapped siding	0.81	0.81
3. Rigid foam insulating sheathing	4.0	4.0
4. Mineral fiber ball insulation, 3.5 in.	13.0	—
5. Wood stud, nominal 2 × 4	—	4.38
6. Gypsum wallboard, 0.5 in.	0.45	0.45
7. Inside surface, still air	0.68	0.68
	$R_1 = 19.11$	$R_2 = 10.49$

Table 2-4. Calculation of the Resistance (*R* Value) of Various Materials
(Copyright 1993 by the American Society of Heating, Refrigerating and Air-Conditioning, from Fundamentals Handbook. Used by permission.)

Element	*R* (Insulated Cavity)	*R* (Studs and Plates)	*R* (Headers)
1. Outside surface, 15 mph wind	0.17	0.17	0.17
2. Wood bevel lapped siding	0.81	0.81	0.81
3. Wood fiberboard sheathing, 0.5 in.	1.32	1.32	1.32
4. Mineral fiber ball insulation, 3.5 in.	21.0	—	—
5. Wood stud, nominal 2 × 4	—	6.88	—
6. Gypsum wallboard, 0.5 in.	—	—	3.75
7. Inside surface, still air	—	—	0.90
8. Air space, 2 in.	—	—	0.90
9. Gypsum wallboard, 0.5 in.	0.45	0.45	0.45
10. Inside surface, still air	0.68	0.68	0.68
	$R_1 = 24.43$	$R_2 = 10.31$	$R_3 = 8.98$

Since the U-factor is the reciprocal of the R-value, $U_1 = 0.041$, $U_2 = 0.097$, and $U_3 = 0.111$ Btu/hr/ft^2/°F.

Design Temperatures

The outdoor temperature in the different sections of the country change considerably from day to day. As an example, a northern city such as Minneapolis, Minnesota, may have extremely high temperatures during the summer time, but the average outdoor temperature in Minneapolis is lower than that in Dallas, Texas. When estimating the conduction heat load, it is necessary to choose the design temperatures on which to base the calculations. Charts that are available from the Weather Bureau show that the extreme temperatures occur in most localities only about 10'% of the time in both the heating and the cooling seasons. Because of this, the equipment designed for a particular application is not sized to handle either the highest or the lowest temperature that will occur. Rather, it is based on the average maximum and minimum temperatures. It would not be economically feasible to install a unit to handle conditions that would occur only about 10% of the time. Table 2-5 is an approximation of outside design conditions for the various parts of the country.

Table 2-5. Design Temperature Conditions (°F) Used for Calculating Heat Loads for Heating or Cooling for Various Regions of the United States

(Copyright 1993 by the American Society of Heating, Refrigerating and Air-Conditioning, from Fundamentals Handbook. Used by permission.)

		Extreme Temperatures		Mean Temperatures		Design conditions		
						Winter	Summer	
State	City	Low	High	January	July	Dry bulb	Dry bulb	Wet bulb
Ala.	Mobile	–1	103	52	81	15	95	79
Ariz.	Phoenix	16	119	51	90	25	110	75
Ark.	Little Rock	–12	108	41	81	5	96	78
Calif.	San Francisco	27	101	50	58	30	90	65
Colo.	Denver	–29	105	30	72	–10	95	72
Conn.	New Haven	–14	101	28	72	0	95	75
D.C.	Washington	–15	106	33	77	0	96	78
Fla.	Jacksonville	10	104	55	82	30	95	79
Ga.	Atlanta	–8	103	43	78	5	95	78
Idaho	Boise	–28	121	30	73	–10	100	70
Ill.	Chicago	–23	103	25	74	–10	95	75
Ind.	Indianapolis	–25	106	28	76	–5	96	76
Iowa	Dubuque	–32	106	19	74	–15	96	75
Kas.	Wichita	–22	107	31	79	0	100	75
Ky.	Louisville	–20	107	34	79	0	98	77
La.	New Orleans	7	102	54	82	25	96	80
Maine	Portland	–7	105	34	77	0	96	78
Mass.	Boston	–18	104	28	72	–5	95	75
Mich.	Detroit	–24	104	24	72	–10	95	75
Minn.	St. Paul	–41	104	12	72	–20	95	75
Miss.	Vicksburg	–1	104	48	81	15	96	80
Mo.	St. Louis	–22	108	31	79	0	98	79
Mont.	Helena	–42	103	20	66	–20	90	68

Table 2-5. (*Continued*)

State	City	Extreme Temperatures		Mean Temperatures		Design conditions		
						Winter	Summer	
		Low	High	January	July	Dry bulb	Dry bulb	Wet bulb
Neb.	Omaha	–32	111	22	77	–15	100	75
Nev.	Winnemucca	–28	104	29	71	–10	95	70
N.C.	Charlotte	–5	103	41	78	10	96	79
N.D.	Bismarck	–45	108	8	70	–25	98	70
N.H.	Concord	–35	102	22	68	–15	95	75
N.J.	Atlantic City	–7	104	32	72	0	95	75
N.M.	Santa Fe	–13	97	29	69	0	92	70
N.Y.	New York City	–14	102	31	74	0	95	77
Ohio	Cincinnati	–17	105	30	75	0	98	77
Okla.	Oklahoma City	–17	108	36	81	0	100	76
Ore.	Portland	–2	104	39	67	10	95	70
Penn.	Philadelphia	–6	106	33	76	0	95	78
R.I.	Providence	–12	101	29	72	0	95	75
S.C.	Charleston	7	104	50	81	15	96	80
S.D.	Pierre	–40	112	16	75	–20	100	72
Tenn.	Nashville	–13	106	39	79	5	98	79
Texas	Galveston	8	101	54	83	25	95	78
Utah	Salt Lake City	–20	105	29	76	–5	95	70
Vt.	Burlington	–28	100	19	70	–15	92	73
Va.	Norfolk	2	105	41	79	10	98	78
Wash.	Seattle	3	98	40	63	10	90	67
W. Va.	Parkersburg	–27	106	32	75	–5	96	77
Wis.	Milwaukee	–25	102	21	70	–15	94	75
Wyo.	Cheyenne	–38	100	26	67	–15	92	70

This table shows the outside design conditions for various cities throughout the United States. This table should be used when estimating air conditioning heat loads. It includes the design wet bulb for summer and the design dry bulb for winter calculations. Example: the correct outside summer design conditions for Detroit, Michigan, would be 95°F dry bulb and 75°F wet bulb. The winter design conditions would be –10°F.

When the design conditions are properly established, the unit will prove satisfactory for all comfort conditioning applications even if the outside design conditions are exceeded for a short period of time. When the outside conditions are exceeded for a short period of time there will be a rise in either or both of the dry bulb temperature or the relative humidity inside the conditioned space. In most cases this rise will only be temporary. In comfort air conditioning applications, the larger equipment will not usually be justified in either initial cost or operating costs. However, in industrial applications this may not be absolutely true. Industrial units are chosen to handle a load for the benefit of the product or process. Therefore, a temperature other than that required for the design conditions for even a few hours may cause a loss of many dollars' worth of products. In some installations it may cause the equipment to completely shut-down the process. In cases such as these, the extra initial cost and the added costs of operation can be justified. A general rule for industrial applications is to design the system for a 5°F higher outside design dry bulb and a 3°F outside design wet bulb temperature than those used for comfort cooling.

Occupancy

When estimating the heat load for the occupants of a building, one must consider that the heat load for people who are sitting down is quite different from the heat load for people who are active. These are about the only classifications needed for comfort air conditioning applications. Notice that this is the first item that has been divided into sensible and latent heat components (Example 3). To calculate this, simply determine how many of the occupants fit into each of the categories and multiply this factor by the proper figure.

It should be noticed that the amount of heat given off by a human depends on the activity of that individual. Example: People who are seated, or at rest, give off approximately 400 Btu/hr total heat. When they are at a medium activity level, the heat given off may increase to 675 Btu/hr. Part of this heat load is sensible and part is latent. Notice that the sensible heat load remains almost constant.

Example 3

A restaurant has 6 employees and a seating capacity of 35 customers. Compute the body heat load by using the factors given previously.

Solution

Sensible heat	=	41 × 225	= 9225
Latent heat	=	35 × 175	= 6125
		6 × 450	= 2700
Total body heat	=	18050 Btu	

(9225 Btu sensible + 8825 latent heat)

Electric Lights and Appliances

About 95% of the electricity supplied to lights actually goes into heat; only about 5% is used for lighting purposes. Because of this, all the electricity supplied to the lighting system should be applied to the heat load of the structure. Electrical energy is calculated at 3.41 Btu/W. As with all other loads, only the lighting that is used during the peak period should be used in the load estimate. When sufficient light is supplied by skylights or windows during the peak load period, the lighting load should not be included in the estimate. In applications where the lighting would be on because of the unavailability of the sun light, the lighting must be included. Such an instance would be a store or restaurant in a lower floor of a multistory building where the sun would probably never reach to supply the required lighting. In installations where the sun will shine through windows and electric lighting is required at the same time, both loads at the peak period must be included in the estimate.

Incandescent lighting is figured at 3.41 Btu/hr. per watt. Fluorescent lighting is figured at 3.41 Btu/hr. per watt plus the heat given off by the ballast, which is about an additional 25% of the lighting load. Thus, when fluorescent lighting is used, an additional 25% must be added to the lighting heat load estimate.

All electric appliances, such as toasters and hair dryers, are calculated at 3.41 Btu/W. When all these appliances dissipate all their heat into the conditioned space, their total combined wattage is multiplied by 3.41 to calculate the appliance heat gain per hour. The amount of time and the period during which the appliance is used must also be considered. If the appliance is not used during the peak period, and it is fairly small, it may not be necessary to include it in the total heat load calculation. Example: A toaster used in a restaurant would probably be used almost continuously during the breakfast period when the solar load was smaller than it would during the noon period or the peak load period. Thus, it would probably be less of a load during the peak period than when it is actually used. However, all these factors must be considered and nothing taken for granted when estimating the heat load on a building.

According to the size, electric motors give off different amounts of heat per watt of energy used. The larger the motor the more efficient it is and the less heat it will give off during operation. Any electric motors that are operated within the conditioned space must be considered as part of the heat load. The approximate amount of heat given off by different appliances that are operated within a conditioned space are shown in Table 2-6. Notice that electric motors are listed in the table on their Btu per horsepower per hour rating.

Gas and Steam Appliances

These types of appliances usually include food cooking and warming devices, such as steam tables and coffee urns. They are listed in tables showing their heat load in Btu per hour of operation (Table 2-7). Realize that for each pound of steam condensed into water by the coil in steam devices, 960 Btu must be added to the total heat load estimate. It should be calculated on the basis that half is sensible heat and half is latent heat. The latent heat results from the water vapor being released by the food when it is heated. Some appliances, such as coffee urns, are heated directly by a gas flame. When this situation is encountered the heat generated by the burning gas must be added to the calculations. When a gas is burned the products of combustion are mostly carbon dioxide (CO_2) and water vapor (H_2O). The heat load from these types of appliances should also be calcu-

Table 2-6. Heat from Electric Appliances

ITEM	AMOUNT OF HEAT
Lights	3.41 Btu/W
Appliances (toasters, hair dryers, etc.)	3.41 Btu/W
Motors	
1/8-1/2 hp	4250 Btu/hp per hour
3/4-3 hp	3700 Btu/hp per hour
5-20 hp	3000 Btu/hp per hour

Table 2-7. Heat from Gas and Steam Appliances

Item	Btu/Hr Sensible	Latent	Total
Steam tables, per sq ft top surface	1,000	1,000	2,000
Restaurant coffee urns	5,000	5,000	10,000
Natural gas, per cu ft	500	500	10,000
Manufactured gas, per cu ft	275	275	550
Steam condensed in warming coils, per pound	480	480	960

NOTES:
1. If there are exhaust hoods over steam tables and coffee urns, include only 50% of the heat load from these sources.
2. Coffee urns may also be figures on the basis of consuming approximately 1 cu ft of natural gas or approximately 2 cu ft of manufactured gas per hour per gallon rated capacity.

lated on the basis that half is sensible and half is latent heat. Generally, the amount of heat given off by gas appliances such as coffee urns can be calculated by using 1 cubic foot of natural gas and 2 cubic feet of manufactured gas for each gallon of capacity of the appliance.

When exhaust hoods are used over these types of appliances, the sensible heat and latent heat loads can usually be reduced by 50%.

Infiltration

The following is a discussion of the infiltration through the various components and systems used in buildings. The following discussion concerns both residential and commercial buildings.

Walls—(18 to 50%; 35% ave.). The interior walls as well as the exterior walls contribute to the infiltration rate of the structure. Infiltration occurs between the sill plate and the foundation, cracks below the bottom of the gypsum wall board, electrical outlets, plumbing penetrations, and at the top plates of the walls into the attic. Since the interior walls are not filled with insulation, open paths connecting to the attic allow the walls to act like heat exchanger fins within the conditioned space.

Ceiling Details—(3 to 18%; 13% ave.). Infiltration across the top ceiling of the conditioned space is undesirable because it reduces the effectiveness of the insulation in the attic and contributes to the infiltration heat loss. Ceiling infiltration also reduces the effectiveness in buildings that do not have attics. Recessed lighting, plumbing, and electrical penetrations leading into the attic are some particular areas of concern.

Heating System—(3 to 28%; 18% ave.). The installation of the furnace or the ductwork in the conditioned space, the venting arrangement of a gas burning furnace, and the location of the combustion air supply all affect infiltration into the building. It has been shown that the air leakage of ducts is in about the 50% range. However, field repairs can eliminate most of the leakages. The average of 18% contribution of leakage significantly underestimates the impact on total system performance. The pressure differential across the duct leaks are approximately ten times higher than typical pressures across the building envelope leaks.

Windows and Doors—(6 to 22%; 15% ave.). There is more variation in the amount of infiltration among the different types of windows, such as casement versus double-hung, than among new windows of the same type from different manufacturers. Windows that seal by compressing a weather strip, such as casements, have significantly less infiltration ratings than windows having sliding seals.

Fireplaces—(0 to 30%; 12% ave.). When the fireplace is not in use, poor fitting dampers allow conditioned air to escape through the chimney. Glass doors on the front of the fireplace will reduce the excess air while the fire is burning, but they will rarely seal the fireplace structure more tightly than does a closed damper. Chimney caps or fireplace plugs with warning devices that they are in place effectively reduce the infiltration through a cold fireplace.

Vents in a Conditioned Space—(2 to 12%; 5% ave.). Exhaust vents in conditioned spaces are not usually equipped with dampers that close properly.

Diffusion through Walls—(less than 1% ave.). Diffusion, in comparison to infiltration through holes and other openings in the structure, is not important. Typical values for the permeability of building materials at 0.12 in. WC produce an air exchange rate of less than 0.01 air changes per hour by wall diffusion in a typical residence. Table 2-8 shows the effective leakage areas for a variety of residential building components.

Table 2-8. Effective Leakage Areas (Low-Rise Residential Applications). (Copyright 1993 by the American Society of Heating, Refrigerating and Air Conditioning Engineers, Inc., from Fundamentals Handbook. Used by permission.)

	Units (see note)	Best Est.	Mini-mum	Maxi-mum
Ceiling				
General	in^2/ft^2	0.026	0.011	0.04
Drop	in^2/ft^2	0.0027	0.00066	0.003
Chimney	in^2/ea	4.5	3.3	5.6
Ceiling penetrations				
Whole house fans	in^2/ea	3.1	0.25	3.3
Recessed lights	in^2/ea	1.6	0.23	3.3
Ceiling/Flue vent	in^2/ea	4.8	4.3	4.8
Surface-mounted lights	m/ea	0.13		
Crawl space				
General (area for exposed wall)	in^2/ft	0.144	0.1	0.24
8 in. by 16 in. vents	in^2/ea	20		
Door frame				
General	in^2/ea	1.9	0.37	3.9
Masonry, not caulked	in^2/ft^2	0.07	0.024	0.07
Masonry, caulked	in^2/ft^2	0.014	0.004	0.014
Wood, not caulked	in^2/ft^2	0.024	0.009	0.024
Wood, caulked	in^2/ft^2	0.004	0.001	0.004
Trim	$in^2/lftc$	0.5		
Jamb	$in^2/lftc$	4	3 6	5
Threshold	$in^2/lftc$	1	0.6	12
Doors				
Attic/crawl space, not weatherstripped	in^2/ea	4.6	1.6	5 7
Attic/crawl space weatherstripped	in^2/ea	2.8	1.2	2 9
Attic fold down, not weatherstripped	in^2/ea	6.8	3.6	13
Attic fold down, weatherstripped	in^2/ea	3.4	2.2	6.7
Attic fold down with insulated box	in^2/ea	0.6		
Attic from unconditioned garage	in^2/ea	0	0	0

Units (see note)	Best Est.	Mini-mum	Maxi-mum	
Sole plate, floor/wall, caulked	$in^2/lftc$	0.4	0.04	0.61
Top plate, band joist	$in^2/lftc$	0.05	0.04	0.19
Piping/Plumbing/Wiring penetrations				
Uncaulked	in^2/ea	0.9	0.31	3.7
Caulked	in^2/ea	0.3	0.16	(1.3
Vents				
Bathroom with damper closed	in^2/ea	1.6	0.39	3.1
Bathroom with damper open	in^2/ea	3.1	0.95	3.4
Dryer with damper	in^2/ea	0.46	0.45	1.1
Dryer without damper	in^2/ea	2.3	1.9	5.3
Kitchen with damper open	in^2/ea	6.2	2.2	11
Kitchen with damper closed	in^2/ea	08	0.16	1.1
Kitchen with tight gasket	in^2/ea	0.16		
Walls (Exterior)				
Cast-in-place concrete	in^2/ft^2	0.007	0.0007	0.026
Clay brick cavity wall, finished	in^2/ft^2	0.0098	0.0007	0.033
Precast concrete panel	in^2/ft^2	0.017	0.0004	0.024
Lightweight concrete block, unfinished	in^2/ft^2	0.05	0.019	0.058
Lightweight concrete block, painted or stucco	in^2/ft^2	0.016	0.0075	0.016
Heavyweight concrete block, unfinished	in^2/ft^2	0.0036		
Continuous air infiltration barrier	in^2/ft^2	0.0022	0.0008	0.003
Rigid sheathing	in^2/ft^2	0.005	0.0042	0.006
Window framing				
Framing, masonry, uncaulked	in^2/ft^2	0.094	0.082	0.148
Framing, masonry, caulked	in^2/ft^2	0.019	0.016	0.03
Framing, wood, uncaulked	in^2/ft^2	0.025	0.022	0.039
Framing, wood, caulked	in^2/ft^2	0.004	0.004	0.007
Windows				
Awning not weather tripped	in^2/ft^2	0.023	0.011	0.035

(Continued)

Units	Best (see note)	Mini- Est.	Maxi- mum	Units mum
Double,not weather stripped	in^2/ft^2	0.16	0.1	0.32
Double, weather stripped	in^2/ft^2	0.12	0.04	0.33
Elevator (passenger)	in^2/ea	0.04	0.022	0.054
General, average	in^2/lftc	0.16	0.12	0.23
Interior (pocket, on top floor)	in^2/ea	2.2		
Interior (stairs)	in^2/lftc	0.5	0.13	0.76
Mail slot	in^2/lftc	2		
Sliding exterior glass patio	in^2/ea	3.4	0.46	9.3
Sliding exterior glass patio	in^2/ft^2	0.079	0.009	0.22
Storm (difference between with and without)	in^2/ea	0.9	0.46	0.96
Single, not weatherstripped	in^2/ea	3.3	1.9	8.2
Single, weatherstripped	in^2/ea	1.9	0.6	4.2
Vestibule (subtract per each location)	in^2/ea	1.6		
Electrical outlets/Switches				
No gaskets	in^2/ea	0.38	0.08	0.96
With gaskets	in^2/ea	0.023	0.012	0.54
Furnace				
Sealed(or no)combustion	in^2/ea	0	0	0
Retention head or stack damper	in^2/ea	4.6	3.1	4.6
Retention head and stack damper	in^2/ea	3.7	2.8	4.6
Floors over crawl spaces				
General	in^2/ft^2	0.032	0.006	0.071
Without ductwork in crawl space	in^2/ft^2	0.0285		
With ductwork in crawl space	in^2/ft^2	0.0324		
Fireplace				
With damper closed	in^2/ft^2	0.62	0.14	1.3
With damper open	in^2/ft^2	5.04	2.09	5.47
With glass doors	in^2/ft^2	0.58	0.06	0.58
With insert and damper closed	in^2/ft^2	0.52	0.37	0.66
With insert and damper open	in^2/ft^2	0.94	0.58	1.3
Gas water heater	in^2/ea	3.1	2.3	3.9
Joints				
Ceiling-wall	in^2/lftc	0.76	0.081	1.3
Sole plate, floor/wall, uncaulked	in^2/lftc	2	0.2	2.8

Best (see note)	Mini- Est.	Maxi- mum	mum	
Awning with weatherstripping	in^2/ft^2	0.012	0.006	0.017
Casement with weatherstripping	in^2/lftc	0.12	0.05	1.5
Casement without weather-stripping	in^2/lftc	0.14		
Double horizontal slider without weatherstripping	in^2/lftc	0.56	0.01	1.7
Double horizontal slider, wood with weatherstripping	in^2/lftc	0.28	0.076	0.87
Double horizontal slider, aluminum with weatherstripping	in~/lftc	0.37	0.29	0.4
Double hung without weather-stripping	in^2/lftc	1.3	0.44	3.1
Double hung with weather-stripping	in^2/lftc	0.33	0.1	0.97
Double hung without weather-stripping, with storm	in^2/lftc	0.5	0.25	0.86
Double hung with weather - stripping, with storm	in^2/lftc	0.4	0.22	0.5
Double hung with weatherstrip-ping, with pressurized track	in~ tc	u.24	0.2	0.28
Jalousie	in^2/louver	0.524		
Lumped		in^2/lfts	0.2 i	0.0046
Single horizontal slider, weatherstripped		in^2/lfts	0.34	0.1
Single horizontal slider, aluminum	2/lfts	0.4	0.14	1
Single horizontal slider, wood	in^2/lfts	0.22	0.14	0.5
Single horizontal slider, wood clad	in^2/lfts	0.33	0.27	0.41
Single hung, weatherstripped	in^2/lfts	0.44	0.32	0.63
Sill	in^2/lftc	0.11	0.071	0.11
Storm inside, heat shrink	in^2/lfts	0.009	0.0046	0.009
Storm inside, rigid sheet with magnetic seal	in^2/lfts	0.061	0009	0.12
Storm inside, flexible sheet with mechanical seal	in^2/lfts	0.078	0.009	0.42
Storm inside, rigid sheet with mechanical seal	in^2/lfts	0.2	0.023	0.42
Storm outside, pressurized track	in^2/lftc	0.27		
Storm outside, 2 track	in^2/lftc	0.63		
Storm outside, 3 track	in^2/lftc	1.25		

Multifamily Building Leakage—The leakage distribution is particularly important in multifamily apartment buildings. These types of buildings cannot be treated as single-zone residences because of the internal resistance between the apartments. The leakage between apartments varies widely, tending to be small in newer construction, and ranging to as high as 60% of the total apartment leakage in older buildings. Little data on interzonal leakage has been reported because of the expense and difficulty of making the necessary measurements.

Commercial Building Envelope Leakage—The building envelopes of commercial buildings are generally considered to be very tight. The National Association of Architectural Metal Manufacturers specifies a maximum leakage per unit of wall area of 0.060 cfm/ft^2 at a pressure difference of 0.30 in. of water column (WC) exclusive of leakage through operable windows. However, tests have shown that the commercial type buildings are not as tight as first thought.

Air Leakage through Internal Partitions—In large buildings the air leakage associated with internal partitions becomes very important. Elevator shafts, stairs, and service shaft walls, floors, and other interior partitions are of major concern in these types of buildings. Their leakage characteristics are needed to determine the infiltration through exterior walls and air flow patterns within the building. These internal resistances are also important in the event of fire to predict smoke movement patterns and to evaluate smoke control systems.

Table 2-9 shows the air leakage areas for different internal partitions of commercial buildings.

Figure 2-3 shows examples of measured air leakage rates of elevator shaft walls.

Consult the ASHRAE Handbook—Applications for models of performance of smoke control systems and their application. Leakage openings at the top of elevator shafts are equivalent to office areas of 620 to 1550 in^2. Air leakage rates through stair shaft and elevator doors are shown in Figure 2-4.

These are based on the average crack around the doors. The leakage areas associated with other openings in commercial buildings are also important for air movement calculations. These include partitions, suspended ceilings in buildings where the space above the ceiling is used in the air distribution system, and other components of the air distribution system.

Table 2-9. Leakage Areas for Internal Partitions in Commercial Buildings (at 0.3 in. of water and C_D = 0.65)

Construction Element	Wall Tightness	Area Ratio
		A/A_w
Stairwell walls	Tight	0.14×10^{-4}
	Average	0.11×10^{-3}
	Loose	0.35×10^{-3}
Elevator shaft walls	Tight	0.18×10^{-4}
	Average	0.84×10^{-3}
	Loose	0.18×10^{-2}
		A/A_f
Floors	Average	0.52×10^{-4}

A = leakage area A_w = wall area A_f = floor area

Air Leakage through Exterior Doors—Door infiltration depends on the type of door, room, and building. In residences and small buildings where the doors are used infrequently, the air exchange associated with a door can be estimated based on the air leakage through the cracks between the door and the frames. A frequently opened single door, as in a small retail store, has a much larger amount of airflow.

Infiltration Calculations—Natural infiltration will occur in all buildings to some extent. The amount will depend upon the interior pressure, wind velocity, tightness of the structure, and other construction details. In residences or small office buildings where smoking is not allowed or other source of fumes to produce objectionable odors are prohibited, natural infiltration could possibly provide an ample quantity of fresh air if the building has window or skylight openings that are at least equal to or exceed 5% of the total floor square foot area. The room must be large enough to allow at least 50 square feet of space and at least 500 cubic feet of internal volume for each individual in the building. Under these conditions, the infiltration rate is sufficient to meet the fresh air requirement for the building. Forced ventilation will not be necessary unless it is required

Figure 2-3. Air-Leakage rate of elevator shaft walls.
(Copyright 1993 by the American Society of Heating, Refrigerating and Air Conditioning Engineers, Inc., from Fundamentals Handbook. Used by permission)

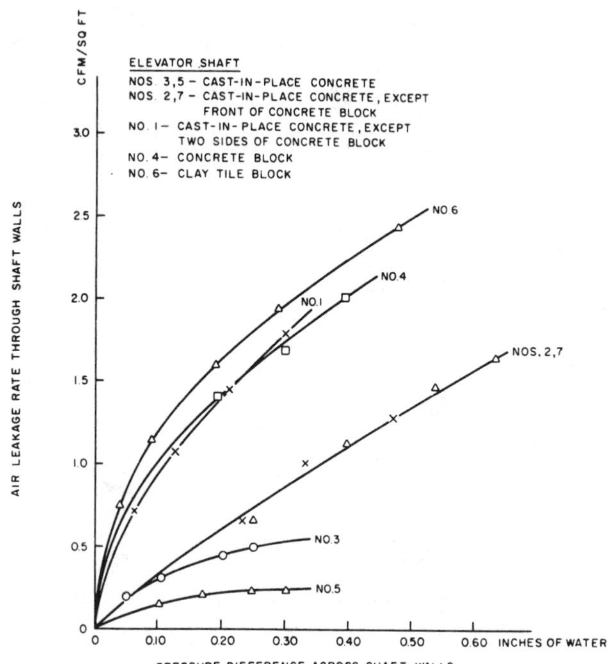

by local ordinances (Table 2-10).

Properly designed installations do not usually depend on infiltration to supply the necessary ventilation requirements.

The infiltration through doors and windows has been placed in tables for convenience (Table 2-11).

When there are two exposed walls to be considered, use the wall with the greatest amount of leakage. When more than two walls are exposed, use either the wall with the greatest amount of leakage or half of the total of all the walls, whichever quantity is the largest.

Example 4

Determine the infiltration into a building with three exposed walls. One of the walls has three average-fit, weather-stripped, double-hung windows 2 ft. by 2 ft. One of the other walls has one industrial-type, steel sash window, 3 ft. by 3 ft. The third wall has a poorly fitted, weather-

Figure 2-4. Air-Leakage rate of door versus average crack width.
(Copyright 1993 by the American Society of Heating, Refrigerating and Air Conditioning engineers, Inc., from Fundamentals handbook. Used by permission)

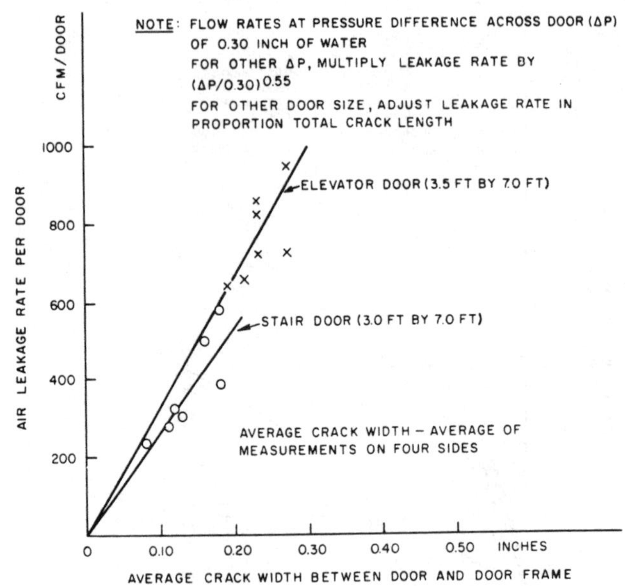

stripped door, measuring 3 ft by 6 ft. The average wind velocity is 5 mph during the cooling season.

Solution

Determine the total length of the crack around the double-hung windows. Use the following procedure:

Cracks at the top, bottom, and center of the window = 2 ft × 3 = 6 ft.
Vertical cracks on each side of the window = 4 ft × 2 = 8 ft.
Total crack length for one double-hung window = 8 + 6 = 14 ft.
Total crack for three windows = 14 ft × 3 = 42 ft.

When the wind velocity is 5 mph, an average-fit weather-stripped double-hung window has an infiltration of 5 cu ft per hour per foot of crack (Table 2-11).

The infiltration would then be 42 ft × 5 = 210 cu ft of air per hour in

Table 2-10. Minimum Outdoor Air Required for Ventilation (Subject to Local Code Regulations)

Application	*CFM per person*
Apartment or residence	10-15
Auditorium	5-7-1/2
Barber shop	10-15
Bank or beauty parlor	7-1/2-10
Broker's board room	25-40
Church	5-7-1/2
Cocktail lounge	20-30
Department store	5-7-1/2
Drugstore	7-1/2-10
Funeral parlor	7-1/2-10
General office space	10-15
Hospital rooms (private)	15-25
Hospital rooms (wards)	10-15
Hotel room	20-30
Night clubs and taverns	15-20
Private office	15-25
Restaurant	12-15
Retail shop	7-1/2-10
Theater (smoking permitted)	10-15
Theater (smoking not permitted)	5-7-1/2

this wall.

Calculate the steel sash window as follows: 3 ft × 4 = 12 ft.

The infiltration rate for an industrial steel sash window at 5 mph wind velocity is 50 cu ft per hour (Table 2-11).

The total crack around the steel sash industrial window is 12 ft. Thus, 12 ft × 50 = 600 cu ft per hour in the second wall.

The infiltration around the door in the third wall is calculated as follows:

Total crack length = (6 ft × 2) + (3 ft × 2) = 18 ft.

The infiltration for a poorly fitted weather-stripped door is 30 cu ft per hour with a wind velocity of 5 mph (Table 2-11).

The infiltration around the door then is 18 ft × 30 = 540 cu ft per hour in the third wall.

The total infiltration for all three walls is 1350 cu ft per hour one-half

Table 2-11. Infiltration per Foot of Crack per Hour in Cubic Feet

Type of Opening	Condition	Wind Velocity (mph)					
		5	10	15	20	25	30
Double-hung wood window	Average fit, not weather-stripped	6	20	40	60	80	100
Double-hung wood window	Average fit, weather-stripped	5	15	25	35	50	65
Double-hung wood window	Poor fit, weather-stripped	6	20	35	50	70	90
Double-hung metal window	Not weather-stripped	20	45	70	100	135	170
Double-hung metal window	Weather-stripped	6	18	20	44	58	75
Steel sash	Casement, good fit	6	18	30	44	58	75
Steel sash	Casement, average fit	12	30	50	75	100	125
Steel sash	Industrial type	50	110	175	240	300	375
Doors	Good fit, not weather-stripped	15	35	55	75	100	125
Doors	Poor fit, not weather-stripped	55	140	225	310	400	500
Doors	Poor fit, weather-stripped	30	70	110	155	200	250

NOTE: Use of storm sash permits a 50% infiltration reduction for poorly fitting windows only.

Table 2-12. Heat Transfer Factors for Ducts

Duct	*Btu/Sq Ft per Hour I°F Difference*
Sheet metal, not insulated	1.13
Average insulation, 1/2 in. thick	0.41

of the infiltration would be 675 cu ft per hour. Since this quantity is greater than the infiltration through any single wall, it is used to determine if there is a sufficient amount of fresh air being supplied to the space by infiltration. The air conditioning equipment must condition all infiltration air; therefore, this air must be added to the heat load estimate. If the quantity of infiltration is not large enough to supply the required amount of fresh air, the necessary air must be brought in through the conditioning equipment so that the fresh air requirements will be met. When the fresh air is brought into the building by the conditioning equipment, the building is placed under a positive pressure and the infiltration air is reduced, if not completely eliminated, requiring that all fresh air be brought in by the equipment. This is a far superior design because all the fresh air is treated by the conditioning equipment.

Infiltration Heat Load—The air that enters a conditioned space places an internal heat load on the space. Because of this it must be divided into its sensible and latent heat components. Ventilation air is not to be included along with any infiltration air introduced into the building. Ventilation air is a load on the equipment rather than the space. Only the total heat load is used when calculating the heat load caused by ventilation air.

Example 5

The total infiltration air calculated in Example 4 above was 675 cu ft per hour. Since the psychrometric chart is based on air by the pound we must convert cu ft per hour into pounds per hour. What is the sensible and latent components of the infiltration air?

Solution

Let us assume that the outside air is at 95°F dry bulb and 77°F wet bulb, the design conditions for New York City (see Table 2-5). Let's assume that the inside design conditions are 80°F dry bulb and 76°F wet bulb. We can find on the psychrometric chart that dry air at 95°F has a weight of 0.0715 lb per cu ft (Figure 2-5).

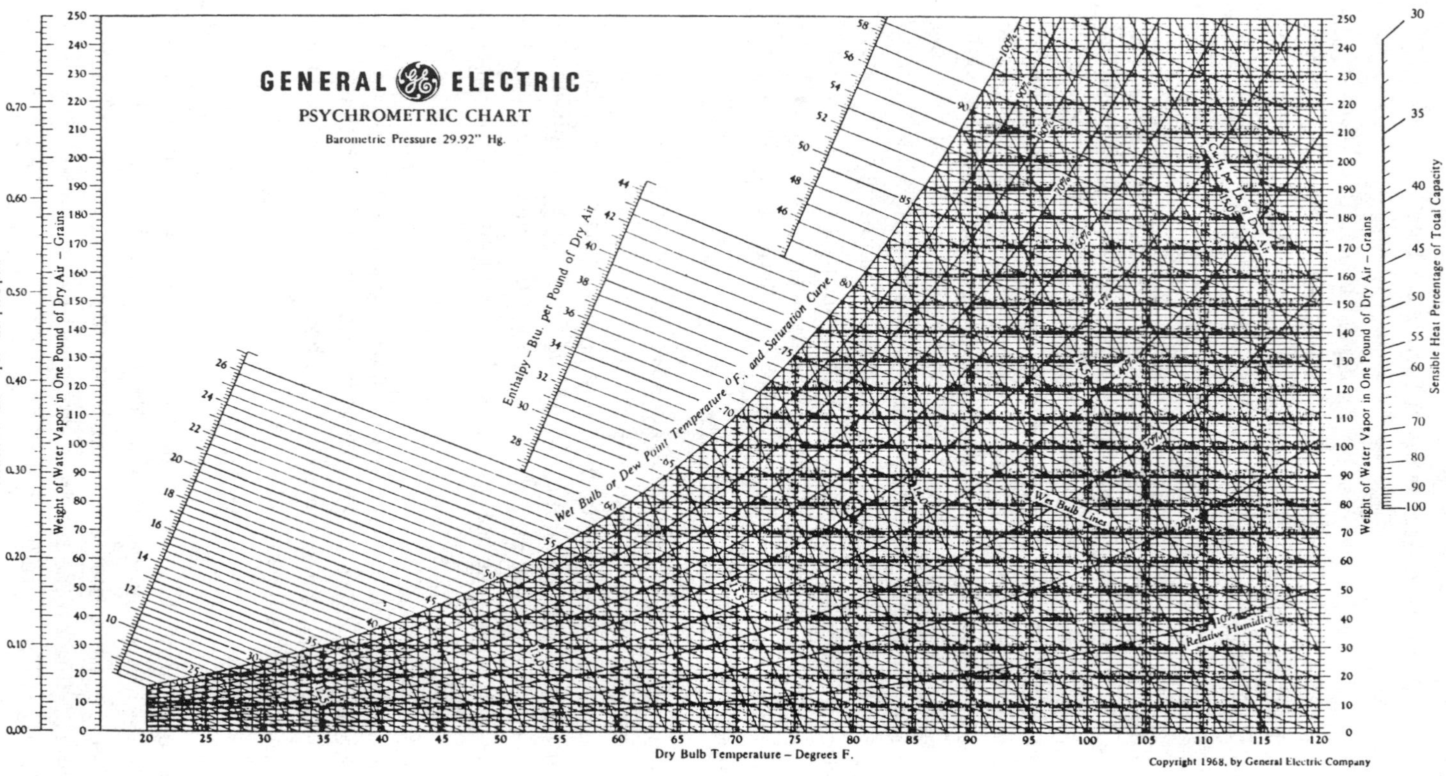

Figure 2-5. Psychrometric chart. (Courtesy of The Trane Company)

Thus

$$675 \times 0.0715 = 48.62 \text{ lb of dry air per hour}$$
$$\text{Total heat at 75°F wet bulb} = 38.45 \text{ Btu per pound of dry air}$$
$$\text{Total heat at 67°F wet bulb} = 31.85 \text{ Btu per pound of dry air}$$
$$\text{Total heat removed} = 6.6 \text{ Btu per pound of dry air}$$

Then, 48.62 × 6.6 = 320.89 Btu per hour total heat removed from the infiltration air. The sensible and latent heat must be separated into their component heat loads.

$$\text{Sensible heat content at 95°F} = 22.95 \text{ Btu per pound of dry air}$$
$$\text{Sensible heat content at 80°F} = 19.32 \text{ Btu per pound of dry air}$$
$$\text{Sensible heat removed} = 3.63 \text{ Btu per hour}$$

Therefore, 48.62 × 3.63 = 179.09 Btu per hour sensible heat that must be removed from the air. The total heat less the sensible heat is equal to the latent heat. Thus, 328.57 – 179.09 = 149.48 Btu per pound of latent heat to be removed from the air.

A review of Chapter 1 is helpful at this point.

Ventilation Heat Load—The minimum outside air requirements should be compared with the amount of air that is admitted through infiltration (Table 2-11). If it is determined that the infiltration air will not provide sufficient fresh air for the space, then ventilation air must be provided. When ventilation is brought into the space through the equipment, a positive pressure is built up inside the space which will prevent infiltration air from entering. Thus, both types are not included in the heat load estimate. In Example 4 it was determined that 675 cu ft of air entered the space by infiltration. When the space is an office having two occupants, 1200 cu ft of air per hour is required because each occupant requires 10 cu ft per min. (cfm) (Table 2-10). Then, 2 × 10 × 60 = 1200 cu ft per hour. In our example, infiltration would not supply a sufficient amount of fresh air to the space to meet these conditions. The heat load would need to include the 1200 cu ft per hour of fresh air.

When using ventilation air it is not required that the heat be broken down into sensible and latent heat. Only the total heat load is calculated for the estimate. To determine this heat load the cu ft of air per hour would need to be converted to pounds of dry air per hour. The conversion is then multiplied by the difference in total heat content between the two samples of air as determined from the psychrometric chart (Figure 2-5).

Duct Heat Load—In a normal installation the amount of heat lost through the ducts is very small when compared to the total heat load of the space and can usually be forgotten. However, in installations where the ducts must be installed in kitchens or other warm or hot areas, the heat load must be calculated and added to the estimate. Remember that any heat added to or taken from the air from the time it leaves the equipment until the time it returns to the equipment must be removed or added to the air. Duct heat load factors are provided in tabular form and are shown in Table 2-12.

Estimating the duct heat load is usually done after the system has been estimated because the ducts have not yet been sized. When this calculation is required it usually causes a redesign of the system. When the supply duct is located within the conditioned space it is not necessary to include the duct heat load in the design because any heat lost or gained from the space will be added to the air. Duct heat gain in these types of installations becomes a plus for the equipment, because the exact amount of heat gained or lost through the ducts is gained by the space.

The sizes of the ducts are directly related to the amount of sensible heat that must be added to or removed from the air because the amount of air that must be supplied to the space is determined by the amount of sensible heat that must be removed or added to the space. Therefore, the duct heat loss is actually a percentage of the sensible heat load for the space. Usually this percentage will vary from a minimum of 0% to a maximum of about 5%. When attempting to determine the percentage of duct heat load, the length of the duct that passes through the unconditioned space as well as the temperature of the surrounding air must be considered in this estimation. The above mentioned percentages are made assuming that the ducts are insulated in the unconditioned space.

The procedure using Table 2-12 to calculate the duct heat load is the same as that for any other heat flow calculation. The area of the exposed duct is determined and then multiplied by the proper factor selected from Table 2-12. The total is then multiplied by the difference between the dry bulb temperature of the air inside the duct and the temperature of the air outside the duct.

Fan Heat Load—The air is caused to flow through the conditioning equipment by a fan. The fan uses a motor to push the air through the system. When the motor is operating heat is produced which is transferred to the air flowing over it. This heat must be removed by the refrigeration system used for cooling purposes. The fan also adds heat during the heating season which is a plus for the equipment. The horsepower of the fan may not be known when the equipment load is being estimated, but there is also an allowance that can be estimated depending upon the amount of air to be circulated, which depends on the total sensible heat load of the space. Thus, the fan sensible heat load is taken as a percentage of the sensible heat load for the space. This amount will usually be about 3% to 4% of the total sensible heat load estimated. It is usually customary to add about 5% as a total for both the duct heat load and the fan motor heat load. For most installations this percentage will be sufficient; however, when special conditions are encountered some allowances must be made to cover this load. When the system is purchased from the manufacturer as a package, the fan heat load is considered in the Btu rating of the equipment.

Safety Factor—Due to the large number of variables that may be encountered, and some of them unknown, heat load estimation is not an exact science. As an example, the heat transfer coefficients can be determined very accurately; however, the design engineer cannot be positive that the construction of the space being considered is in every detail identical to the test panels from which the coefficients were established. Other factors such as the internal loads, lighting, people, equipment, appliances, shading factors, and the exterior color of the building could possibly all vary from the conditions on which the design was based. Because of these many variables, it is proper and reasonable to use a safety factor when estimating heat loads. Usually, this safety factor will vary from a maximum on very small installations (because any variation could possibly affect the performance of the equipment) to a minimum on very large installations (where there is greater chance of variations cancelling out each other). Most design engineers use a safety factor of about 10%. However, the same safety factor should be applied to both sensible and latent heat loads to prevent upsetting the relationship between these two factors. The importance of this relationship will be more apparent with experience in estimating heat loads for spaces. The safety factor should not be used to cover negligence in estimating the heat load.

SUMMARY

The person who is estimating the size of an air conditioning system must have a working knowledge of the different heat sources or heat loss through the structure.

The three primary functions of air conditioning equipment are to maintain indoor air conditions that (1) aid in human comfort, (2) are required by a product, or (3) are required by a process performed in the conditioned space.

The size of the equipment is determined by the actual peak heat load requirements.

The first step when making an estimate is to make a comprehensive survey of the building and the surroundings to make certain that an accurate evaluation of the heat loads is made.

Use a survey and checklist to make certain that a thorough and accurate survey is completed.

The two headings for heat loads are (1) those that result in an internal heat load on the conditioned space, and (2) those that result in an external heat load.

Many of the heat loads are not at their peak at the same time.

When estimating the heat load for a space we are looking for one hour during the day when the sum of the heat loads is at a peak, and not the sum of the heat load peaks.

When estimating the conduction heat load on a space, each type of construction material must be separated and figured alone.

When estimating the heat load on a space, it is necessary to determine the proper design temperature upon which to base the calculations.

The design temperature of comfort conditioning is based on the average maximum temperature for the area in which the building is located.

Industrial conditioning equipment is sized according to the maximum temperature that will be encountered during the process.

The peak sun load on a single-story building is usually at the peak when the sun effect is on the roof.

Usually, the areas on the east and west side of the building and to a certain extent the south facing walls and glass are the governing factors when estimating the heat load of a building.

The amount of heat given off by people at rest is different from that given off by active people.

Electrical lighting within the conditioned space is calculated by the full amount of electricity multiplied by 3.41 Btu per watt. Use only the lights that are on during the peak load in the estimate.

In installations where the sun will shine into the windows and the electric lights are also needed, both are used in the estimation process.

Electric appliances are calculated on the basis of 3.41 Btu per watt of power used during operation.

The amount of heat given off by electric motors will vary according to the motor size. Smaller motors are less efficient than the larger motors and therefore give off more heat.

Steam tables and coffee urns and other such devices that are used for food warming also add heat to the conditioned space and must be included in the heat load estimate.

Infiltration will be present in almost all spaces unless a positive pressure is maintained inside the space.

Natural infiltration may supply a sufficient quantity of fresh air to the space, especially when there are skylights and operable windows that are equal to at least 5% of the floor area in square feet.

The heat load caused by ducts can usually be omitted from the estimate, unless they are placed in hot or cold spaces. Then the heat load must be added to the estimate.

A 10% safety factor is proper and reasonable for most types of installations.

Chapter 3

Residential Heat Load Calculation Procedure

We will now calculate the heat load for a residential building. The calculation will use the heat gain and heat loss factors introduced in Chapter 2.

INTRODUCTION

During this calculation example we will estimate the heat load for a single-story residential building. The window and door schedule is listed on the floor plan for the structure.

While computing this heat load calculation, please refer to the paragraphs, figures, and tables indicated. The information found there will help you to better understand each step as it is undertaken. The assumed design conditions, building orientation, type of construction and so on should be indicated on the survey and check list.

This example is presented only as an illustration for the use of data and the procedures used in estimating the building heat load. Therefore, it is not to be considered as a recommendation for construction details.

In the following step by step procedure we will use the method and forms used by TU Electric Company. The utility company in your area will probably have similar data for your area. They are usually glad to share if you will just ask them. There are more tedious forms available, however, they will generally produce the same results only after hours of labor in making the calculations. Step 1. Refer to Figure 3-1, TU Electric Form 219R and Figure 3-2 the TU Electric Weather Chart.

The cooling design conditions for the Fort Worth, Texas area are usually selected as 100°F outside temperature and 75°F inside temperature. Place this information in the proper space at the top of the form in Figure 3-1 (Figure 3-3).

Subtract the two to obtain the temperature difference between the two (25°F).

Figure 3-1. Residential Heating and Cooling Estimate Form. (Courtesy of TU Electric Co.)

HEATING AND COOLING ESTIMATE
RESIDENTIAL

Customer__________
Address__________
Town__________ District__________
Accepted by__________ Organization__________

DESIGN CONDITIONS
COOLING D.B. °F
Outside Temp.______
Inside Temp.______
Temp. Diff.______

HEATING D.B. °F
Inside Temp.______
Outside Temp.______
Temp. Diff.______

INSULATION
Ceiling______
Wall______
Floor______
Window______

HOUSE FACES OVERHANG OR SHADING
☐ North______
☐ East______
☐ South______
☐ West______

CONSTRUCTION		COOLING FACTOR							HEATING FACTOR	QUANTITY	COOLING Btuh	HEATING Btuh
1. GLASS-SOLAR GAIN Windows-Doors-Skylights (Sq.Ft.) (Note A)	SC	No Shading	Inside Shading & Overhang 0′	1′	2′	3′	4′	Outside Shades				
North Northeast		32 50	14 23	0 21	0 19	0 19	0 17	0 17				
East Southeast		50 80	23 36	20 26	19 19	17 8	14 0	13 21				
South Southwest		100 155	45 70	33 51	10 39	0 17	0 0	26 42				
West Northwest		190 128	86 58	82 54	80 51	67 51	57 49	57 37				
Horizontal		185										

	U	DESIGN D.B. TEMP. DIFFERENTIAL 20	22	25	30	65	60	55	50	QUANTITY	COOLING Btuh	HEATING Btuh
2. GLASS TRANSMISSION (Sq.Ft)		U X TD				U X TD						
Standard Single Glazing Insulating-Double Glazing	1.13 .78	22.6 15.6	24.9 17.2	28.2 19.5	33.9 23.4	73.4 50.7	67.8 46.8	62.2 42.9	56.5 39.0			
Storm Window Storm or Insulating Glass with Thermal Break	.67 .56	13.4 11.2	14.7 12.3	16.8 14.0	20.1 16.8	43.6 36.4	40.2 33.6	36.8 30.8	33.5 28.0			
3. DOORS--(Sq.Ft.)		U X TD				U X TD						

Solid Wood or Hollow Core	.55	11.0	12.1	13.8	16.5	35.8	33.0	30.3	27.5			
Wood with Storm Door	.34	6.8	7.5	8.5	10.2	22.1	20.4	18.7	17.0			
Metal with 1 1/2" Urethane	.11	2.2	2.4	2.8	3.3	7.2	6.7	6.1	5.6			
4a. FLOOR: SLAB (Linear Ft. Exposed Edge)		U X TD				U X TD						
No Edge Insulation	.81					52.6	48.6	44.6	40.5			
R-4 Edge Insulation	.68					44.2	40.8	37.4	34.0			
R-7 Edge Insulation	.55					35.8	33.0	30.2	27.5			
4b. FLOORS: ENCLOSED CRAWL SPACE (Sq. Ft.)		U X Zero				U X (TD - 20)						
No Insulation	.270					12.2	10.8	9.4	8.1			
R-7 Insulation	.093					4.2	3.7	3.2	2.8			
R-11 No Insulation	.073					3.3	2.9	2.6	2.2			
R-13 Insulation	.060					2.7	2.4	2.1	1.8			
R-19 Insulation	.046					2.1	1.8	1.6	1.4			
R-22 Insulation	.039					1.8	1.6	1.4	1.2			
4c. FLOORS: OPEN CRAWL SPACE (Sq. Ft.)		U X (TD - 5)				U X TD						
No Insulation	.374	5.6	6.4	7.5	9.4	24.3	22.4	2-.6	18.7			
R-7 Insulation	.103	1.5	1.8	2.1	2.6	6.7	6.2	5.7	5.2			
R-11 Insulation	.073	1.1	1.2	1.5	1.8	4.7	4.4	4.0	3.6			
R-13 Insulation	.064	1.0	1.1	1.3	1.6	4.2	3.8	3.5	3.2			
R-19 Insulation	.046	0.7	0.8	0.9	1.2	3.0	2.8	2.5	2.3			
R-22 Insulation	.041	0.6	0.7	0.8	1.0	2.7	2.5	2.3	2.0			
5. Walls--(Sq.Ft.) Net		U X TD				U X TD						
No Insulation--Solid Masonry	.389	7.8	8.6	9.7	11.7	25.3	23.3	21.4	19.5			
No Insulation--Wood Siding	.320	6.4	7.0	8.0	9.6	20.8	19.2	17.6	16.0			
No Insulation--Brick Veneer	.240	4.8	5.3	6.0	7.2	15.6	14.4	13.2	12.0			
R-5 Insulation	.128	2.6	2.8	3.2	3.8	8.3	7.7	7.0	6.4			
R-7 Insulation	.109	2.2	2.4	2.7	3.3	7.1	6.5	6.0	5.5			
R-11 Insulation	.075	1.5	1.7	1.9	2.3	4.9	4.5	4.1	3.8			
R-13 Insulation	.065	1.3	1.4	1.6	2.0	4.2	3.9	3.6	3.3			
R-16 Insulation	.054	1.1	1.2	1.4	1.6	3.5	3.2	3.0	2.7			
R-19 Insulation	.047	0.9	1.0	1.2	1.4	3.1	2.8	3.6	3.3			
R-24 Insulation	.038	0.8	0.8	1.0	1.1	2.5	2.3	2.1	1.9			

(Continued)

Figure 3-1. (*Concluded*)

CONSTRUCTION	U	COOLING FACTOR — DESIGN D.B. TEMP. DIFFERENTIAL 20	22	25	30	HEATING FACTOR — DESIGN D.B. TEMP. DIFFERENTIAL 65	60	55	50	QUANTITY	COOLING Btuh	HEATING Btuh
6a. CEILINGS WITH ATTIC (Sq. Ft.) (Note B)		U X (TD + 40)				U X TD						
No Insulation	.598	35.9	37.1	38.9	41.9	38.9	35.9	32.9	29.9			
R-4 Insulation	.176	10.6	10.9	11.4	12.3	11.4	10.6	9.7	8.8			
R-7 Insulation	.114	6.8	7.1	7.4	8.0	7.4	6.8	6.3	5.7			
R-11 Insulation	.079	4.7	4.9	5.1	5.5	5.1	4.7	4.3	4.0			
R-19 Insulation	.048	2.9	3.0	3.1	3.4	3.1	2.9	2.6	2.4			
R-22 Insulation	.042	2.5	2.6	2.7	2.9	2.7	2.5	2.3	2.1			
R-26 Insulation	.036	2.2	2.2	2.3	2.5	2.3	2.2	2.0	1.8			
R-30 Insulation	.032	1.9	2.0	2.1	2.2	2.1	1.9	1.8	1.6			
R-33 Insulation	.029	1.7	1.8	1.9	2.0	1.9	1.7	1.6	1.5			
R-38 Insulation	.025	1.5	1.6	1.6	1.8	1.6	1.5	1.4	1.2			
6b. CEILINGS *NO ATTIC (Sq. Ft.) (Note B)		U X (TD + 45)				U X TD						
No Insulation	.470	30.6	31.5	32.9	35.3	30.6	28.2	25.9	23.5			
R-4 Insulation	.160	10.4	10.7	11.2	12.0	10.4	9.6	8.8	8.0			
R-5 Insulation	.130	8.5	8.7	9.1	9.8	8.5	7.8	7.2	6.5			
R-7 Insulation	.109	7.1	7.3	7.6	8.2	7.1	6.5	6.0	5.5			
R-11 Insulation	.076	4.9	5.1	5.3	5.7	4.9	4.6	4.2	3.8			
R-19 Insulation	.047	3.1	3.1	3.3	3.5	3.1	2.8	2.6	2.4			
R-26 Insulation	.035	2.3	2.1	2.2	2.3	2.3	2.1	1.9	1.8			
R-30 Insulation	.031	2.0	2.1	2.2	2.3	2.0	1.9	1.7	1.6			
7. INFILTRATION: VOL. METHOD (Cu. Ft.) SENSIBLE	q_s = .018	q_s X w_c X TD (w_c @ 7.5 m/h wind = 0.72)				q_s X w_c X TD (w_c @ 15.0 m/h = 1.00)						

	Air Changes Per Hour Saved	Improvement to Structure
	0.3564 0.2808	Soleplate Sealed Wiring @ Plumbing Holes and Furrdowns Sealed
	0.0756	Exterior Doors @ Windows Weather Stripped
	0.1836	Exterior Doors @ Windows Rough opening Caulked
	0.0216	Attic Access Outside Conditioned Space or Weather Stripped

	0.0756	Outside Sheathing Holes Sealed and Polyethylene Film Installed					
	0.0864	Ventless or Dampered Range Hood Installed (or no range vent)					
1.75 ______ = ______ Air Changes/hr X (*Note C)			0.26 0.29 0.32 0.39	1.17 1.08 0.99 0.90			
8. PEOPLE--SENSIBLE HEAT (Ave. No.)			250				
9. SUB TOTAL---SENSIBLE HEAT (Total Lines 1 through 8) Btuh							
10. TOTAL SENSIBLE-INCLUDING DUCT LOSS (Multiply Line 9 SubTotal by Duct Factor (From Table Below)						X______	X______
11. UNITARY COOLING EQUIPMENT REQUIREMENT (Btuh): Line 10 X 1.25							
12. PEOPLE--LATENT HEAT (Ave. No.)			200				
13. TOTAL LOAD Btuh (Total Lines 11 and 12)							

DUCTWORK LOCATION	INSUL. THICKNESS	COOLING FACTOR	HEATING FACTOR
Attic--vented	1"	1.15	1.25
--vented	2"	1.10	1.15
--unvented	1"	1.20	1.20
--unvented	2"	1.15	1.15
Crawlspace--vented	2"	1.10	1.15
--unvented	1"	1.05	1.10
Within conditioned area	1"	1.00	1.00
Within slab	0"	1.20	1.25

*Note A 1. Use only the one largest gain
2. Multiply the factors by shading coefficient (SC) of .90 for plate glass, .85 for double glass
3. No solar gain used if overhang exceeds 4', or if glass is shaded by permanent structure

*Note B For light colored roof, Multiply COOLING Factor by .75

*Note C Sum of air changes per hour saved

*Note D 12,000 Btuh/ton; 3413 Btuh/kW

*Note E See Ft. Worth H.U.D. circular Letter No. 87-2.
The Warranter of the HVAC equipment accepts the calculated loads as his/her own calculated loads as his/her own calculations

Figure 3-2. TU Electric Weather Chart. (Courtesy of TU Electric Co.)

	Outside Temperature Design (°F)*		Heating Degree Days (65°F) Base**	Annual Heating (Consumption) Factor***
	Summer	Winter		
Dallas Div. Ft. Worth & Central Region: Arlington, Dallas, Duncanville, Euless, Ft. Worth Garland, Grand Prairie, Grapevine, Farmers Branch, Irving, Lancaster, Mesquite, Plano, Richardson	100	20	2,400	310
Eastern Region: Athens, Corsicana, Crokett, Lufkin, Nacogdoches, Palestine, Terrell, Tyler	100	20	2,300	330
Northern Region: Decatur, Denison, Gainesville, McKinney, Mineral Wells, Paris, Sherman, Sulphur Springs	100	20	2,750	344
Northwest Region: Archer City, Breckenridge, Burkburnett, DeLeon, Eastland Electra, Graham, Henrietta, Iowa Park, Wichita Falls	100	20	2,875	355
Southern Region: Brownwood, Cleburne, Hillsboro, Killeen, Round Rock, Stephenville, Taylor, Temple, Waco, Waxahachie	100	25	2,225	321
Western Region: Andrews, Big Spring, Colorado City, Crane, Lamesa, Midland, Monahans, Odessa, Snyder, Sweetwater	100	20	2,750	344

*Generalized from ASHRAE Fundamentals Handbook 1985, Chapter 24
**Accuracy 10%
***Factors for maintaining 75°F indoors at design conditions. Units = kWh/MBtuh

Figure 3-3. Entering Design Conditions on Heat Load Form. (Courtesy of TU Electric Co.)

DESIGN CONDITIONS	
COOLING D.B. °F	
Outside Temp.	________
Inside Temp.	________
Temp. Diff.	________

Step 2. The heating design is usually calculated on 75°F inside temperature. From the TU Electric Weather Chart, the outside design temperature is selected for the geographical area involved. For example Fort Worth has a 20°F outside winter design temperature (Figure 3-4).

Figure 3-4. Section of TU Electric Weather Chart. (Courtesy of TU Electric Co.)

Outside Temperature Design (°F)* Summer	Outside Temperature Design (°F)* Winter	Heating Degree Days (65°F) Base**	Annual Heating Consumption Factor***
100	20	2,400	310

Step 3. Insert the inside and the outside design temperatures on the form and subtract the two to find the temperature differential (Figure 3-5).

Figure 3-5. Entering Inside and Outside Design Conditions on Heat Load Form. (Courtesy of TU Electric Co.)

DESIGN CONDITIONS			
COOLING D.B. °F		HEATING D.B. °F	
Outside Temp.	100	Inside Temp.	75
Inside Temp.	75	Outside Temp.	20
Temp. Diff.	25	Temp. Diff.	55

Step 4. Refer to the set of house plans to find the additional design criteria. In addition to the floor plan, reference should be made to other

parts of the plans, such as the roof plan, foundation plan, cross sections and elevations. If some of the needed information cannot be found on the plans, it can be obtained by consulting with the architect, builder, or owner.

See the floor plan of the house (Figure 3-6). Much of the design information can be found on it, but more is needed. The following also applies:

1. The house is for a family of four people.
2. It will be built on a slab without perimeter insulation.
3. The house will have a hip roof with a 2 foot overhang.
4. The windows will be insulating glass with a thermal break. The doors will be metal with 1-1/2" Urethane.
5. Inside shading will be used over the glass areas.
6. The walls will have R-16 insulation. The ceilings will have R-30 insulation.
7. The attic will be ventilated. The duct insulation will be 2 inches thick.
8. The ceilings will be 8 feet high.
9. A heat pump with an SEER of 11 will be used for both heating and cooling the house.
10. As seen from the arrow to the north on the floor plan (Figure 3-6) the front of the house will face the west.

Step 5. Use the data which applies to the house construction to fill in the remainder of the data block at the top of the form (Figure 3-7).

Step 6. Item 1. Glass Solar Gain: Locate the correct column under the cooling factor heading. In this example, the overhang is 2 feet and there is inside shading. Circle the 2' under the Inside Shading and Overhang Column. The factors directly below the 2' figure are the number of Btus of radiant heat gain per square foot of glass for each direction of exposure (Figure 3-8).

Step 7. Item 1: Circle the solar gain factors under the 2' column which apply to each direction of glass exposure (Figure 3-9).

The glass is exposed on the north, east, south, and west elevations. The solar gain factors which apply are circled under the 2' overhang column opposite to the direction of the exposure listed on the form.

Step 8. Item 1. Quantity: Under the quantity column heading, insert the square feet of glass for the corresponding direction of exposure after first determining the glass area for exposure. Refer to the Floor Plan and the Window Schedule (Figure 3-6).

WINDOW SCHEDULE				
MARK	TYPE	SIZE	MFG. NO.	MATL
A	RANCH WALL	9'0" x 6'1"	9061	ALMN
B	" "	4'0" x 6'1"	4061	"
C	SINGLE HUNG	3'0" x 4'5"	3045	"
D	" "	3'0" x 3'1"	3031	"
E	" "	2'0" x 3'1"	2031	"
F	HORIZ. SLIDER	5'11 5/8" x 1'11 5/8"	6020SV	"

DOORS	
MARK	SIZE
1	3'0" x 6'8"
2	2'6" x 6'8"
12	6'0" x 6'8"

Figure 3-6. Floor Plan. (Courtesy of TU Electric Co.)

Figure 3-7. Entering Data that Applies to the Building Construction. (Courtesy of TU Electric Co.)

Insulation		House Faces	Overhang or Shading
Ceiling	R-19	__ North	2'
Wall	R-11	__ East	2'
Floor	single	__ South	2'
Window	single	__ West	2

Figure 3-8. Entering Overhang and Inside Shading on Heat Load Form. (Courtesy of TU Electric Co.)

Construction		Cooling Factor						
1. Glass—Solar Gain Windows-Doors-Skylights (Sq. Ft.) (Note A)	SC	No Shading	Inside Shading & Overhang					Outside Shades
			0'	1'	(2')	3'	4'	
North		32	14	0	0	0	0	0
Northeast		50	23	21	19	19	17	17
East		50	23	20	19	17	14	13
Southeast		80	36	26	19	8	0	21
South		100	45	33	10	0	0	26
Southwest		155	70	51	39	17	0	42
West		190	86	82	80	67	57	57
Northwest		128	58	54	51	51	49	37
Horizontal		185						

Figure 3-9. Entering Solar Gain Factors on Heat Load Form. (Courtesy of TU Electric Co.)

Construction		Cooling Factor						
1. Glass—Solar Gain Windows-Doors-Skylights (Sq. Ft.) (Note A)	SC	No Shading	Inside Shading & Overhang					Outside Shades
			0'	1'	(2')	3'	4'	
North		32	14	0	(0)	0	0	0
Northeast		50	23	21	19	19	17	17
East		50	23	20	(19)	17	14	13
Southeast		80	36	26	19	8	0	21
South		100	45	33	(10)	0	0	26
Southwest		155	70	51	39	17	0	42
West		190	86	82	(80)	67	57	57
Northwest		128	58	54	51	51	49	37
Horizontal		185						

The glass area is found for each window by multiplying the window width in feet by the window height in feet. The glass door area is found in a like manner. The total area of glass exposure for each direction is the sum of the individual window areas and glass door areas located in that direction. As follows:

Glass facing north:

2-F	2(6 × 2) = 24 sq. ft.
1-sliding door	(6 × 7) = 42 sq. ft.
	total = 66 sq. ft.

Glass facing east:

1-F	(6 × 2) = 12 sq. ft.
1-E	(2 × 3) = 6 sq. ft.
	total = 18 sq. ft.

Glass facing south:

1-D	$(3 \times 3) = 9$ sq. ft.
1-C	$(3 \times 41/2) = 13\ 1/2$ sq. ft.
	total = 22 1/2 sq. ft.

Glass facing west:

2-B	$2(4 \times 6) = 48$ sq. ft.
1-E	$(2 \times 3) = 6$ sq. ft.
1-A	$(9 \times 6) = 54$ sq. ft.
	total = 108 sq. ft.

Fill in the square feet of glass area in the quantity column opposite the direction of exposure, and under the appropriate shading category. Insert the value for the SC (shading coefficient) to compensate for plateglass or multiple layers as shown on Form 219R (Figure 3-10).

Step 9. Item 1. Cooling Factors: Multiply the cooling factors times the quantity of glass for each direction of exposure and insert these figures in the cooling Btuh Column (Figure 3-10). Note A gives the shading factors to modify the cooling Btuhs should other than standard single-pane window glass be used. For instance, the Btuh for double glass would be the product of 0.85 times the cooling factor times the area of glass (Figure 3-11).

For all practical purposes, the sun can impose a maximum solar gain from only one vertical wall or exposure at any given time; therefore, only the one largest vertical calculated Btuh solar gain will be included in the total cooling load estimate. Identify the largest cooling Btuh figure by marking it in some manner, such as an asterisk, or by crossing-out the other amounts.

A skylight was not used in this particular house, so the "Horizontal" area of glass category listed under "Glass-Solar Gain" is ignored. If horizontal glass (i.e., skylight) was used, then the total square feet (area) of horizontal quantity should be multiplied by the solar gain cooling factor to obtain the cooling Btuh for solar gain from horizontal glass. This should be marked with an asterisk also and added along with the largest vertical area solar gain Btuh to obtain the total cooling Btuh load when calculating the subtotal for Item 9.

Step 10. Item 1. Design Conditions: From the established design conditions (Step 1 and Step 3), determine the temperature differentials for heating and cooling. In the last line of Item 1, circle the heating and cooling temperature differentials for a cooling temperature differential (TD) of 25°F and a heating TD of 55°F (Figure 3-12).

Figure 3-10. Entering Quantity of Glass on Heat Load Form. (Courtesy of TU Electric Co.)

Construction	Cooling Factor							Heating Factor	Quantity	Cooling Btuh	Heating Btuh
1. Glass—Solar Gain Windows-Doors-Skylights (Sq. Ft.) (Note A) SC	No Shading	Inside Shading & Overhang 0'	1'	2'	3'	4'	Outside Shades				
North	32	14	0	0	0	0	0		66		
Northeast	50	23	21	19	19	17	17				
East	50	23	20	19	17	14	13		18		
Southeast	80	36	26	19	8	0	21				
South	100	45	33	10	0	0	26		23		
Southwest	155	70	51	39	17	0	42				
West	190	86	82	80	67	57	57		108		
Northwest	128	58	54	51	51	49	37				
Horizontal	185										

Figure 3-11. Indicating the Largest Glass Solar Gain on Heat Load Form. (Courtesy of TU Electric Co.)

Construction		Cooling Factor							Heating Factor	Quantity	Cooling Btuh	Heating Btuh
1. Glass—Solar Gain Windows-Doors-Skylights (Sq. Ft.) (Note A)	SC	No Shading	Inside Shading & Overhang					Outside Shades				
			0'	1'	(2')	3'	4'					
North		32	14	0	(0)	0	0	0		66		
Northeast		50	23	21	19	19	17	17				
East		50	23	20	(19)	17	14	13		18		
Southeast		80	36	26	19	8	0	21				
South		100	45	33	(10)	0	0	26		23		
Southwest		155	70	51	39	17	0	42				
West		190	86	82	(80)	67	57	57		108	8640	
Northwest		128	58	54	51	51	49	37				
Horizontal		185										

Figure 3-12. Indicating the Heating and Cooling Temperature Differentials on Heat Load Form. (Courtesy of TU Electric Co.)

Construction	U	Cooling Factor	Heating Factor	Quantity	Cooling Btuh	Heating Btuh
		Design D.B Temp. Differential				
		20 22 (25) 30	65 60 (55) 50			

Step 11. Items 2 through 7: (Figure 3-13).

Circle all of the heating and cooling factors which apply and are used in Items 2 through 6. These values are located directly under the circled design dry bulb (D.B.) temperature differential for both heating and cooling and directly across from the type of construction used in the building. Item 7 has only one heating and one cooling factor for each Design D.B. temperature differential. See Step 12 for additional instructions on Item 7.

The following characteristics apply to the structure. (Factors which apply are circled under the 25°F cooling and 55°F heating differential columns for Items 2, 3, 4a, 6a, and 7 on Form 219R (Figure 3-6).

1. Insulating glass with a thermal break (Item 2).
2. Metal windows with 1 1/2" Urethane (Item 3).
3. Slab floor with no insulation (Item 4a).
4. Wall insulation of R-16 (Item 5).
5. Ceiling with R-30 insulation and attic space above (Item 6a).
6. Doors and windows are weather-stripped and the attic access is in the garage (unconditioned space). The range has a dampered range hood.

Step 12. Item 7: For Item 7 determine the construction features of the house structure that have resulted in reduced infiltration. Identify these on the form by an arrow or by circling the reduction factor. After making the air change reduction determinations, add the marked numbers to obtain a total and place it in the blank provided beneath the "Air Changes Per Hour Saved" column. This figure is subtracted from 1.75 (the number of air changes in an unimproved structure) to find the "Total Air Changes Per Hour." Figure 3-14 indicates the amount of infiltration of unconditioned air and the air changes saved per hour with each method.

Figure 3-13. Indicating All of the Heating and Cooling Factors that Apply on Heat Load Form. (Courtesy of TU Electric Co.)

CONSTRUCTION	U	COOLING FACTOR				HEATING FACTOR				QUANTITY	COOLING Btuh	HEATING Btuh
		DESIGN D.B TEMP. DIFFERENTIAL										
		20	22	(25)	30	65	60	(55)	50			
2. **GLASS TRANSMISSION** (Sq. Ft.)		U X TD				U X TD						
Standard--Single glazing	1.13	22.6	24.9	(28.2)	33.9	73.4	67.8	(62.2)	56.5			
Insulating--Double Glazing	.78	15.6	17.2	19.5	23.4	50.7	46.8	42.9	39.0			
Storm Window	.67	13.4	14.7	19.5	20.1	43.6	40.2	36.8	33.5			
Storm or Insulating Glass with Thermal Break	.56	11.2	12.3	14.0	16.8	36.4	33.6	30.8	28.0			
3. **DOORS**--(Sq. Ft.)												
Solid Wood or Hollow Core	.55	11.0	12.1	(13.8)	16.5	35.8	33.0	(30.3)	27.5			
Wood with Storm Door	.34	6.8	7.5	8.5	10.2	22.1	20.4	18.7	17.0			
Metal with 1 1/2" Urethane	.11	2.2	2.4	2.8	3.3	7.2	6.7	6.1	5.6			
4A. **FLOOR: SLAB** (Linear Ft. Exposed Edge)		U X Zero				U X TD						
No Edge Insulation	.81					52.6	48.6	(44.6)	40.5			
R-4 Edge Insulation	.68					44.2	40.8	37.4	34.0			
R-7 Edge Insulation	.55					35.8	33.0	30.2	27.5			
4b. **FLOORS: ENCLOSED CRAWL SPACE** (Sq. Ft.)		U X Zero				U X TD						
No Edge Insulation	.270					12.2	10.8	9.4	8.1			
R-7 Insulation	.093					4.2	3.7	3.2	2.8			

	U											
R-11 Insulation	.073					3.3	2.9	2.6	2.2			
R-13 Insulation	.060					2.7	2.4	2.1	1.8			
R-19 Insulation	.046					2.1	1.8	1.6	1.4			
R-22 Insulation	.039					1.8	1.6	1.4	1.2			
4c. FLOORS: OPEN CRAWL SPACE (Sq. Ft)		U X (TD - 5)				U X TD						
No Insulation	.374	5.6	6.4	7.5	9.4	24.2	22.4	20.6	18.7			
R-7 Insulation	.103	1.5	1.8	2.1	2.6	6.7	6.2	5.7	5.2			
R-11 Insulation	.073	1.1	1.2	1.5	1.8	4.7	4.4	4.0	3.6			
R-13 Insulation	.064	1.0	1.1	1.3	1.6	4.2	3.8	3.5	3.2			
R-19 Insulation	.046	0.7	0.8	0.9	1.2	3.0	2.8	2.5	2.3			
R-22 Insulation	.041	0.6	0.7	0.8	1.0	2.7	2.5	2.3	2.0			
5. WALLS--(Sq. Ft.) Net		U X TD				U X TD						
No Insulation-Solid Masonry	.389	7.8	8.6	9.7	11.7	25.3	23.3	21.4	19.5			
No Insulation--Wood Siding	.320	6.4	7.0	8.0	9.6	20.8	19.2	17.6	16.0			
No Insulation--Brick Veneer	.240	4.8	5.3	6.0	7.2	15.6	14.4	13.2	12.0			
R-5 Insulation	.128	2.6	2.8	3.2	3.8	8.3	7.7	7.0	6.4			
R-7 Insulation	.109	2.2	2.4	2.7	3.3	7.1	6.5	6.0	5.5			
R-11 Insulation	.075	1.5	1.7	1.9	2.3	4.9	4.5	4.1	3.8			
R-13 Insulation	.065	1.3	1.4	1.6	2.0	4.2	4.5	4.1	3.8			
R-16 Insulation	.054	1.1	1.2	1.4	1.6	3.5	3.2	3.0	2.7			
R-19 Insulation	.047	0.9	1.0	1.6	2.0	3.1	2.8	2.6	2.4			
R-24 Insulation	.038	0.8	0.8	1.0	1.1	2.5	2.3	2.1	1.9			

(Continued)

Figure 3-13 (*Concluded*)

CONSTRUCTION	U	COOLING FACTOR				HEATING FACTOR				QUANTITY	COOLING Btuh	HEATING Btuh
		DESIGN D.B. TEMP. DIFFERENTIAL										
		20	22	25	30	65	60	55	50			
6a. CEILINGS-WITH WITH ATTIC(Sq. Ft.) (*Note B)		U X (TD + 40)				U X TD						
No Insulation	.598	35.9	37.1	38.9	41.9	38.9	35.9	32.9	29.9			
R-4 Insulation	.176	10.6	10.9	11.4	12.3	11.4	10.6	9.7	8.8			
R-7 Insulation	.114	6.8	7.1	7.4	8.0	7.4	6.8	6.3	5.7			
R-11 Insulation	.079	4.7	4.9	5.1	5.5	5.1	4.7	4.3	4.0			
R-19 Insulation	.048	2.9	3.0	3.1	3.4	3.1	2.9	2.6	2.4			
R-22 Insulation	.042	2.5	2.6	2.7	2.9	2.7	2.5	2.3	2.1			
R-26 Insulation	.036	2.2	2.2	2.3	3.4	2.3	2.2	2.0	1.8			
R-30 Insulation	.032	1.9	2.0	2.1	2.2	2.1	1.9	1.8	1.6			
R-33 Insulation	.029	1.7	1.8	1.9	2.0	1.9	1.7	1.6	1.5			
R-38 Insulation	.025	1.5	1.6	1.6	1.8	1.6	1.5	1.4	1.2			
6b. CEILINGS--NO ATTIC (Sq. Ft.)(*Note B)		U X (TD + 45)				T X TD						
No Insulation	.470	30.6	31.5	32.9	35.3	30.6	28.2	25.9	23.5			
R-4 Insulation	.160	10.4	10.7	11.2	12.0	10.4	9.6	8.8	8.0			
R-5 Insulation	.130	8.5	8.7	9.1	9.8	8.5	7.8	7.2	6.5			
R-7 Insulation	.109	7.1	7.3	7.6	8.2	7.1	6.5	6.0	5.5			
R-11 Insulation	.076	4.9	5.1	5.3	5.7	4.9	4.6	4.2	3.8			
R-19 Insulation	.047	3.1	3.1	3.3	3.5	3.1	2.8	2.6	2.4			
R-26 Insulation	.035	2.3	2.1	2.2	2.3	2.3	2.1	1.9	1.8			
R-30 Insulation	.031	2.0	2.1	2.2	2.3	2.0	1.9	1.7	1.6			
7. INFILTRATION: VOL. METHOD (Cu. Ft.)- SENSIBLE	q_s = .018	q_s X w_c X TD (w_c@ 7.5 m/h wind= 0.72)				q_s X w_c X TD (w_c@ 15.0m/h wind= 1.00)						

	Air Changes Per Hour Saved	Improvement to Structure					
	0.3564	Soleplate Sealed					
	0.2808	Wiring & Plumbing Holes and Furrdowns sealed					
	0.0756	Exterior Doors & Windows Weather Stripped					
	0.1836	Exterior Doors & Windows Rough Opening Caulked					
	0.0216	Attic Access Outside Conditioned Space or Weather Stripped					
	0.0756	Outside Sheathing Holes Sealed and Polyethylene Film Installed					
	0.0864	Ventless or Dampered Range Hood Installed (or no Range vent)					
1.75 - .1836 = 1.57 Air Changes/hr x (*Note C)			0.26 0.29 (0.32) 0.39	1.17 1.08 (.99) .90	13264	6664	20616

Figure 3-14. Method of Reducing Infiltration of Unconditioned Air (Ranges from 0.67 to 1.75 Air Changer Per Hour). (Courtesy of TU Electric Co.)

Infiltration Reduced By	Air Changes Saved Per Hour
A. Soleplate Sealed (33%)	0.3564
B. Wiring & Plumbing Holes and Furrdowns Sealed (26%)	0.2808
C. Exterior Doors & Windows Weather Stripped	0.0756
D. Exterior Door & Window Rough Openings Caulked (17%)	0.1836
E. Attic Access Outside Conditioned Space or Weather Stripped (2%)	0.0216
F. Outside Sheathing Holes Sealed and Polyethylene Film Installed (9%)	0.0756
G. Ventless or Dampered Range Hood Installed or No Range Vent (8%)	0.0864
TOTAL	1.0800

Step 13. Determine the quantity of Items 2 through 7: Items 2 and 3 are in square feet. Item 4a is in linear feet. Item 7 is in cubic feet. (The units for each quantity are indicated). Insert these figures on the form under the quantity column and directly across from the items to which they apply.

The following calculations were made after reference to the floor plan and to additional information supplied in the example in Step 4 (Figure 3-15).

Item 2. Glass Transmission: The total of the glass area found under quantity of solar gain, Item 1.

$66 + 18 + 23 + 108 = 215$ sq. ft.

Item 3. Doors: The area of the doors is found by totaling the areas of all exterior doors. The area for each door is found by multiplying the length in feet times the width in feet of each door (Figure 3-16).

$2\text{-}1/2 \times 7 = 18$
$3 \times 7 = 21 = 39$ sq. ft.

Figure 3-15. Entering Calculated Quantities on Heat Load Form. (Courtesy TU Electric Co.)

CONSTRUCTION	U	COOLING FACTOR				HEATING FACTOR				QUANTITY	COOLING Btuh	HEATING Btuh
		DESIGN D.B. TEMP. DIFFERENTIAL										
		20	22	25	30	65	60	55	50			
2. GLASS TRANSMISSION (Sq. Ft.)		U X TD				U X TD						
Standard--Single Glazing	1.13	22.6	24.9	28.2	33.9	73.4	67.8	62.2	56.5	215		
Insulation--Double Glazing	.78	15.6	17.2	19.5	23.4	50.7	46.8	42.9	39.0			
Storm Window	.67	13.4	14.7	16.8	20.1	43.6	40.2	36.8	33.5			
Storm or Insulating Glass with Thermal Break	.56	11.2	12.3	14.0	16.8	36.4	33.6	30.8	28.0			

Figure 3-16. Entering Door Information on Heat Load Form. (Courtesy of TU Electric Co.)

3. DOORS--(Sq. Ft.)		U X TD				U X TD						
Solid Wood or Hollow Core	.55	11.0	12.1	13.8	16.5	35.8	33.0	30.3	27.5	39	109	238
Wood with Storm Door	.34	6.8	7.5	8.5	10.2	22.1	20.4	18.7	17.0			
Metal with 1 1/2" Urethane	.11	2.2	2.4	2.8	3.3	7.2	6.7	6.1	5.6			

Item 4a. Slab: The quantity is in linear feet of exposed edge. It is the number of feet measured around the house, but does not include the garage in this example (Figure 3-17).

2(37) + 2(54) = 182 feet.

Item 5. Walls: The net wall area = (total length of outside wall in feet × ceiling height in feet) minus (door and window area in square feet) (Figure 3-18).

Total area = 182 × 8 = 1456 sq. ft.
Net area = 1456 - 39 - 215 = 1202 sq. ft.

Item 6a. Ceiling: The length of the structure in feet times the width of the structure in feet. See Figure 3-19.

31 × 29 = 899
33 × 23 = 759
Total = 1658 sq. ft.

Item 7. Infiltration Volume: The area of the structure in square feet times the ceiling height in feet = volume (Figure 3-20).
1658 × 8 = 13,264 cubic feet.
Insert these calculated quantities on the form opposite the corresponding description of construction. Refer to Figure 3-21 for additional quantities.

Step 14. Cooling Btuh: Determine the cooling Btuh for each Item (2 through 6) by multiplying the circled cooling factor times the quantity and inserting this figure in the COOLING Btuh column. Determine the heating Btuh for each Item (2 through 6) by multiplying the circled heating factor times the quantity and inserting this figure in the HEATING Btuh column. Do the same for Item 7, except in addition multiply the air changes figure. Btuh calculations are shown in Item 2 and Item 7. Others can be seen in Figure 3-21.

Step 16. Item 9: Obtain a subtotal for the cooling requirement by totaling Items 1 through 8 which appear under the COOLING Btuh column. Be sure to use only the single largest vertical solar gain PLUS any horizontal solar gain in Item 1. Make a subtotal for heating Btuh by totaling Items 2 through 7 under the HEATING Btuh column.

Figure 3-17. Entering Floor Information on Heat Load Form. (Courtesy of TU Electric Co.)

	U	U X Zero				U X TD						
4a. FLOOR: SLAB (Linear Ft. Exposed Edge)												
No Edge Insulation	.81					52.6	48.6	(44.6)	40.5	182		8117
R-4 Edge Insulation	.68					44.2	40.8	37.4	34.0			
R-7 Edge Insulation	.55					35.8	33.0	30.2	27.5			
4b. FLOOR: ENCLOSED CRAWL SPACE (Sq. Ft.)		U X Zero				U X (TD - 20)						
No Insulation	.270					12.2	10.8	9.4	8.1			
R-7 Insulation	.093					4.2	3.7	3.2	2.8			
R-11 Insulation	.073					3.3	2.9	2.6	2.2			
R-13 Insulation	.060					2.7	2.4	2.1	1.8			
R-19 Insulation	.046					2.1	1.8	1.6	1.4			
R-22 Insulation	.039					1.8	1.6	1.4	1.2			
4c. FLOORS: OPEN CRAWL SPACE (Sq. Ft.)		U X (TD - 5)				U X TD						
No Insulation	.374	5.6	6.4	7.5	9.4	24.3	22.4	2.6	18.7			
R-7 Insulation	.103	1.5	1.8	2.1	2.6	6.7	6.2	5.7	5.2			
R-11 Insulation	.073	1.1	1.2	1.5	1.8	4.7	4.4	4.0	3.6			
R-13 Insulation	.064	1.0	1.1	1.3	1.6	3.0	2.8	2.5	2.3			
R-19 Insulation	.046	0.7	0.8	0.9	1.2	3.0	2.8	2.5	2.3			
R-22 Insulation	.041	0.6	0.7	0.8	1.0	2.7	2.5	2.3	2.0			

Figure 3-18. Entering Wall Information on Heat Load Form. (Courtesy of TU Electric Co.)

5. WALLS--(Sq.Ft.) Net		U X TD				U X TD						
No Insulation-Solid Masonry	.389	7.8	8.6	9.7	11.7	25.3	23.3	21.4	19.5			
No Insulation-Wood Siding	.320	6.4	7.0	8.0	9.6	20.8	19.2	17.6	16.0			
No Insulation-Brick Veneer	.240	4.8	5.3	6.0	7.2	15.6	14.4	13.2	12.0			
R-5 Insulation	.128	2.6	2.8	3.2	3.8	8.3	7.7	7.0	6.4			
R-7 Insulation	.109	2.2	2.4	2.7	3.3	7.1	6.5	6.0	5.5			
R-11 Insulation	.075	1.5	1.7	1.9	2.3	4.9	4.5	4.1	3.8			
R-13 Insulation	.065	1.3	1.4	1.6	2.0	4.2	3.9	3.6	3.3	1202	1683	3606
R-16 Insulation	.054	1.1	1.2	(1.4)	1.6	3.5	3.2	(3.0)	2.7			
R-19 Insulation	.047	0.9	1.0	1.2	1.4	3.1	2.8	2.6	2.4			
R-24 Insulation	.038	0.8	0.8	1.0	1.1	2.5	2.3	2.1	1.9			

Figure 3-19. Entering Ceiling Information on Heat Load Form. (Courtesy of TU Electric Co.)

CONSTRUCTION	U	COOLING FACTOR				HEATING FACTOR				QUANTITY	COOLING Btuh	HEATING Btuh
		DESIGN D.B. TEMP. DIFFERENTIAL										
		20	22	25	30	65	60	55	50			
6a. CEILINGS--WITH ATTIC (Sq. Ft.) (*Note B)		U X (TD + 40)				U X TD						
No Insulation	.598	35.9	37.1	38.9	41.9	38.9	35.9	32.9	29.9			
R-4 Insulation	.175	10.6	10.9	11.4	12.3	11.4	10.6	9.7	8.8			
R-7 Insulation	.114	6.8	7.1	7.4	8.0	7.4	6.8	6.3	5.7			
R-11 Insulation	.079	4.7	4.9	5.1	5.5	5.1	4.7	4.3	4.0			
R-19 Insulation	.048	2.9	3.0	3.1	3.4	3.1	2.9	2.6	2.4			
R-22 Insulation	.042	2.5	2.6	2.7	2.9	2.7	2.5	2.3	2.1			
R-26 Insulation	.036	2.2	2.2	2.3	2.5	2.3	2.2	2.0	1.9	1658	3482	2984
R-30 Insulation	.032	1.9	2.0	2.1	2.2	2.1	1.9	1.8	1.6			
R-33 Insulation	.029	1.7	1.8	1.9	2.0	1.9	1.7	1.6	1.5			
R-38 Insulation	.025	1.5	1.6	1.6	1.8	1.6	1.5	1.4	1.2			
6b. CEILINGS--NO-ATTIC (Sq. Ft.) (*Note B)		U X (TD + 45)				U X TD						
No Insulation	.470	30.6	31.5	32.9	35.3	30.6	28.2	25.9	23.5			
R-4 Insulation	.160	10.4	10.7	11.2	12.0	10.4	9.6	8.8	8.0			
R-5 Insulation	.130	8.5	8.7	9.1	9.8	8.5	7.8	7.2	6.5			
R-7 Insulation	.109	7.1	7.3	7.6	8.2	7.1	6.5	6.0	5.5			
R-11 Insulation	.076	4.9	5.1	5.3	5.7	4.9	4.6	4.2	3.8			
R-19 Insulation	.047	3.1	3.1	3.3	3.5	3.1	2.8	2.6	2.4			
R-26 Insulation	.035	2.3	2.3	2.4	2.6	2.3	2.1	1.9	1.8			
R-30 Insulation	.031	2.0	2.1	2.2	2.3	2.0	1.9	1.7	1.6			

Figure 3-20. Entering Infiltration Information on Heat Load Form. (Courtesy of TU Electric Co.)

	Air Changes Per Hour Saved	Improvement to Structure					
→	0.3564	Soleplate Sealed					
→	0.2808	Wiring & Plumbing Holes and Furrdowns sealed					
→	0.0756	Exterior Doors & Windows Weather Stripped					
→	0.1836	Exterior Doors & Windows Rough Opening Caulked					
→	0.0216	Attic Access Outside Conditioned Space or Weather Stripped					
→	0.0756	Outside Sheathing Holes Sealed and Polyethylene Film Installed					
→	0.0864	Ventless or Dampered Range Hood Installed (or no Range vent)					

Figure 3-21. Calculating Infiltration Load on Heat Load Form. (Courtesy of TU Electric Co.)

7. **INFILTRATION: VOL. METHOD (Cu. Ft.) - SENSIBLE**	q_s = .018	q_s X w_c X TD (w_c@ 7.5 m/h wind= 0.72)	q_s X w_c X TD (w_c@ 15.0 m/h wind= 1.00)			
1.75 1.08 = 0.67 Air Changes/hr x (*Note C)		0.26 0.29 (0.32) 0.39	1.17 1.08 (.99) .90	13264	2844	8798

Step 17. Item 10: From the Ductwork Location Chart in the lower left section on the back of the form (Figure 3-22) determine the heating and cooling duct loss factors for the appropriate location and insulation level of the duct system. Circle the duct loss factors. In the example we are referring to the conditioned air supply ductwork, not the return air ductwork. The factors are 1.10 and 1.15 respectfully for a ventilated attic and duct insulation of 2" thickness.

Figure 3-22. Entering Ductwork Information on Heat Load Form. (Courtesy of TU Electric Co.)

Ductwork Location	Insul. Thickness	Cooling Factor	Heating Factor
Attic —vented	1"	1.15	1.25
—vented	2"	1.10	1.15
—unvented	1"	1.20	1.20
—unvented	2"	1.15	1.10
Crawl space —vented	2"	1.10	1.15
—unvented	1"	1.05	1.10
Within Conditioned Area	1"	1.00	1.00
Within Slab	0"	1.20	1.25

Step 18. Item 10: Place the duct loss factors found above in the appropriate blanks in line 10. Determine the total sensible cooling Btuh of the structure by multiplying line 9 cooling Btuh by the cooling duct loss factor. Determine the total sensible heating Btuh of the structure by multiplying line 9 heating Btuh by the heating duct loss factor.

Step 19. Item 11: Multiply the sensible cooling Btuh in Item 10 by 1.25 latent heat factor to obtain the structure cooling Btuh.

Step 20. Item 12: Insert the average number of people under the quantity column. It is the same number that is in Item 8. Multiply this figure by 200 to obtain the latent heat cooling Btuh.

Step 21. Item 13: The total cooling Btuh is obtained by adding the cooling Btuh in Items 11 and 12. The total heating Btuh is the same as that shown in Item 10.

Step 22. Tons of cooling is found by dividing the cooling Btuh in Item 13 by 12,000 (See Note D).

Step 23. The Kilowatt capacity of electric heating is found by dividing the heating Btuh in Item 13 by 3.413 (See Note D).

SUMMARY

The size of heating and cooling equipment is determined by the actual peak load requirement of the building.

The peak sun load will occur, in most cases, at one single time. However, if an equal load occurred at two different times it would not matter which load is used.

When determining the design conditions for your area from a chart, use a town or city near your locality if your particular locality is not listed.

When calculating heat loads, use only the net area of the wall, windows, and doors.

Chapter 4

Residential Equipment Sizing

Heating and cooling equipment must be properly sized if it is to be expected to operate as designed. The estimating procedure must be accurate if the properly sized equipment is to be selected. If the equipment is sized too small, the temperature inside the space will not be maintained properly on warmer days. Likewise, when the equipment is sized too large, humidity control and operating costs will be poor.

INTRODUCTION

There first must be an accurate heat load estimation before the equipment can be properly sized. The correct procedure was described in Capters 2 and 3. The heat load calculation form will indicate the proper size for equipment installed in a particular situation.

The proper sizing of equipment cannot be overemphasized. A unit that is sized too small may operate more economically and the original cost will be less than one that is properly sized. However, the too-small unit will operate continuously during warmer weather and will also result in poor temperature control and customer dissatisfaction. The life of the unit will be shortened when it must operate continuously. When the unit is sized too large, the initial cost will be higher than when properly sized. There will be poor humidity control inside the conditioned space. The equipment operating expense will be higher than normal, which will also cause a dissatisfied customer.

HEATING EQUIPMENT SIZING

It is just as important to properly balance the equipment to the load as it is to properly balance any other system components. Not only is the balance between the different components of the system vital, but the balance between the system and the load, the balance between cost and performance, and many other factors must be given careful consideration.

Equipment that will meet all of the infinite number of design conditions will be difficult to locate. Because of this single factor, the selection of the proper equipment becomes a problem for the design engineer. It generally becomes a matter of choosing the equipment that meets the most requirements for a particular installation, including equipment first cost, operating costs, and efficiency. It usually becomes a problem of which considerations can, and which considerations cannot, be compromised. Just as important is learning how the capacity of the equipment is determined and the performance characteristics of different models and sizes of equipment, and the different combinations of equipment that are available.

It is a great help that the manufacturer has made many of these decisions during the manufacturing design of the equipment. Usually, the engineers have completely analyzed any problems with the unit and have compiled their answers in charts, tables, and formulas which are all easy to use in the field. Because of this, selecting equipment is a matter of understanding and using all of the selection aids that the manufacturer has made available to us (Figure 4-1). We will start our discussion by choosing units that are relatively easy to select, then gradually move to the more difficult equipment selections.

GENERAL INFORMATION

Furnaces and other packaged equipment are fairly simple to select because the individual components have already been chosen by the equipment manufacturer. Thus, balancing the unit to the heating load, and determining such things as the cubic feet per minute (cfm) that the equipment will deliver when installed, and selecting the required electrical characteristics remain to be chosen.

Data Required for Heating Unit Selection

When selecting the unit for a particular application there are definite types of information that are needed for the proper selection. The following list contains some of the information required to properly make the selection of furnaces for residential and light commercial applications:

1. The indoor and outdoor design temperatures
2. The total heat loss load in Btuh

Figure 4-1. Specifications — Downflow U.S. Models. (Courtesy of NORDYNE)

Model Numbers	G2RL-045A12	G2RL-050A13	G2RL-075A14	G2RL-090A18	G2RL-105A18	G2RL-120A18
Input Btuh (a)	45.000	60.000	75.000	90.000	105.000	120.000
Heating Capacity Btuh	41.000	54.000	68.000	83.000	95.000	110.000
AFUE (b)	90.1	90.5	90.3	91	90	90.5
Heat External Static Pressure - in W.C.	.10	.12	.12	.15	.20	.20
Motor H.P. - Speed - Type	1/4-4-PSC	1/3-3-PSC	1/2-4-PSC	3/4-4-PSC	3/4-4-PSC	3/4-4-PSC
Motor Full Load Amps	6.5	8.2	9.6	11.0	11.0	11.0
Heating Speed	Low	Med	Med-Low	Low	Low	Med-Low
Cooling Speed	High	High	High	High	High	High
Cooling CFM @.5" ESP	1240	1330	1390	1800	1785	1765
Maximum External Static Pressure (in.)	.50	.50	.50	.50	.50	.50
Temperature Rise Range °F	50-80	40-70	50-80	50-80	50-80	50-80
Filter Size - (c)	20 × 22 × 1	20 × 22 × 1	24 × 22 × 1	20 × 22 × 1	24 × 22 × 1	24 × 22 × 1
Approximate Shipping Weight (lbs.)	150	155	160	189	192	194
Combustion Flr. Base (Opt.) RH-RXGC-	B21	B21	B24	B24	B24	B24

All models are 115V, 60Hz - Gas Connections are 1/2" N.P.T.

(a) Ratings up to 2,000 ft. above sea level. Over 2,000 ft. ratings should be reduced 4% for each 1,000 ft. above sea level.

(b) Isolated combustion - per D.O.E. test procedures.

(c) Cleanable filter.

Figure 4-1. (*Continued*)

SPECIFICATIONS - Downflow Canadian Models

Model Numbers (a)	G2RL-045C12	G2RL-060C13	G2RL-075C14	G2RL-090C18	G2RL-105C18	G2RL-129C18
Input Btuh (b)	45,000	60,000	75,000	90,000	105,000	120,000
Heating Capacity Btuh	42,300	57,000	70,500	83,700	97,600	112,800
Output Efficiency (c)	94	95	94	93	93	94
Heat External Static Pressure - in W.C.	.10	.12	.12	.15	.20	.20
Blower (D × W)	11 × 7	11 × 7	11 × 7	11 × 10	11 × 10	11 × 10
Motor H.P. - Speed - Type	1/4-4 -PSC	1/3-3-PSC	1/2-4-PSC	3/4-4-PSC	3/4-4-PSC	3/4-4-PSC
Motor Full Load Amps	6.5	8.2	9.6	11.0	11.0	11.0
Heating Speed	Low	Med	Med-Low	Low	Low	Med-Low
Cooling Speed	High	High	High	High	High	High
Cooling CFM @.5" ESP	1240	1330	1390	1800	1785	1765
Maximum External Static Pressure (in.)	.50	.50	.50	.50	.50	.50
Temperature Rise Range °F	50-80	40-70	50-80	50-80	50-80	50-80
Filter Size - (d)	20 × 22 × 1	20 × 22 × 1	20 × 22 × 1	24 × 22 × 1	24 × 22 × 1	24 × 22 × 1
Approximate Shipping Weight (lbs.)	150	155	160	189	192	194
Combustion Flr. Base (Opt.) RH-RXGC-	B21	B21	B21	B24	B24	B24

All models are 115V, 60Hz - Gas Connections are 1/2" N.P.T.
(a) For Canada only
(b) Ratings shown are for elevations up to 2,000 ft. Derate 10% for elevations of,500 ft. to 4,500 ft.
(c) Steady state efficiency per Canadian Gas Association
(d) Cleanable filter.

Figure 4-1. (*Concluded*)

BLOWER PERFORMANCE

Model No.	Blower Size	Motor HP **	Blower Speed	CFM Air Delivery External Static Pressure - Inches Water Column 0.1	0.2	0.3	0.4	0.5	0.6	0.7	0.8
G2RL-045(_)12	11 × 7	1/4	High	1335	1332	1313	1282	1243	1299	1156	—
			Med-High	978	964	946	926	904	881	859	—
			Med-Low	884	866	851	837	819	792	754	—
			Low*	629	620	610	597	582	561	536	—
G2RL-060(_)13	11 × 7	1/3	High	1481	1445	1408	1369	1328	1282	1232	1177
			Med*	1018	1006	995	984	970	952	928	897
			Low	686	689	693	697	699	695	686	668
G2RL-075(_)14	11 × 7	1/2	High	1555	1515	1475	1434	1389	1339	1282	1215
			Med-High	1307	1279	1254	1231	1206	1176	1139	1092
			Med-Low*	1021	1006	995	985	975	959	937	904
			Low	892	875	860	847	832	814	792	762
G2RL-090(_)18	11 × 10	3/4	High	2059	2010	1947	1876	1800	1721	1644	—
			Med-High	1782	1815	1762	1708	1651	1586	1510	—
			Med-Low	1563	1526	1495	1464	1430	1386	1330	—
			Low*	1267	1252	1237	1220	1195	1161	1112	—
G2RL-105(_)18	11 × 10	3/4	High	2046	1990	1926	1858	1786	1714	1643	—
			Med-High	1881	1833	1781	1724	1663	1597	1525	—
			Med-Low	1651	1613	1565	1514	1460	1401	1339	—
			Low*	1370	1327	1296	1271	1241	1200	1139	—
G2RL-120(_)18	11 × 10	3/4	High	2065	2009	1939	1857	1765	1665	1561	—
			Med-High	1900	1847	1784	1711	1628	1535	1431	—
			Med-Low*	1646	1605	1558	1502	1434	1350	1249	—
			Low	1342	1321	1287	1241	1183	1113	1030	—

*Heating speed.
Air conditioning models are certified at a maximum ESP of 0.5" W.C.
**1/4-3/4 Permanent split capacitor motor type.

3. The type of heating unit to be used
4. The required cfm for the application
5. The external static pressure created by the ductwork

The Indoor Design Temperatures: Only the dry bulb temperatures are required when considering these temperatures. Usually the indoor design conditions are between 70°F and 75°F.

The Total Heat Loss in Btuh: The total heat loss is determined by proper use of the heat load estimate form. There are two methods of calculating the amount of heat loss from ducts located in unconditioned areas. They are: (1) use the method included on the heat loss estimate form, and (2) multiply the calculated heat loss by 0.95%. This step will indicate the actual amount of heat that will be delivered to the conditioned space by the furnace.

Type of Heating Equipment: There are several factors that will determine the type of furnace chosen. Some of these factors are: type of available (or desired) energy and the type of air flow needed (upflow, downflow, or horizontal). These are discussed as follows:

Type of Energy Used: This means the energy type that will be used—gas, oil, or electricity. Other factors include: the availability and cost of each source of energy. The dependability and convenience of each of the energy sources, and consideration of areas where the desired type of energy may be available in the near future. It may be desirable to install a furnace of the type desired but using an alternate type of energy, e.g., an LP gas furnace installed when natural gas will soon be available. This will reduce the cost of changing to the types of energy desired.

Type of Air Flow: Upflow type furnaces are the type most often chosen for residential applications (Figure 4-2).

During operation of the upflow furnace, air is drawn in through the bottom of the furnace, pushed around the heat exchanger, and discharged out the top. Upflow furnaces are ideally suited for installations in the basement using an under the floor duct system along with floor-level or high-wall diffusers. They are also popular in installations where the furnace is connected to an air distribution system that is installed in the attic of the building and using overhead or high-wall diffusers.

Downflow, or counterflow, type furnaces are designed with the heat exchanger and the blower sections reversed (Figure 4-3).

During operation, the air is drawn in through the top of the furnace, forced through the heat exchanger, and discharged out the bottom of the

Figure 4-2. Upflow gas furnace. (Courtesy of NORDYNE)

furnace. These types of furnaces are popular when the building does not have a basement where the furnace can be installed. They may also be used in commercial buildings that are built on a slab or when the furnace is to be installed over a crawl space. Usually the furnace is installed over a hole in the floor. The discharge air plenum and the air distribution system are installed underneath the floor (Figure 4-4).

Usually when the air distribution system is located below the floor, floor-type diffusers are used.

Horizontal furnaces are popular for installations in buildings when floor space is limited. These types of furnaces are generally installed in an attic or crawl space. Sometimes they are suspended from the ceiling in

Figure 4-3. Blower location in downflow furnace. (Courtesy of NORDYNE)

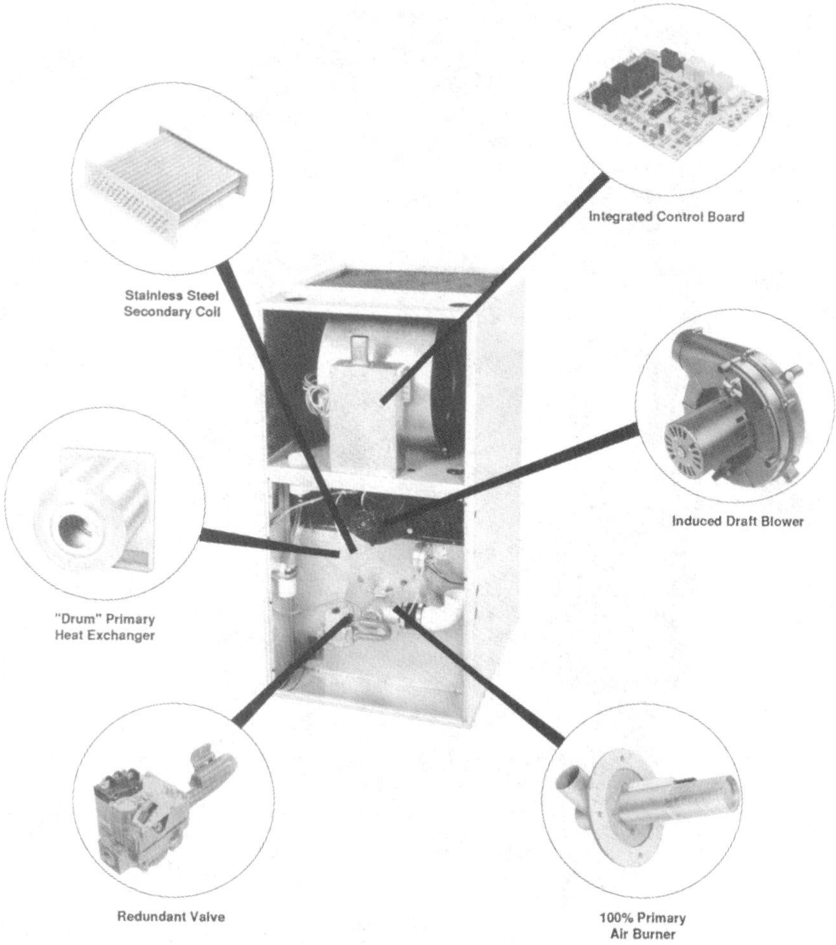

commercial installations (Figure 4-5).

During operation, the air is drawn in through one end of the furnace, forced through the heat exchanger, then forced out the other end of the unit. The air is then forced through the air distribution system to the heated space (Figure 4-6).

Before manufacturers' tables can be used to choose the type of furnace to use, the direction of air flow and the type of energy available must be known. Also, when the furnace is to be a part of a year-round conditioning system, the type and size of the cooling equipment must also be known because of the different air requirements for cooling applications.

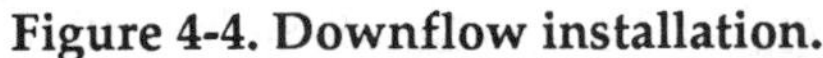

Figure 4-4. Downflow installation.

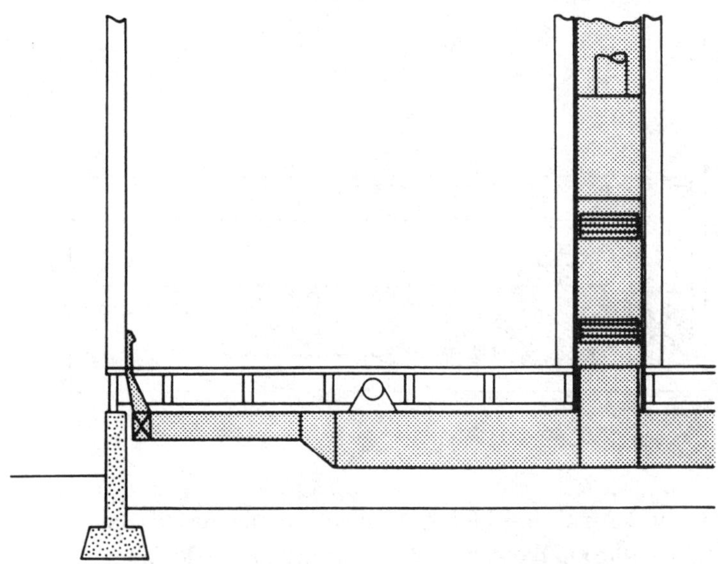

Figure 4-5. Horizontal furnace. (Courtesy of NORDYNE)

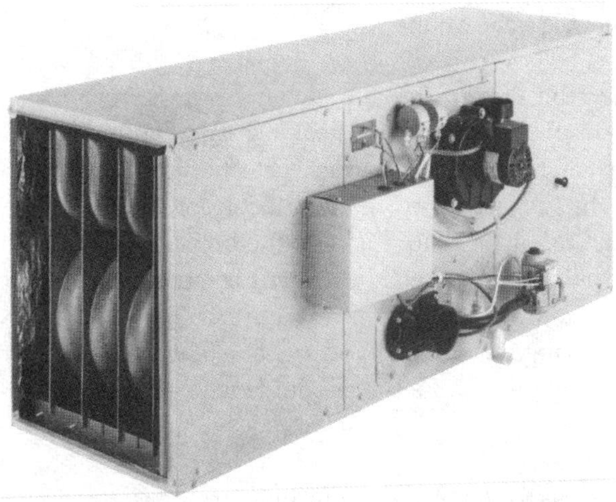

The Required cfm: In order for the heated air to be properly distributed within the conditioned space, the proper cfm must be calculated and the air distribution system designed to meet this requirement. For residential and light commercial installations there are several different methods that can be used to design the air distribution system. The different methods are produced by different organizations, such as the National Warm

Figure 4-6. Horizontal installation.

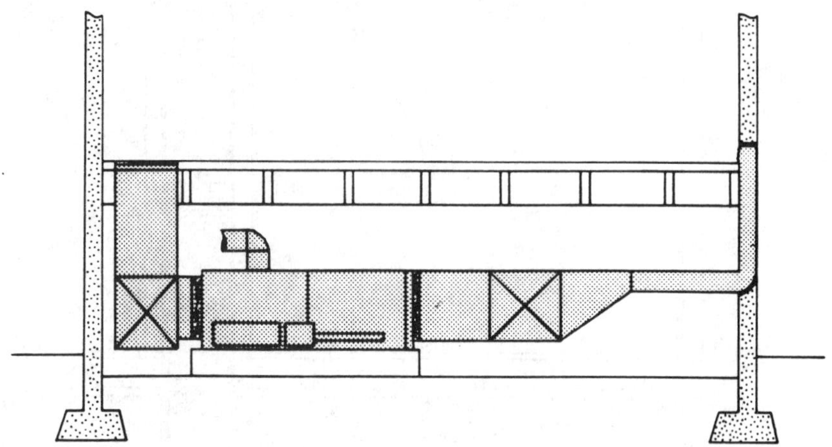

Air Heating and Air Conditioning Association and ASHRAE. The one to use is dependent upon the personal preference of the designer.

External Static Pressure Created by the Ductwork: The external static pressure created at the required cfm is a very important component of the duct system design calculations. The static pressure must be included in the duct system specifications.

Furnace Selection Procedure

After completion of the total heat loss calculation, only two steps are remaining in the heating-only furnace selection procedure. The steps are: (1) selecting the furnace from the manufacturers' tables, and (2) determining the correct required heating cfm of the furnace.

When making this selection we will use the building for which the heat load estimation was completed (Figure 4-7).

The following information and steps will be used to select a gas-fired up-flow furnace for use in the model house. Construction details for Figure 4-7 are:

1. The house is for a family of four people.
2. It will be built on a slab without perimeter insulation.
3. The house will have a hip roof with a 2-foot overhang.
4. The windows will be of insulating glass with a thermal break. The doors will be metal with 1-1/2" Urethane insulation.
5. Inside shading will be used over the glass areas.
6. The walls will have R-16 insulation; ceilings will have R-30.

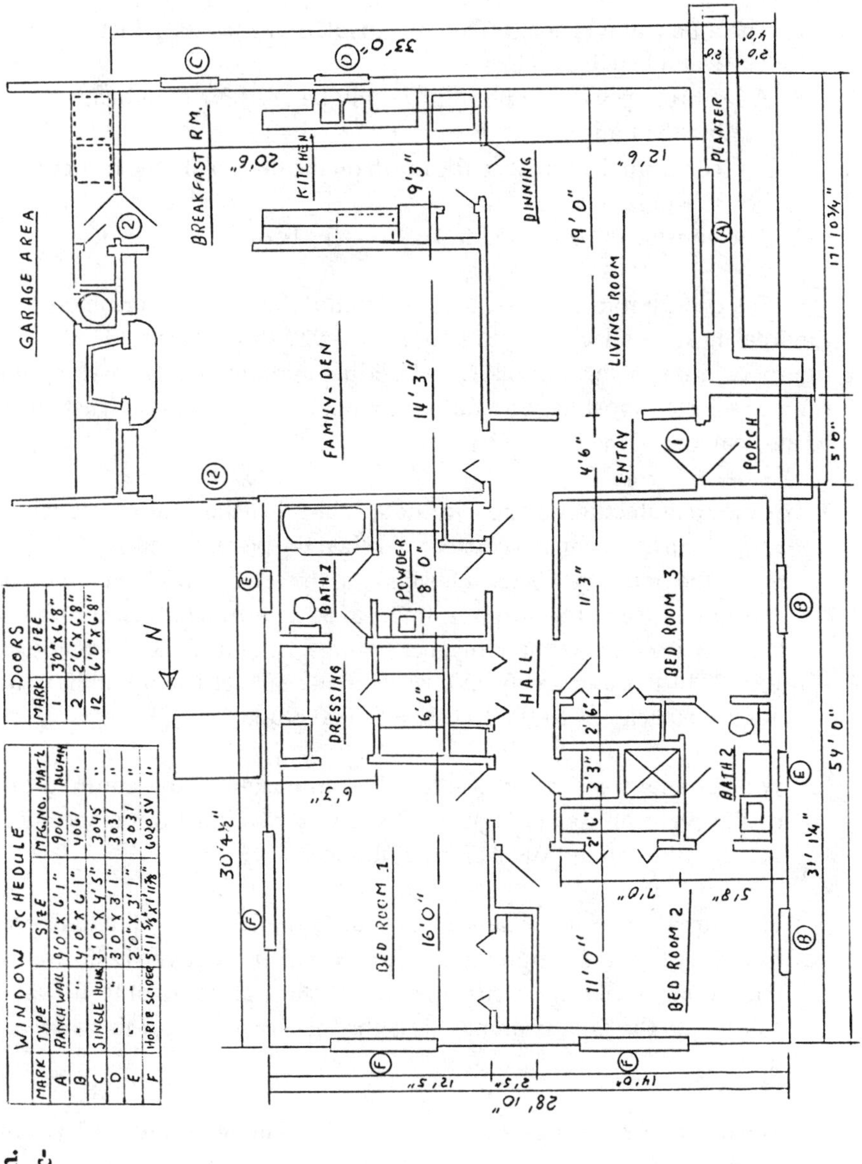
GARAGE AREA
BREAKFAST RM.
KITCHEN
DINING
LIVING ROOM
FAMILY-DEN
ENTRY
PORCH
PLANTER
BATH 1
POWDER
DRESSING
HALL
BED ROOM 1
BED ROOM 2
BED ROOM 3
BATH 2
N
33'0"
54'0"
28'10"
30'4½"
17'10¾"
31'1¼"
WINDOW SCHEDULE
MARK | TYPE | SIZE | MFG. NO. | MAT'L
A | RANCH WALL | 9'0" x 6'1" | 9061 | ALUMN
B | " | 4'0" x 6'1" | 4061 | "
C | SINGLE HUNG | 3'0" x 4'5" | 3045 | "
D | " | 3'0" x 3'1" | 3031 | "
E | " | 2'0" x 3'1" | 2031 | "
F | HORIZ SLIDER | 5'11 5/8" x 1'11 7/8" | 6020 SV | "
DOORS
MARK | SIZE
1 | 3'0" x 6'8"
2 | 2'6" x 6'8"
12 | 6'0" x 6'8"

Figure 4-7. Floor Plan. (Courtesy of TU Electric Co.)

7. The attic will be vented. The duct insulation will be 2" thick.
8. The ceilings will be 8 feet.
9. A heat pump with an SEER of 11 will be used for both heating and cooling the house.
10. As seen from the arrow to the north on the floor plan, the front of the house will face west.
11. The house will be located in Fort Worth, Texas.

Selection: From the heat load estimation form we determined that the total heat loss for the building was 34,920 Btuh (Figure 3-21). This figure includes the heat loss for the building and the duct. Therefore, the furnace selected must have an output (bonnet) rating that is at least equal to the heat loss of the house or greater.

1. Some manufacturer's capacity data tables list the output capacities and others list the input capacities. When the output capacity is listed, this is the figure that is used for making the selection. When the input capacity is listed, the output rating can be determined by multiplying the input rating by the efficiency rating of the furnace. Example: A 100,000 Btuh input gas-fired furnace that is 80% efficient will have an output capacity of $100{,}000 \times 0.80 = 80{,}000$ Btuh.

2. We can now select a furnace that has an output rating equal to our requirement or above. Figure 4-8 shows a manufacturer's specification chart for an upflow gas-fired furnace.

3. This furnace has an air handling capacity of 820 cfm with a temperature rise of 75°F through the furnace. It will also deliver 1235 cfm for the cooling application with an external static pressure of 0.50" water column on the duct system. This is the furnace to use for this application.

4. The height of the furnace is 51 in.; it is 27-1/4 in. wide and is 27-1/4 in. deep. (See dimensions in Figure 4-9.)

   If a furnace of these dimensions will fit into the designated space, the installation will be simple. However, should the opening be smaller than the furnace, either the size of the opening must be increased to accommodate the furnace, or the furnace must be installed in another location.

Figure 4-8. Specifications

Model Numbers	G1RS-050A12	G1RS-075A13	G1RS-100A16	G1RS-125A20
Input Btuh (a), (b)	50,000	75,000	100,000	125,000
Output Btuh (a)	47,000	69,750	92,000	115,000
AFUE	**94.0**	**93.0**	**84.6**	**85.6**
California Seasonal Efficiency (%)	84.1	83.3	.12	.15
Motor H.P. - Speeds	1/3-3	1/3-3	3/4-3	3/4-3
High Speed 0.5 ESP (cfm)	1235	1300	1600	1975
Low Speed 0.5 ESP (cfm)	820	800	1165	1315
Vent Diameter Size (inches)	2	2	3	3
Air Filter Size (inches)	13 × 23 × 1/2	13 × 23 × 1/2	16-1/2 × 23 × 1/2	20 × 23 × 1/2
Weight - Shipping (pounds)	168	180	201	221

(a) Capacity and efficiency ratings in accordance with D.O.E. ICS test procedures.
(b) Ratings shown are for elevation up to 2,000 ft. Ratings should be reduced by 4% per 1,000 ft. above sea level.
AFUE = Annual Fuel Utilization Efficiency

Figure 4-9. Typical furnace dimension chart.

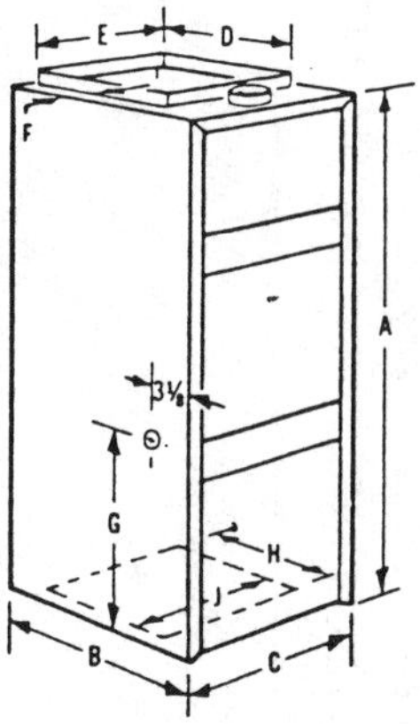

dimensions (in inches)

Model No.	HL50-D HL75-DD HL80-D2	HL75-B3 HL100-D HL100-B3	HL100-B4 HL125-B3 HL125-B4	HL125-B5 HL150-B4	HL150-B5 HL150-B75
A	51	51	51	51	51
B	27¼	27¼	27¼	27¼	27¼
C	15¼	18½	23	27½	31
D	19⅝	19⅝	19⅝	19⅝	19⅝
E	13¼	16½	21	25½	29
F	1	1	1	1	1
G	28	28	28	28	28
H	23¼	23¼	23¼	23¼	23¼
J	10¾	14	18½	23	23

5. The required vent diameter in inches. This must be taken into account when figuring the installation cost because an error here would be very costly if double wall type B vent pipe is to be used.

   This type of vent pipe is very expensive. Also, if replacing a furnace, the existing venting system must be of the proper size or the vent transitions must be determined and figured into the installation costs.

6. The furnace weighs 190 lb. The flooring must be sufficiently strong to support this as a minimum weight. The ductwork must also be figured into this weight so that the supporting platform can be properly selected.

Electric Furnace Selection

The selection of an electric furnace is a relatively simple process. These types of furnaces are rated in kW per hour. To convert the kW input to Btuh divide the kW by 3,413. The result will be the Btuh of the furnace. Since electric furnaces are theoretically 100% efficient there is no conversion factor needed to learn the Btuh rating of the furnace. The rating of the furnace may be changed by simply adding more elements to the furnace or removing them and changing to the correct air flow. Heat strips are available in a variety of Btuh capacities. Check with the furnace manufacturer for the correct strip to use for the application being considered. When cooling is to be installed also check to make certain that the blower will furnish the desired volume of air to satisfy the needs of the cooling unit selected.

Heat Pump Selection

The selection of a heat pump for the heating cycle is relatively simple. Generally heat pump systems are sized according to the cooling capacity of the application. Therefore, simply choose a heat pump that will provide the required amount of cooling Btuh, then add heat strips to the air handling section to obtain the desired amount of heating Btuh required for the application.

Example

In our example house the cooling load estimate indicated that 27,574 Btuh were needed to satisfy the cooling demand for the house.

We can see (Table 4-1) that with a 10 SEER unit the outdoor unit model number would be T1BA-030K with coil model number C1BL-030C/U (providing 29,800 Btuh cooling). This unit would have a heating rating of 28,800 Btuh at 47°F and 15,200 Btuh at 17°F. Since there are two units that are basically equal the choice would probably be determined by their price. Since either one would do the job, choose the less expensive unit. There would need to be heat strips added to supply 19,720 Btu to meet the outside design conditions. Since one 5 kW heat strip will supply approximately 17,065 Btuh there would need to be two 5 kW or one 7.5 kW heat strip added to handle the heating demand for the application.

Cooling Equipment Sizing

The proper balancing of the equipment to the load is very important. The balancing of the system components is also important for trouble-free operation with a long life expectancy of the equipment. It is especially important to have a properly designed air distribution system for the unit so that maximum efficiency and comfort can be realized.

General Information

The proper selection of cooling equipment is relatively simple, whether the selection is of a split system, remote system, or a packaged system. The simplicity is because the manufacturers have developed tables to be used in selecting their equipment. Always use the tables so that the correct unit can be chosen for the application. The tables make only the balancing of the unit capacity to the load, determining the cfm to be delivered, and the required electrical characteristics necessary. The components have already been balanced.

Data Required for Cooling Unit Selection: There is certain information that must be determined before the selection of the equipment is completed. The following is a list of this needed information:

Table 4-1. System heating and cooling capacities. (Courtesy of NORDYNE)

Outdoor Unit	With this coil						
Model Number	Model Number	Cooling Btuh	SEER	Heating Btuh @47	Heating Btuh @17	HSPF	Nominal cfm
T1BA-018K	C1BL-018C/U	17,500	10.0	17,500	10,000	6.8	675
T1BA-024K	C1BA-024C/U	23,800	10.0	22,000	12,000	6.8	900
T1BA-024K	C1BL-024C/U	23,400	10.0	21,600	11,600	6.8	900
T1BA-030K	C1BA-030C/U	30,000	10.0	29,200	15,600	6.8	1,125
T1BA-030K	C1BL-030C/U	29,800	10.0	28,800	15,200	6.8	1,125
T1BA-036K	C1BA-036C/U	35,600	100	35,600	20,000	7.0	1,350
T1BA-042K	C1BA-042C/U-13	40,500	10.0	40,500	23,400	7.0	1,400
T1BA-042K	C1BA-042C/U-14	40,500	10.0	40,500	23,400	7.0	1,450
T1BA-048K	C1BA-048C/U-16	46,000	10.0	46,000	26,600	7.0	1,600
T1BA-048K	C1BA-048C/U-15	45,000	10.0	45,500	26,000	7.0	1,500
T1BA-060K	C1BA-060C/U	56,000	10.0	57,500	35,000	7.0	1,700

1. The summer design outdoor dry bulb (DB) and wet bulb (WB) temperatures.
2. The design wet bulb (WB) and dry bulb (DB) temperatures of the conditioned space.
3. The required amount of ventilation air in cfm.
4. The total cooling load in Btuh.
5. The cfm of the supply air.
6. The condensing medium (air or water) and the temperature of the medium entering the condenser.
7. The duct external static pressure.

The Design Outdoor Air Temperatures: There are two reasons for knowing the design outdoor air conditions: (1) to complete the cooling (heat gain) load estimate, and (2) to establish the conditions of the air that enters the evaporator coil when positive ventilation is used.

The design outdoor air temperatures used should be those that are usually used in the geographical area of the house. The design outdoor air temperatures for most large U.S. cities can be found in tables (Figure 2-5). When working in an area that is not listed in the table, use the closest city within 100 miles. When there is no listing available within a suitable distance, use the accepted outdoor air design temperatures from a heating, ventilating, and air conditioning guide, a local engineer, a local weather bureau, or from the local power company.

The Conditioned Space Design Dry Bulb (DB) and Wet Bulb (WB) Temperatures: There are two reasons for knowing the space design conditions: (1) the heat load estimate, and (2) the evaporator entering air temperature calculation. In some installations the air entering the evaporator is a mixture of outside air and return air. Therefore, when the temperatures of the return air (indoor design) and the ventilation air (outdoor design), together with the percentage of ventilation air are known, the temperature of the air entering the evaporator can be easily determined.

When considering bid-specification jobs, the indoor air conditions will be specified by the design engineer. When other types of jobs are considered, the design indoor conditions are established which are acceptable to the owner and which are consistent with local experience and recommendations by the industry.

The Required Ventilation Air in cfm: The ventilation air cfm must be known before attempting to determine the temperature of the air entering the evaporator or attempting to determine the sensible heat capacity of the cooling system. The ventilation air requirement is determined from the cooling load estimate form and can be obtained from the load calculation form.

The Total Cooling Load in Btuh: The total cooling load (heat gain) is the basis for the complete air conditioning unit selection. The tentative and final equipment selections are made on the basis of this calculation.

The heat gain is located on the heat load estimate form. There are several short forms available for making this estimation. In cases when special conditions must be met, when the construction features are considerably different from those described on the heat load estimation form, a more detailed load estimating form should be used.

The cfm Air Delivery Required: When the long-form load estimation form is used, the required cfm can be calculated from the total sensible heat load and the rise in the DB temperature of the supply air. Some load estimation forms do not have a cfm requirement listed. When this occurs, the cooling unit is chosen according to the Btuh requirement of the structure only. It is then assumed that the rated air delivery is 400 cfm per ton of refrigeration. This 400 cfm per ton figure is used when designing the duct system and making static pressure calculations.

The Condensing Medium (Air or Water) and the Temperature of the Medium Entering the Condenser: When water-cooled condensers are used with well water or city water, the entering water temperature must be determined so that the gallons per minute (gpm) that must flow through the condenser to reach the desired design condensing temperatures can be calculated. When a cooling tower is used, the discharge water temperature at the tower must be known so that the gpm flow and the final condensing temperature can be properly calculated.

The city or well water temperature can be measured at the source of supply. The cooling tower condenser water temperatures are obtained from the tower manufacturer's literature (Table 4-2).

The capacity of an air-cooled condenser is determined by the difference between the condensing temperatures and the design outdoor air DB temperatures. Most equipment manufacturers publish this data in tables (Table 4-3).

The entering air temperatures for air cooled condensers are the various outdoor design air temperatures.

The Duct External Static Pressure: When an air conditioning unit is to be connected to a duct system, the external static pressure on the duct system must be calculated to determine the fan speed and the fan motor

Table 4-2. Cooling Tower Rating Chart (Courtesy of The Marley Company)

No.	Capacity Tons-Refrig. 78° WB	Overall Dimensions Wide	Overall Dimensions Deep	Overall Dimensions High	Pump Head In Feet	H.P.	Motor Voltage	Cold Water Outlet	Ship. Wt.	Max. Oper. Wt.	FOB Louisville Ea.	Part No.	Basin Cover Ea.	Part No.	Air Inlet Screen Ea.
91000	5	32½"	69"	60⁵⁄₁₆"	4.5	¼	115/230/60/1	2" FIPT	460 Lbs.	960 Lbs.	$1066.00			91055	$ 38.25
91001	7	32½"	69"	60⁵⁄₁₆"	4.5	⅓	115/230/60/1	2" FIPT	460 Lbs.	960 Lbs.	1079.00			91055	38.25
91002	10	32½"	69"	61⅛"	4.5	⅓	115/230/60/1	2" FIPT	480 Lbs.	1060 Lbs.	1158.00			91056	40.75
91003	15	32½"	69"	70³⁄₁₆"	5.3	½	115/230/60/1	2" FIPT	520 Lbs.	1100 Lbs.	1342.00				
91004	15	32½"	69"	70³⁄₁₆"	5.3	½	200/60/3	2" FIPT	520 Lbs.	1100 Lbs.	1342.00	91050	12.00	91057	52.75
91005	15	32½"	69"	70³⁄₁₆"	5.3	½	230/460/60/3	2" FIPT	520 Lbs.	1100 Lbs.	1342.00				
91006	20	32½"	72¾"	80½"	6.5	½	115/230/60/1	4" IPT	580 Lbs.	1020 Lbs.	1500.00				
91007	20	32½"	72¾"	80½"	6.5	½	200/60/3	4" IPT	580 Lbs.	1020 Lbs.	1500.00			91058	58.00
91008	20	32½"	72¾"	80½"	6.5	½	230/460/60/3	4" IPT	580 Lbs.	1020 Lbs.	1500.00				
91009	25	46½"	75⅜"	80½"	6.5	½	115/230/60/1	4" IPT	750 Lbs.	1410 Lbs.	1645.00				
91010	25	46½"	75⅜"	80½"	6.5	½	200/60/3	4" IPT	750 Lbs.	1410 Lbs.	1645.00				
91011	25	46½"	75⅜"	80½"	6.5	½	230/460/60/3	4" IPT	750 Lbs.	1410 Lbs.	1645.00				
91012	30	46½"	75⅜"	80½"	6.5	1	115/230/60/1	4" IPT	750 Lbs.	1410 Lbs.	1776.00				
91013	30	46½"	75⅜"	80½"	6.5	1	200/60/3	4" IPT	750 Lbs.	1410 Lbs.	1776.00	91051	19.75	91059	72.25
91014	30	46½"	75⅜"	80½"	6.5	1	230/460/60/3	4" IPT	750 Lbs.	1410 Lbs.	1776.00				
91016	40	46½"	75⅜"	80½"	8.5	2	200/60/3	4" IPT	750 Lbs.	1410 Lbs.	2053.00				
91017	40	46½"	75⅜"	80½"	8.5	2	230/460/60/3	4" IPT	750 Lbs.	1410 Lbs.	2053.00				
91018	55	64"	83¹⁄₁₆"	91⅜"	9.0	2	200/60/3	4" IPT	1170 Lbs.	2210 Lbs.	2539.00				
91019	55	64"	83¹⁄₁₆"	91⅜"	9.0	2	230/460/60/3	4" IPT	1170 Lbs.	2210 Lbs.	2539.00				
91020	65	64"	83¹⁄₁₆"	91⅜"	8.0	3	200/60/3	4" IPT	1190 Lbs.	2230 Lbs.	2829.00	91052	27.75	91060	108.00
91021	65	64"	83¹⁄₁₆"	91⅜"	8.0	3	230/460/60/3	4" IPT	1190 Lbs.	2230 Lbs.	2829.00				
91022	75	76"	89¹⁄₁₆"	91⅜"	8.0	3	200/60/3	6" IPT	1400 Lbs.	2770 Lbs.	3158.00	91053	30.25	91061	117.00
91023	75	76"	89¹⁄₁₆"	91⅜"	8.0	3	230/460/60/3	6" IPT	1400 Lbs.	2770 Lbs.	3158.00	91053	30.25	91061	117.00
91024	90	88"	92¾"	91⅜"	8.0	5	200/60/3	6" IPT	1610 Lbs.	3250 Lbs.	3671.00	91054	33.00	91062	138.00
91025	90	88"	92¾"	91⅜"	8.0	5	230/460/60/3	6" IPT	1610 Lbs.	3250 Lbs.	3671.00	91054	33.00	91062	138.00

4-3. System cooling capacities. (Courtesy of NORDYNE)

WITH THIS BLOWER COIL

Model Number	Btuh	SEER	Nom. Cfm
PBMC-024 K	18,000	10.00	675
N/A	N/A	N/A	N/A
PBMC-024 K	24,000	10.00	900
N/A	N/A	N/A	N/A
N/A	N/A	N/A	N/A
PBMC-030 K	30,000	9.30*	1125
PBMD-030 K	30,000	9.30*	1125
N/A	N/A	N/A	N/A
PBMC-042 K	36,000	9.30*	1300
PBMD-042 K	36,000	9.30*	1300
N/A	N/A	N/A	N/A
PBMC-042 K	41,000	9.30*	1500
PBMD-042 K	41,000	9.,30*	1500
N/A	N/A	N/A	N/A
PBMC-048 K	46,500	9.20	1700
N/A	N/A	N/A	N/A
N/A	N/A	N/A	N/A
PBMC-060 K	57,000	9.00	1900
N/A	N/A	N/A	N/A
N/A	N/A	N/A	N/A

horsepower required. The maximum external static pressure for the equipment is determined by the manufacturer and is listed on the equipment nameplate. The external static pressure for the duct system is calculated when the ductwork is designed and should be included in the duct system specifications. In any case it should never exceed the manufacturer's specifications.

Split-System Equipment Selection Procedure

The same basic procedure is used for both residential and commercial split-system equipment installations. The heat gain estimation must be completed before attempting to actually select any type of system. After the estimate has been completed, equipment can then be chosen to meet the following application and installation considerations:

1. The temperatures of the air entering both the evaporator and the condenser are determined. The outdoor design DB temperature is the condenser-entering air temperature. In residential applications and other types where no ventilation air is brought in through the equipment, the conditioned space WB and DB design temperatures are used as the evaporator-entering air temperatures. When the ventilation air is introduced through the evaporator, the entering air temperatures may be calculated or obtained from a table (Table 4-4).

2. A split system having sufficient capacity is now selected using the manufacturers' tables. When commercial units are sized, the total capacity should be equal to or greater than the calculated heat gain load, and the sensible capacity should be great enough to handle the sensible heat load. For residential applications, the unit capacity may be a little less than the calculated heat gain on the estimate form and should be only a very small amount more.

3. A comparison of the air delivery of the unit versus the external static pressure must be determined. When the cooling cfm requirement is quite a bit more than the heating requirement, the furnace should be equipped with a two-speed fan motor.

Example

Select a split-system cooling unit to cool the house that we used in selecting the furnaces earlier in this chapter. We can assume that a year-around air conditioning system is to be installed. The heat gain calculation was completed in Chapter 3. We will use the same construction details that were used in the selection of the furnace.

Selection

It was determined that the total heat gain for the house was 27,574 Btuh. This is the total heat including both latent and sensible heat, and the duct heat gain. Therefore, the unit selected must have a total Btuh capacity of 27,574 Btuh or greater.

The heating unit must have an output capacity (bonnet) of at least 34,920 Btuh.

Use the following steps to select the unit:

1. Some manufacturers' data list the output capacities at different outdoor ambient air temperatures; others just list them at one temperature (Table 4-3).

Table 4-4. Resulting Temperatures of a Mixture of Return Air and Outdoor Air (to be used only for return air having an 80°F DB temperature and 50% humidity.)

	Dry Bulb Temperature of the Air Mixture			*Wet Bulb Temperature of the Air Mixture*							
	Outside Dry Bulb Temp.			*Outside Wet Bulb Temperature*							
Percentage of outside air	95°	100°	105°	74°	75°	76°	77°	78°	79°	80°	*Percentage of Outside Air*
2%	80.3	80.4	80.5	67.2	67.2	67.2	67.2	67.3	67.3	67.3	2%
4	80.6	80.8	81.0	67.3	67.3	67.4	67.4	67.5	67.5	67.6	4
6	80.9	81.2	81.5	67.5	67.5	67.6	67.7	67.7	67.8	67.9	6
8	81.2	81.6	82.0	67.6	67.7	67.8	67.9	68.0	68.1	68.2	8
10	81.5	82.0	82.5	67.8	67.9	68.0	68.1	68.2	68.4	68.5	10
12%	81.8	82.4	83.0	67.9	68.0	68.2	68.3	68.5	68.6	68.8	12%
14	82.1	82.8	83.5	68.1	68.2	68.4	68.5	68.7	68.9	69.1	14
16	82.4	83.2	84.0	68.2	68.4	68.6	68.8	69.0	69.2	69.4	16
18	82.7	83.6	84.5	68.3	68.5	68.8	69.0	69.2	69.4	69.6	18
20	83.0	84.0	85.0	68.5	68.7	69.0	69.2	69.4	69.7	69.9	20
22%	83.3	84.4	85.5	68.6	68.9	69.1	69.4	69.7	69.9	70.2	22%
24	83.6	84.8	86.0	68.8	69.1	69.3	69.6	69.9	70.2	70.5	24
26	83.9	85.2	86.5	68.9	69.2	69.5	69.8	70.1	70.4	70.8	26
28	84.2	85.0	87.0	69.1	69.4	69.7	70.0	70.4	70.7	71.0	28
30	84.5	86.0	87.5	69.2	69.6	69.9	70.2	70.6	70.9	71.3	30

2. The next step is to select a condensing unit with the capacity that is equal to or greater than 27,574 Btuh. Using Table 4-3, locate the heading BTUH. This indicates that condensing unit model number ACZD-030 KC when matched with coil model number AAZB-030 XC upflow coil, will provide 31,000 Btuh capacity. If the unit was to be a horizontal application the coil model number would be AHCA-042 XB, providing 30,000 Btuh. Each model coil is designed to have 1125 cfm through it at these capacities.

3. The coil must match the furnace outlet air opening as closely as possible. The evaporator coil configurations and dimensions are shown in Figure 4-8.

The furnace dimensions are shown in Figure 4-9. The figure shows the dimensions for a gas furnace. For a 45,000 input furnace at 90% efficiency the air outlet dimensions are A and B, which are A = 21 in. and B = 19-1/2 in. Coil model number AAZB030 XC has dimensions A = 19-1/2 in. and is 18 in. wide. We will need to have a metal flange made to convert the furnace outlet opening to the coil inlet opening. The rated air flow of the coil is within the range listed for the cooling unit.

Vertical Packaged (Self-Contained) Unit Selection Procedure

These units are sometimes referred to as PTAC (packaged terminal air conditioning units). Equipment manufacturers generally recommend selection procedures for their equipment. The procedures may even vary for different models of the same manufacturer. However, when the general procedure is learned, it can be used on just about any brand of equipment. The following is a general procedure that is very popular in the field.

Before the selection process can be started, the heat load estimate form must be completed and the seven items of necessary data, discussed previously under the heading "Data Required for Cooling Unit Selection," must be gathered. When this information is gathered the procedure for selecting a packaged unit can be started.

Selection Procedure: Use the following steps in the selection procedure:

1. The first step is to make a tentative selection based on the heat gain calculation. This selection is made from the manufacturer's nominal unit capacity (Table 4-5).

Table 4-5. Water-cooled unit data.

UF Series Condensed Specifications

Model	UF036W	UF060W	UF090W	UF120W	UF180W	UF240W	UF300W	UF360W
Dimensions (in.): Height	78	78	78	78	78	79½	79½	79½
Width	30	42	54	65	89	76¼	96½	96½
Depth	20	20	20	28	28	38½	38½	38½
Ship Wt. (lbs.)	370	510	672	1020	1267	1445	1865	2025
Filters— †Disposable ‡Permanent, cleanable	†(1)20×25×1	†(1)20×25×1 †(1)16×25×1	†(3)16×25×1	‡(3)19×27×1	‡(8)15×20×1	‡(4)20×25×1 (2)20×20×1	‡(6)25×30×1	‡(6)25×30×1
Performance: Cooling, std rating (Btuh)	36,000	60,000	86,000	120,000	176,000*	240,000*	300,000*	360,000*
Power input (watts)	3,950	6,700	9,700	13,100				

*Units above 120,000 Btuh not ARI rated

UR Series Condensed Specifications

Model	UR030	UR036	UR048	UR060
Dimensions (in.): Height	26 11/32	26 11/32	30	30
Width	42 13/16	42 13/16	52⅜	52⅜
Depth	45 3/16	45 3/16	52⅜	52⅜
Ship Wt. (lbs.)	320	340	450	510
Filters (recommended), dimensions (in.) not included in unit	(2)16×20×1	(2)16×20×1	(2)20×20×1	(2)20×20×1
Performance: Cooling, standard rating (Btuh)	30,000	38,000	46,000	61,000
Power input (watts)	4,300	5,100	5,400	8,300

Electrical data:	UR030		UR048		
Phase/Cycle	1/60		1/60	3/60	3/60
Rated voltage	208-230		230	230	460
	UR036		UR060		
Phase/Cycle	1/60	3/60	1/60	3/60	3/60
Rated voltage	230	208-230	230	208-240	460

2. Compare the rated cfm air delivery for the unit selected, when provided in the table, to the air delivery requirement for the building. If the rating is within 20% of the required cfm, the selection should be made on the basis of the required cfm. If this difference is greater than 20%, a special built-up system may be needed.

3. Calculate the entering dry bulb and wet bulb temperatures, or get them from a table (Table 4-4). Regardless of which method is used, the indoor and the outdoor design temperatures and the ventilation requirements must be known.

4. When water-cooled condensers are used, the condensing temperature should be determined. A tentative condensing temperature of approximately 30°F higher than the entering water temperature is normally satisfactory.

5. Determine the total unit capacity and the sensible heat capacity of the unit at the conditions given for the unit selected. This information is usually listed in the manufacturer's equipment tables.

   a. When selecting water-cooled units, the evaporator entering air dry bulb and wet bulb temperatures are required.

   b. When considering air-cooled units, the outdoor design dry bulb temperature, and the dry bulb and wet bulb temperatures of the air entering the evaporator are required. Use the smallest condenser recommended at this point of the selection procedure. When the capacity of the unit selected falls between the ratings listed in the table, it will be necessary to interpolate to obtain the exact capacities of the unit. The unit capacities listed in these types of tables are for standard air delivery only. When quantities other than standard are to be used, correction factors must be applied to the listed capacities. These correction factors are listed in Table 4-6.

6. The total capacity of the unit selected should be equal to or greater than the total cooling load unless the design temperatures can be compromised. When additional capacity is needed, a new selection must be made to handle the load.

   a. When considering a water-cooled condenser, the condensing temperature should be lowered 5 to 10°F and the new unit capaci-

4-6. Capacity Correction Factors. *

CFM Compared to Rated Quantity	–40%	–30%	–20%	–10%	Std.	+10%	+20%
Cooling Capacity Multiplier	88	93	96	98	1.00	1.02	1.03
Sensible Capacity Multiplier	80	85	90	95	1.00	1.05	1.10
Heating Capacity Multiplier	78	83	89	94	1.00	1.06	1.12

*To be applied to cooling and heating capacities.

ties should be considered before selecting a larger unit. However, if the load remains larger than the unit capacity, the next larger unit should be used.

b. When considering an air-cooled condenser and the unit capacity will not meet the load requirements, a larger-capacity condenser should be selected. If the unit capacity is still not large enough, the next-larger-size unit should be selected.

7. After the actual unit capacity has been established, compare the sensible heat capacity of the unit to the heat gain of the building. If the sensible heat capacity of the unit is not equal to or greater than the cooling load, a larger unit should be selected.

8. When considering water-cooled condensing units, determine the gpm water flow required and the water pressure drop through the condenser. Use the final condensing temperature and the total cooling load in tons of refrigeration to determine the gpm required from the condenser water temperature table, when provided for the considered unit. After the condenser gpm requirement is determined, the pressure drop can be read from the table prepared by the equipment manufacturer, if this information is given.

9. Determine the fan speed and the fan motor horsepower required to deliver the required volume of air against the calculated external static pressure caused by the air distribution system. The air volume is read directly from the manufacturer's charts and performance tables. Should the required fan motor horsepower be greater than the recommended limits, a larger fan motor must be used.

Vertical self-contained units are usually available only in fairly large capacity increments. Many situations are encountered when the actual heat gain load falls between two of these capacities. For example, the calculated heat gain load indicates a 12.5 ton load. The available units are either 10 or 15 tons capacity. With the possibility of future expansion, the building owner may authorize a single 15-ton unit. However, if he is economy minded, he may decide to tolerate a few days which are warmer than normal and authorize the smaller 10-ton unit. The contractor has the choice of offering one 7.5-ton unit and one 5-ton unit to match the load exactly. In this case the customer must be willing to pay the higher cost of a two unit installation.

It is always wise, whatever the case may be, to compare the actual sensible heat gain capacities of the unit with the actual calculated heat gain of the building. It is often most desirable, for comfort reasons, that the unit have the capacity to handle the complete sensible heat load.

When the sensible heat percentage (the ratio of sensible heat gain to latent heat gain) is normally high or low, a more detailed check should be made. This check can be made by checking the equipment manufacturer's capacity tables.

Vertical Packaged (Self-Contained)
Equipment Selection Procedure

These units are often referred to as PTAC units (packaged terminal air conditioning units). We will select a self-contained water-cooled condensing unit to be installed in an office building in Fort Worth, Texas. The condensing water will be supplied by a cooling tower. The indoor unit will use a free discharge air plenum. There is no specified cfm requirements. The necessary data are:

1. The outdoor design temperatures are 100°F DB and 78°F WB.
2. The indoor design temperatures are 78°F DB and 65°F WB.
3. The required ventilation is 300 cfm.
4. The total cooling load is estimated at 7-tons (84,000 Btuh).
5. The condensing medium temperature is 85°F.
6. The required cfm is 400 cfm/ton or 2800 cfm.
7. There is no external static pressure because there is no duct.

Selection

1. From the manufacturer's rating chart, select a unit model UF 090W (Table 4-5).

2. When supplied, determine the rated cfm for the model chosen. This will normally be 400 cfm/ton. In either case the amount of air delivered would be 2868 cfm.

3. Calculate the DB and the WB temperatures of the air entering the evaporator:
   a. Ventilation air requirement ÷ Standard cfm delivery = 300 ÷ 2866 = 10.4% or 10% outside air required.
   b. Entering air DB (°F) = (0.10 × 100) + (0.90 × 78) = 82.2°F DB
   c. The WB temperature with a given mixture is found on the psychrometric chart to be 66.4°F WB.

4. Calculate the condensing temperature: 85°F + 30 = 115°F condensing temperature.

5. Because there is no specified latent heat load, the sensible heat percentage is not considered in this example.

6. Determine the gpm water flow required. When the manufacturer's table does not include this information, it is normally figured to be 3 gpm per ton of refrigeration. Thus $7.16 \times 3 = 21.48$ gpm.

7. Determine the air delivery of the unit in cfm. When this information is not included in the manufacturer's table, it is usually figured to be 400 cfm/ton of refrigeration. This is then $7.16 \times 400 = 2864$ cfm. The air can now be properly distributed to the needed areas.

Horizontal Packaged (Self-Contained) Unit Selection Procedure

When considering a horizontal self-contained unit for commercial installations, essentially the same procedure is used as when selecting vertical self-contained units. However, for residential installations, the ventilation air requirements will not be considered.

Before beginning the selection process, the heat gain estimate must be complete and the seven items of necessary data discussed under the heading "Data Required for Cooling Unit Selection," must be gathered. Once these tasks are completed, the packaged unit may be selected.

Selection

Use the following steps to select a horizontal self-contained unit:

1. From the manufacturer's capacity table, select a unit based on the total calculated heat gain (cooling) calculation.

2. Determine the standard cfm rating of the unit selected, if included, and compare it to the cfm required for the building.

3. Determine the WB and DB temperatures of the air entering the evaporator. Use the formula:

$$\text{Entering air DB} = (\%\ \text{outside air} \times \text{outdoor DB}) + (\%\ \text{inside air} \times \text{indoor DB})$$

This is the same procedure as was used in step 3 in the section headed "Vertical Packaged (Self-Contained) Unit Selection Procedure."

4. Determine the total heat capacity and the sensible heat capacity at the various conditions for the unit selected. This information is obtained from the manufacturer's capacity table, when provided, using the design outdoor DB and the return air DB and WB temperatures.

5. When the total unit capacity is not sufficiently close to the calculated heat gain, another unit must be selected.

The capacity of air-cooled units cannot be adjusted as readily as water-cooled units. Nor do air-cooled packaged units offer a choice of evaporator or condenser coil for varying the system capacity. This problem is alleviated, however, because air-cooled packaged units are readily available in a larger number of capacity ranges. When the calculated heat gain cannot be met exactly, the alternate choice of a larger or smaller unit, or a multiple-unit type of installation, will depend on the owner's preference. The considerations for making this decision are basically the same as those listed for vertical self-contained units.

SUMMARY

The proper sizing and selection of air conditioning equipment is just as important as any other step in the estimating process.

The proper sizing and selection of air conditioning equipment can be accomplished only after an accurate heat loss and heat gain analysis is completed.

The selection of equipment is a problem of choosing the unit that is the closest to satisfying the demands of capacity, performance, and cost.

The information that must be determined before attempting to select a warm air furnace includes such items as (1) the inside and outside design temperatures, (2) the total heat loss in Btuh, (3) the type of heating equipment used, (4) the required cfm, and (5) the external static pressure caused by the ductwork.

When considering the design temperatures, only the dry bulb temperatures are required.

The total heat loss figure is determined by completing a heat loss estimate form.

The type of furnace selected will depend on several application variations, such as the type of energy available, or desired; and whether an upflow, downflow, or horizontal airflow is desired.

The direction of airflow and the type of energy chosen must be known before the furnace manufacturer's tables can be used.

To properly distribute the heated air within the conditioned space requires that the cfm be properly calculated before the duct system is designed.

The external static pressure at the required cfm is a vital part of the complete duct system design calculations. It should, therefore, be included in the duct system specifications.

When the total heat loss has been calculated, there are only two steps left in selecting a furnace for a heating-only system: (1) selecting the furnace from the manufacturer's tables, and (2) determining the necessary cfm adjustment for heating.

The balancing of system components, such as the condensing unit and the evaporator, and the air delivery and the duct system, is important if comfort and maximum efficiency are to be maintained.

The selection of cooling equipment, whether it be a remote, split system, or a packaged unit, has been made relatively simple because the equipment manufacturers have already determined what components will produce a given capacity and efficiency.

The following information must be known before the proper selection of cooling equipment for residential or light commercial installation can be completed: (1) the summer design outdoor air dry bulb (DB) and wet bulb (WB) temperatures, (2) the conditioned space design DB and WB temperatures, (3) the required ventilation air in cfm, (4) the total cooling load in Btuh, (5) the cfm delivery required, (6) the condensing medium (air or water) and the temperature of the medium as it enters the condenser, and (7) the duct external static pressure.

The design outdoor air temperatures used should be those commonly used in the geographical area where the building is located.

The conditioned space design temperatures are also necessary for two reasons: (1) the load calculations, and (2) the evaporator entering air-temperature calculation.

The required cfm of ventilation air must be known before attempting to determine the temperature of the air entering the evaporator or attempting to determine the sensible heat capacity of the air conditioning system.

The total cooling load (heat gain) is the basis for the complete air conditioning unit selection.

When the long-form heat gain calculation form is used, the cfm required can be calculated from the total sensible heat load and the rise in DB temperature of the air supply. On some load calculation forms no cfm requirement is specified. In such cases, the unit is chosen on the Btuh capacity alone.

When water-cooled condensers are used with well water or city water, the entering water temperature must be determined to calculate properly the gpm of water that must flow through the condenser to achieve the design condensing temperatures.

The capacity of an air-cooled condenser is determined on the difference between the condensing temperatures and the design outdoor air DB temperatures.

When an air conditioning unit is to be connected to a duct system, the external static pressure must be known to determine the fan speed and the fan motor horsepower required.

After the heat gain calculations have been completed, the split-system equipment can then be chosen according to the appropriate following application and installation considerations: (1) the temperature of the air entering the evaporator and condenser is determined, (2) a split system with sufficient capacity is selected from the manufacturer's tables, and (3) the air delivery by the unit calculated against the external static pressure must be determined.

Chapter 5

Residential Equipment Location

In the preceding four chapters we discussed psychrometrics, heat loss, heat gain, equipment sizing, and equipment selection for a residential application. In this chapter we will discuss some principal considerations regarding the location and installation of the equipment.

INTRODUCTION

A well-planned location for the cooling and heating equipment is one of the most important steps toward making a good installation. The following are some of the considerations that are involved in planning the location of the various types of units.

FURNACES

There are three types of furnaces when classified by their source of energy: (1) gas, (2) electric, and (3) oil. The gas furnaces are further divided according to the direction of air flow through them. The classifications are: upflow, downflow (counterflow), and horizontal (Figure 5-1). Some gas furnaces are now designed to be used in multiple positions rather than a single position for each model.

Electric furnaces are more flexible in installation and they are not subdivided by the direction of airflow through them (Figure 5-2).

Oil furnaces are generally classified by the direction of airflow through them, like the gas furnaces listed above.

GAS FURNACES (GENERAL)

All gas furnaces require some of the same considerations with regard to location and installation. Therefore, we will discuss at this time

Figure 5-1. Upflow, downflow, and horizontal. (Courtesy of BDP Company)

5-2. Electric furnace. (Courtesy of Weatherking Heating and Cooling)

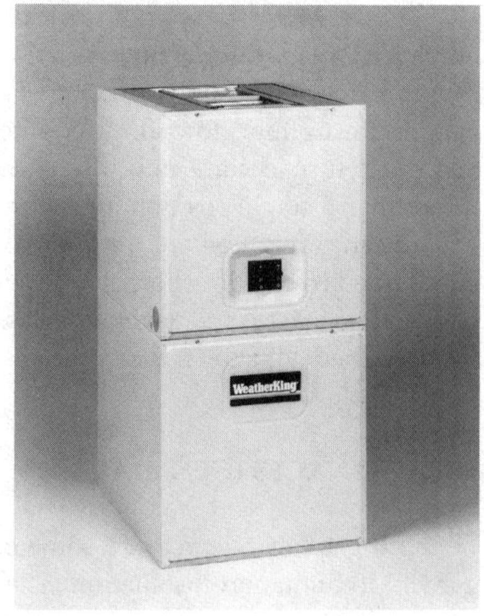

such items as gas pipe connections, venting requirements, return air location, supply air location, minimum clearances, noise, accessibility, length of duct runs, and thermostat location.

Gas Pipe Connections

The closeness of the gas supply pipe to the furnace is very important from a labor viewpoint. When the gas supply pipe is already in the equipment room, a certain amount of labor and materials will be saved. However, if the gas pipe is located on the other side of the building, the cost of labor and materials will be greatly increased, thus increasing the cost of the installation. The gas supply to the building must have the capacity to handle the extra load that the new furnace will place on the supply piping. Be sure to consider any restrictions or requirements that might be required because of local or national building codes.

Venting Requirements

When considering the venting requirements for gas-fired furnaces, any and all local and national codes and ordinances must be followed. An unsafe furnace should not be installed or operated because of the possible hazards to life and property. The furnace location should be so that the vent pipe will provide as straight a path as possible for the flue gases to escape to the atmosphere. Never decrease the size of vent piping from that recommended by the furnace manufacturer. Vent pipe is very expensive; therefore, a short, straight run is not only easier but also more economical to install (Figure 5-3).

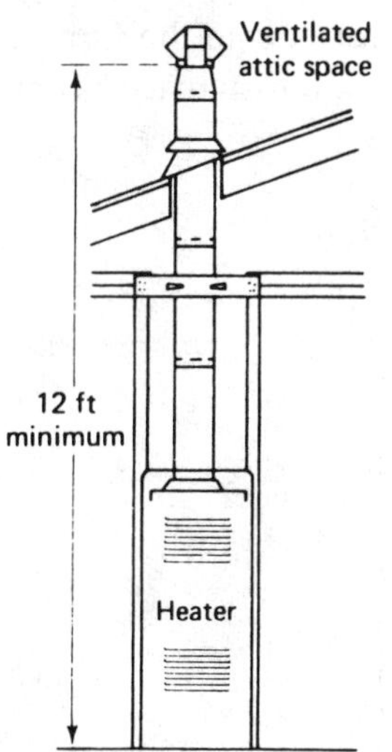

Figure 5-3. Minimum height. (Courtesy of Selkirk Metalbestos, Division of Household International)

When the furnace is to be installed in a closet or other tightly closed room, openings must be provided to ensure adequate combustion air is available to the unit. Be sure that the local and national codes are followed when locating and sizing these openings. It is more desirable from a safety standpoint and for more economical operation if the combustion air can be taken from outside the living space (Figure 5-4, opposite).

Approximately half the air should be introduced near the ceiling and the other half is introduced near the floor (Figure 5-5).

Figure 5-5. Combustion air openings in equipment room wall.

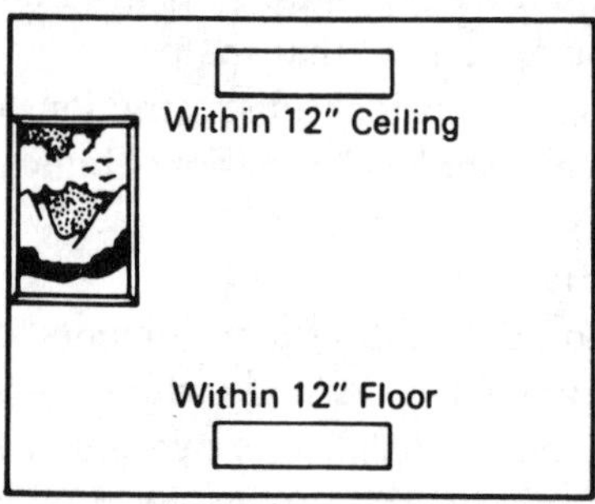

Adequate covering must be placed on the openings to prevent insects from entering through the opening. Be sure to provide extra dimensions to the combustion air intake because of the added restriction caused by the grill or covering.

Return Air Location

There are several methods of getting the return air to the furnace. The most popular are: a central return (Figure 5-6), a return in each room (Figure 5-7), and a combination of the two.

Figure 5-6. Central return location.

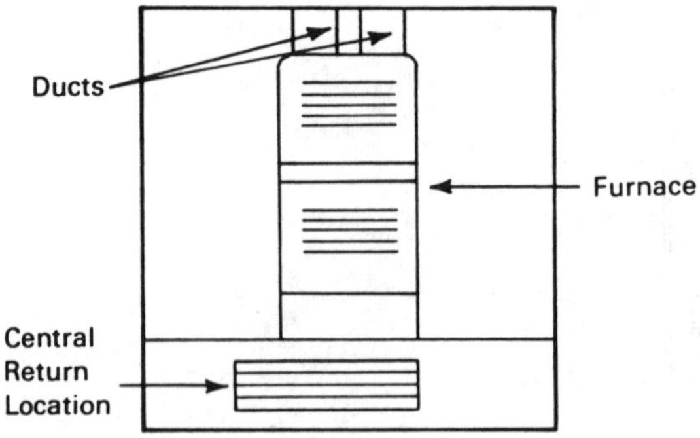

Figure 5-4. Suggested methods of providing air supply. (Courtesy of Selkirk Metalbestos, Division of Household International)

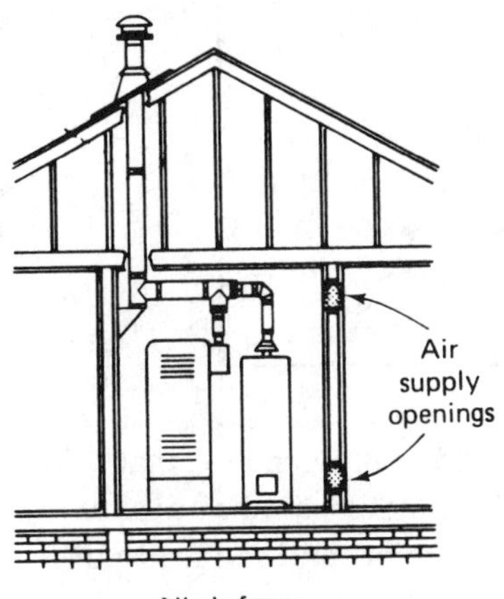

All air from inside building

Free area of each grill = Total input / 1000

(Use 2 grills facing into large interior room)

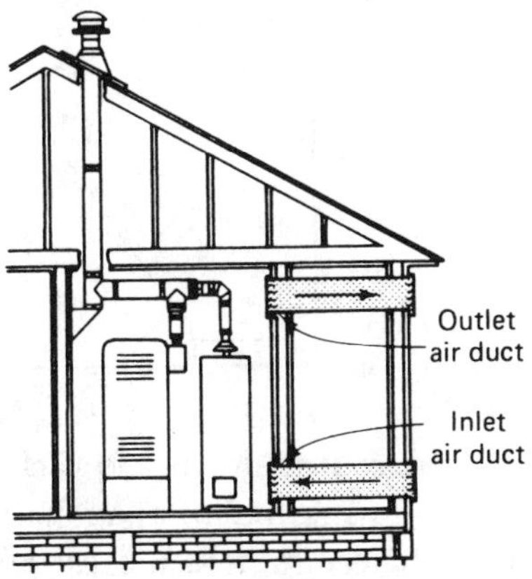

All air from outdoors

Free area of each duct = Total input / 2000

Free area of each grill = Total input / 4000

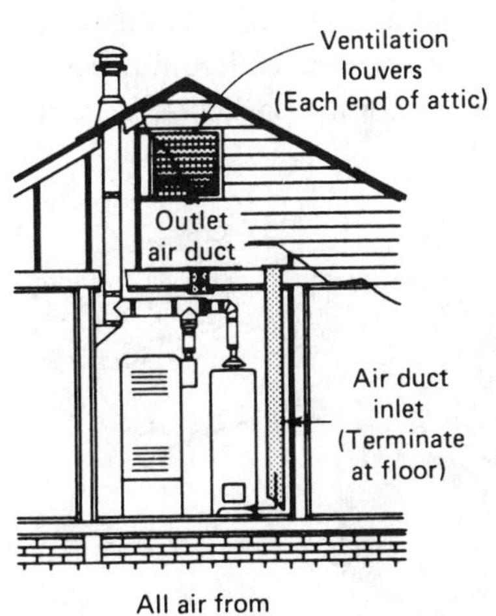

All air from ventilated attic

Free area of each duct or grill = Total input / 4000

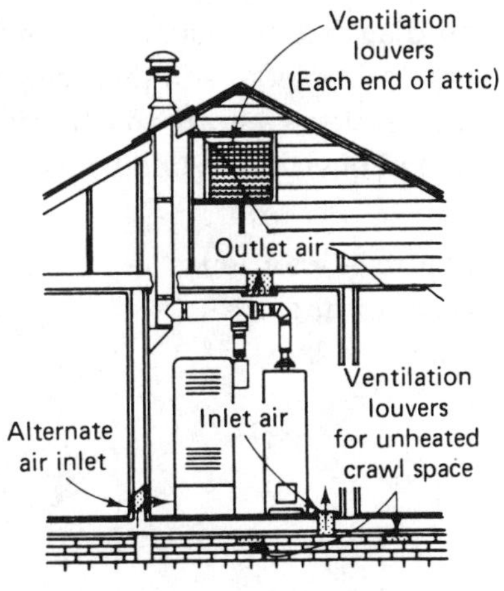

Air in from crawl space, out into attic

Free area of each grill = Total input / 4000

Figure 5-7. A return outlet in each room.

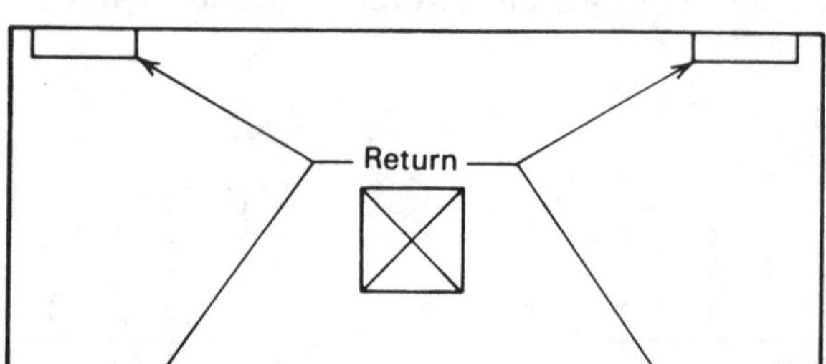

A return air duct to each room is an additional expense that will occur during the installation. However, it is sometimes more desirable than the central return because of the added comfort and reduced noise level.

Under no circumstances should the furnace closet be used as a return air plenum. The return air duct and plenum must be completely isolated from the combustion air in the furnace closet. This precaution is to prevent the products of combustion from entering the living space.

Supply Air Location

The supply air plenum is located on the air outlet of the furnace. The air distribution system is connected to the plenum and directs the air to the desired points in the living area. There are several different types of duct systems and the one that is most desirable for the particular installation in question should be used (Figure 5-8).

Figure 5-8. Various types of duct systems.

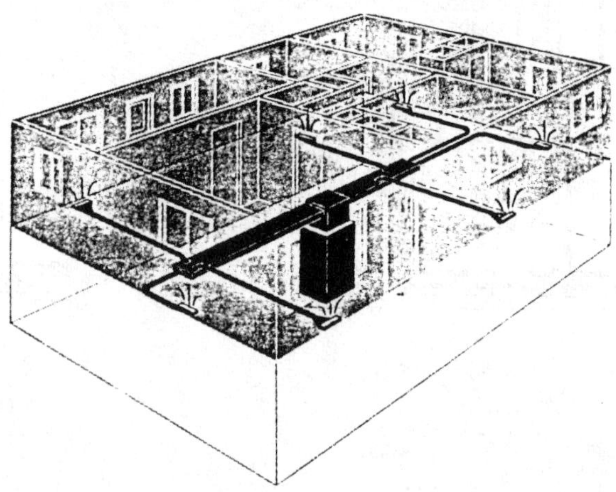

(a) Extended plenum system

Figure 5-8. (*Continued*)

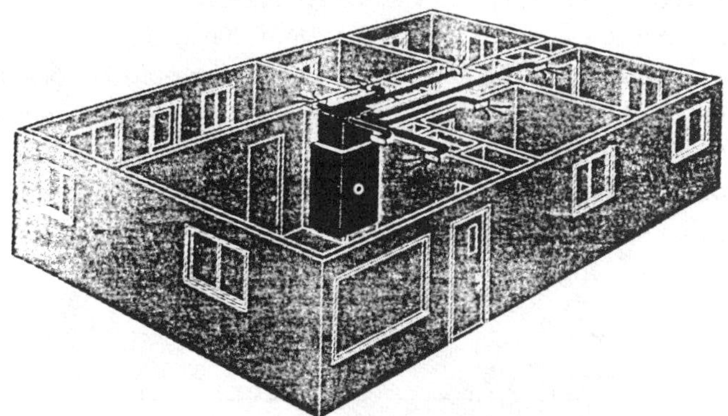

(b) Overhead radial duct system

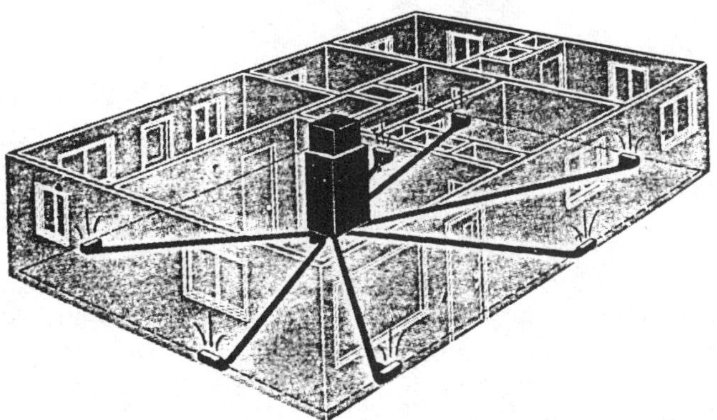

(c) Radial perimeter system

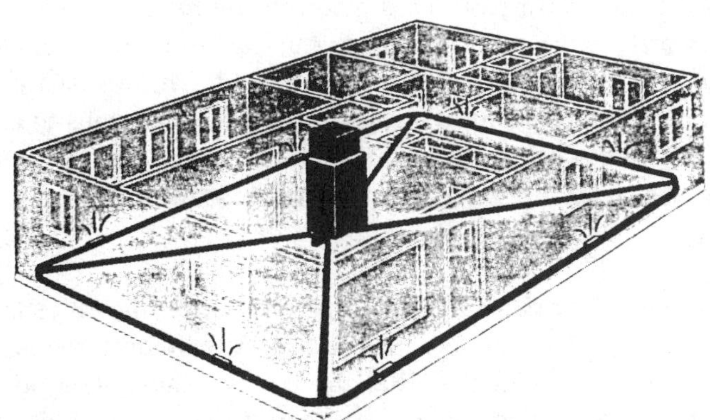

(d) Loop perimeter system

It should be kept in mind that when more material and labor are used, the cost of the installation will also increase. It is usually best to use the most economical air distribution system that will do the job properly.

The supply openings into the living area should be located so that the air will be evenly distributed throughout the space, or where the air will be directed to the point of greatest heat loss or gain. Generally, the air will be directed toward an outside wall or window (Figure 5-9).

Figure 5-9. Supply air outlet location.

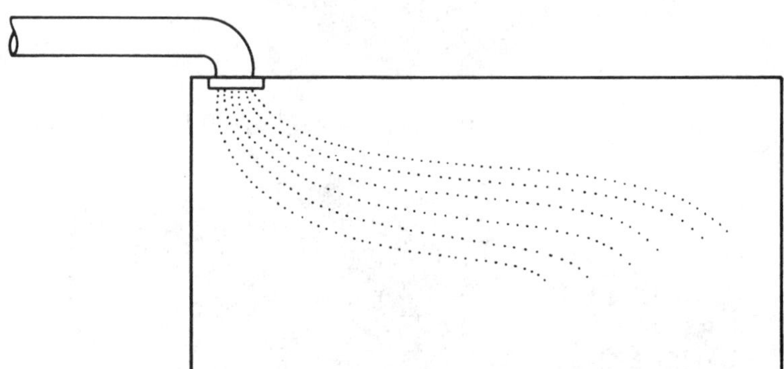

The location of the supply air opening will determine, to a great extent, the length of duct required for that run.

Minimum Clearances

Under no circumstances should combustible material be located within the minimum clearances given in the manufacturer's literature. These clearances are set by the American Gas Association for each model furnace and they must be strictly followed. The furnace dimensions should be correlated with the closet or space dimensions to ensure that sufficient room is available. If not, another location must be found or the room size increased to safely accommodate the furnace.

Noise

Because of air noise and equipment vibration, the unit should not, if at all possible, be located near bedrooms or the family room. However, this will usually increase the installation costs because of the added return air duct required. When such a location cannot be avoided, acoustical materials can be placed in the return air plenum and the air handling section of the equipment to reduce the air noise. Vibration can be reduced, or completely eliminated, by placing the unit on a solid foundation and using vibration-elimination materials.

Accessibility

When considering possible locations for the equipment, sufficient room must be allowed for servicing the unit. A minimum of 24 in. in front of the unit is necessary for changing filters, lubricating bearings, lighting the pilot, and other service requirements.

Length of Duct Runs

The indoor unit should be located so that the length of the duct runs to each room will be as short as possible. The number of turns and elbows must be kept at a minimum. These precautions help reduce heat loss and heat gain through the air distribution system, and allow a simpler design and installation. The shorter duct runs also help to reduce the amount of resistance to the airflow, allowing the air delivery to be balanced much more easily.

Thermostat Location

Satisfactory operation of an air conditioning system depends greatly on careful placement of the thermostat. The following is a list of some major points that must be considered when placing the thermostat.

1. It must be in a room that is conditioned by the unit it is controlling.
2. It must be exposed to normal free air circulation. Do not locate the thermostat where it will be affected by lamps, appliances, fireplaces, sunlight, or drafts from the opening of doors or outside windows.
3. It must be on an inside wall about 4 to 5 ft. from the floor.
4. It must not be located behind furniture, drapes, or other objects that would affect the normal flow of free air over the sensing element.
5. It must not be located where it can be affected by vibrations, such as on a wall of an equipment room, or on the wall in which the door is located.

GAS FURNACE (UPFLOW)

Upflow type furnaces are best suited for closet, garage, or basement installations (Figure 5-10).

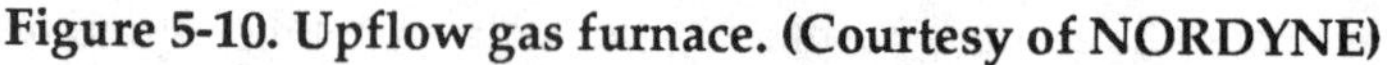

Figure 5-10. Upflow gas furnace. (Courtesy of NORDYNE)

When a cooling unit is to be installed at the same time, there must be a drain closeby so that proper condensate disposal can occur. If not, a condensate pump will need to be considered in the cost of the installation. The floor where the unit will sit should be level, dry, solid, and sufficiently strong to support the weight of the unit. When both the supply and the return airs are to be ducted to the unit, a central location should be chosen to ensure efficient and economical air distribution. Be sure to check the equipment manufacturer's installation information to determine the proper clearances and other pertinent data.

Gas Furnaces (Downflow)

Downflow furnaces are best suited for closet, garage, or second-floor attic installations (Figure 5-11).

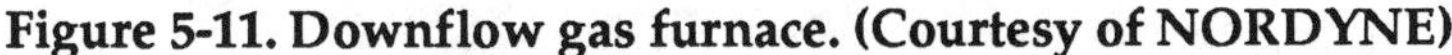

Figure 5-11. Downflow gas furnace. (Courtesy of NORDYNE)

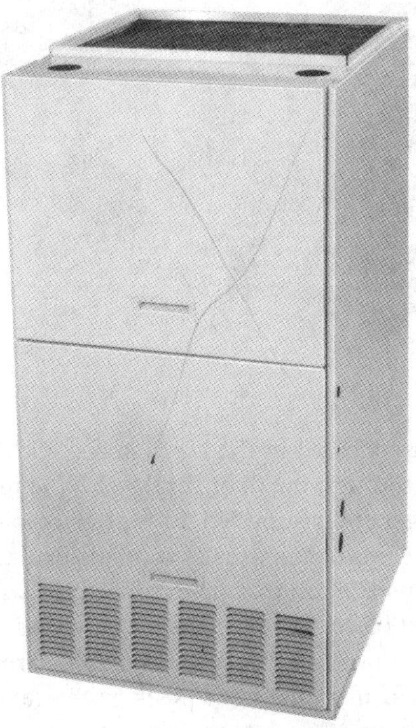

The same basic considerations apply to downflow furnaces as to upflow furnaces. The main exception is that there be sufficient room beneath the furnace and building to install the supply air duct and plenum. These types of systems usually distribute the air from the floor upward (counterflow). However, when installed in a second story attic the air may be delivered to the first floor from the ceiling. The supply ducts are generally placed beneath the floor, in the slab foundation, or between the second-story floor and the first-floor ceiling. Adequate condensate drain facilities should be available or a more expensive drain system will be needed when a cooling unit is also installed.

Gas Furnace (Horizontal)

Horizontal furnaces are best suited for crawl space or attic installations (Figure 5-12).

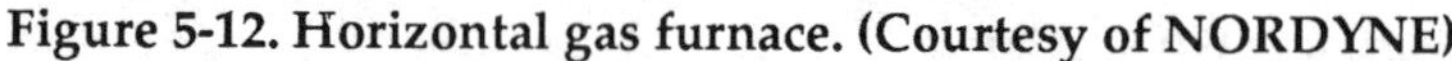

Figure 5-12. Horizontal gas furnace. (Courtesy of NORDYNE)

When they are located in the crawl space, the air is generally supplied to the space through the floor (upflow). When they are installed in the attic, the air is generally supplied through the ceiling (downflow). The installation of the vent pipe is usually more difficult and expensive with crawl space type installation than with other types. Also for this type of furnace, there must be adequate access for servicing the unit. Otherwise, poor service with reduced efficiency and safety will result. These units are generally suspended from the floor joists or the rafters. Horizontal furnaces are usually more difficult to install and service than the other types of furnaces because of where the equipment is located. When they are used along with a cooling system, an auxiliary condensate drain pan and line are required to help prevent water damage to the ceiling should the drain become clogged on attic installations. However, the local codes and ordinances will dictate these requirements.

Condensing Gas Furnaces

These are very high efficiency furnaces which condense the moisture out of the combustion flue gases. This condensate must be properly channeled for disposal. Usually a drain is required to receive the condensate from the furnace. Some municipalities require that this type of condensate be treated before allowing it to enter the city drain system. The local codes and ordinances must be followed to prevent a fine or other undesirable consequences.

Electric Furnaces

Electric furnaces are much more flexible for installation than are gas furnaces. This flexibility is because there is no gas vent to be installed and connected to the furnace (Figure 5-13).

However, it is more economical to locate the furnace as close to the electric service panel as possible. The Btuh rating of electric furnaces can be changed by adding or removing heating elements to the furnace. These furnaces may be installed in virtually any position because there is no vent piping or burner requirements to consider.

Figure 5-13. Electric furnace. (Courtesy of Weatherking Heating and Cooling)

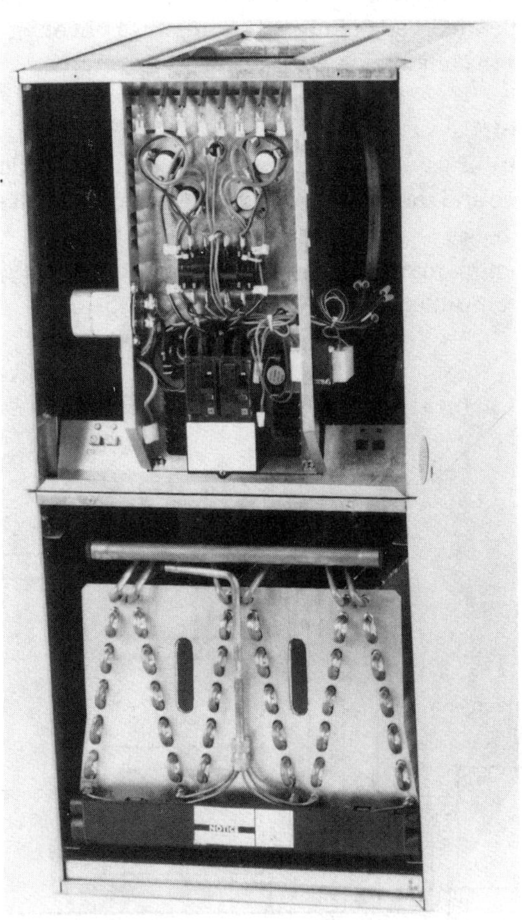

The same general requirements as to the return air location, supply air location, minimum clearances, noise, accessibility for service, length of duct runs, and the location of the thermostat for gas furnaces apply to electric furnaces. When these units are used in combination with a cooling unit, adequate drainage should be provided or provisions made to dispose of the condensate water.

COOLING EQUIPMENT

Basically, there are two types of cooling systems: split-system and self-contained (packaged) units. The split-system evaporator and condensing unit are in different locations. The self-contained unit has the evaporator and the condensing unit in one cabinet. Self-contained systems may be operated as cooling-only units or as heating and cooling units when used in conjunction with a heating apparatus.

Split-System Units

Split-system units usually have the evaporator installed inside the conditioned space and the condensing unit placed outside at some remote location (Figure 5-14).

The evaporator may be located inside a cabinet with a blower or mounted on the air outlet end of a furnace (Figure 5-15).

Figure 5-14. Split-system installation.

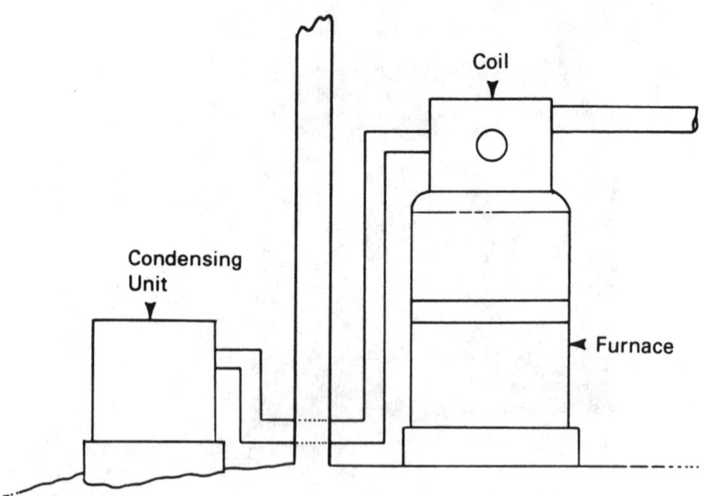

Figure 5-15. Evaporator location.

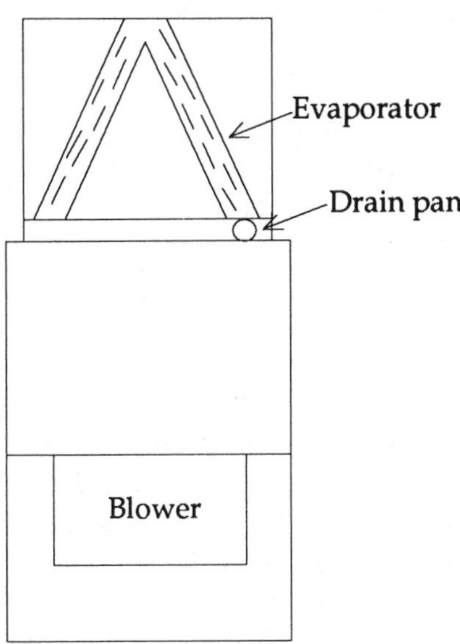

The following considerations should be taken into account when choosing the location of a split-system: support, access, air supply, comfort, length of refrigerant lines and length of the electric lines. These are in addition to those listed for the various types of furnaces discussed earlier in this chapter.

Support

When the condensing unit is placed on the ground a 4-in. concrete slab should be used to support the unit (Figure 5-16).

When the unit is mounted on the roof or other parts of the structure,

Figure 5-16. Concrete slab on ground.

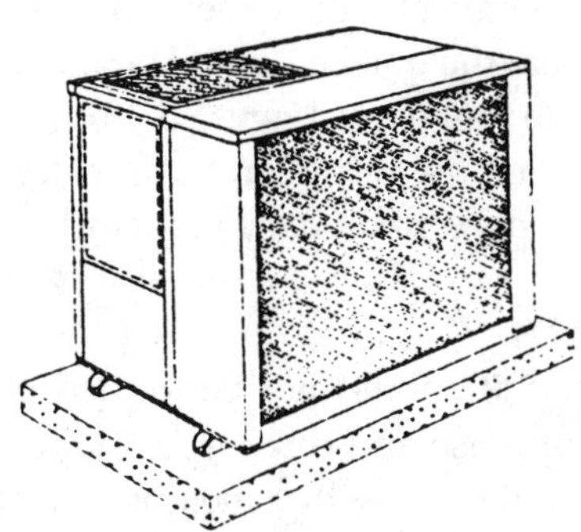

the construction must be checked to make certain that it has sufficient strength to support the weight of the unit.

Access

Two or three feet of clearance should be provided on the side where the access panel is located. This clearance is necessary for routine servicing operations. There should be sufficient clearance around the unit to allow for air flow to the condenser. Check the manufacturer's instructions for the clearance required.

Air Supply

When the desired location offers a restricted air flow, such as in a garage or carport, duct work may be required to allow a sufficient amount of air to flow across the condenser coil and out the fan discharge. When a horizontal discharge unit is installed, it may be desirable to install an air deflector to direct the hot air away from grass, shrubs, flowers, or a neighbor's property (Figure 5-17).

Figure 5-17. Horizontal discharge condenser with an air deflector.

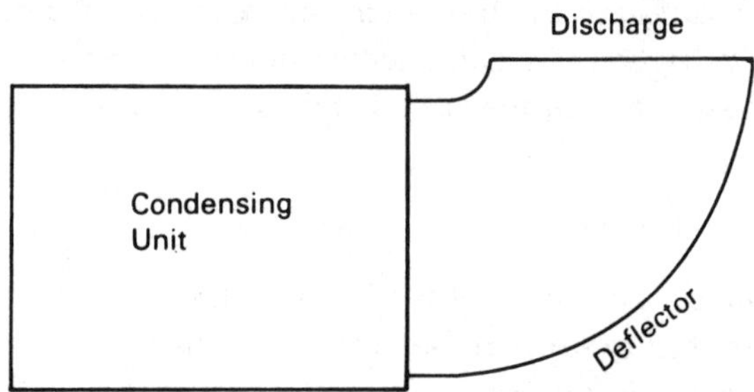

When a vertical discharge condensing unit is used, the unit must be located so that the discharge air will not be recirculated back into the unit. This usually requires that the unit be located far enough from the house so that the air will miss the roof eaves (Figure 5-18).

In such installations the refrigerant and electrical lines should be buried underground to prevent damage to them.

Comfort Considerations

The condensing unit should not be located close to bedrooms or close to a patio, where noise and discharge air would be objectionable. The unit should not interfere with a neighbor's property or personal

Figure 5-18. Vertical discharge condenser.

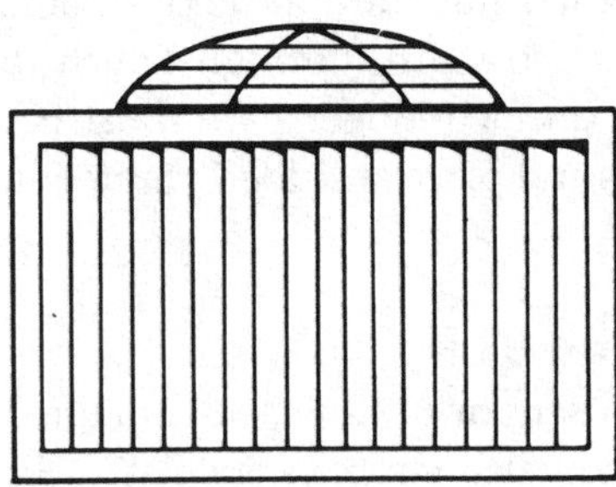

convenience. It is usually preferable to locate the condensing unit by a solid wall or a bathroom.

Length of Refrigerant Lines

The condensing unit should be located as close as possible to the evaporator to reduce the length of the refrigerant lines. Long refrigerant lines are more susceptible to refrigerant leaks than are shorter ones. Also, the cost of the refrigerant lines will be less. With shorter lines the initial refrigerant charge will be smaller, and if a leak should occur, a smaller quantity of refrigerant will be lost. An increased restriction to the flow of refrigerant will be experienced with longer refrigerant lines. This restriction must be kept to a minimum.

Length of Electric Lines

The location of the electric service panel in relation to the unit should be considered. The longer the electric lines, the more costly the installation. However, a long electric line should not outweigh long refrigerant lines. Always check the manufacturer's specifications for the size of wiring for each distance from the electric supply panel.

SELF-CONTAINED (PACKAGED) UNITS

Self-contained cooling equipment is manufactured in basically two types: horizontal and upflow units. The type of design chosen will depend greatly on the installation location. When air-cooled units are located indoors or in through-the-wall installations, a sufficient amount of clearance must be allowed in front of the condenser air intake to allow sufficient air to pass over the condenser coil. A grill or some other type of screen should be installed on the air inlet to prevent debris and foreign matter entering the coil and other internal components of the unit.

When water-cooled units are located indoors, there should be an accessible point in the structure through which the water lines can be installed. Otherwise, a much longer path must be taken, which will increase the initial costs and place an added restriction on the water flowing through the pipes.

Upflow Self-Contained Units

The location chosen for upflow self-contained units must have adequate electrical and water supplies available. It is desirable to have drainage facilities close to the unit to prevent the extra expense of having them installed or that of installing a drain pump. The floor or the foundation where the unit will be located should be strong enough to support the weight of the unit. It should be level, dry, and solid. When supply and/or return ducts are to be installed, the unit should be placed in a central location to ensure that efficient and economical air distribution can be obtained. There should be adequate clearance around the unit to permit routine maintenance. The manufacturer's installation literature usually will provide the weight, necessary clearances, and other pertinent data about the unit.

Horizontal Self-Contained Units

When horizontal self-contained units are under consideration, many of the location considerations listed for upflow packaged units, split-system units, and furnaces will apply. These types of units are generally used on mobile homes, in crawl spaces, and for through-the-wall installations. Remember that a central location is best for air distribution purposes. In addition, electrical power and drain facilities, together with the strength of the structural support, should be considered. Sufficient clearance must be maintained around the unit for servicing procedures. Avoid locating the equipment close to sleeping quarters.

Example Location

In locating the unit for our example, the architect designed the indoor unit location in bathroom 2 between the closets for bedroom 2 and bedroom 3 (Figure 5-19).

The cooling unit design engineer can place the condensing unit in the best location, keeping in mind the facts discussed earlier in this chapter. If this is new construction, there is not much of a problem because the refrigerant lines and electric wires can be installed without any problem during the construction of the house.

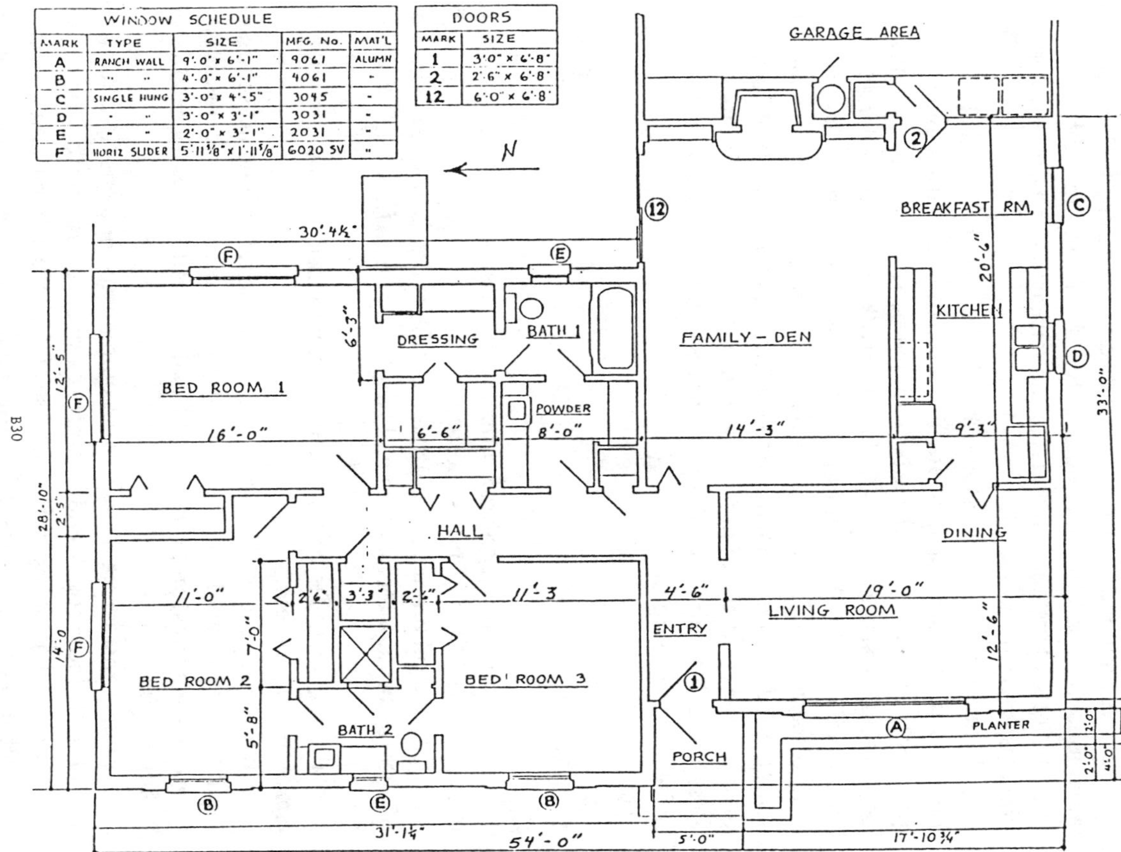

Figure 5-19. Indoor unit location. (Courtesy of TU Electric Co.)

However, if the unit is being installed in an existing house, consideration must be given to such things as types of floors, amount of room in the attic, ease of installation of ducts and refrigerant lines, easy access to a condensate drain, and so on.

In our example let us assume that the unit is to be installed when the house is under construction. In this case, the refrigerant lines would probably be installed in a chase beneath the slab floor so that the lines will be shorter and easier to install. In keeping with the aforementioned installation practices, the best location for the condensing unit is on the west wall of bathroom 2. The noise level would be at a minimum and the refrigerant lines would not be unnecessarily long. It would be a good idea to locate the condensing unit away from the bathroom window, but not close enough to either of the bedroom windows to allow unnecessary noise into the bedroom.

SUMMARY

There are three types of furnaces as classified by their source of energy: (1) gas, (2) oil, and (3) electric.

Gas and oil furnaces are further classified according to the direction of airflow through them: upflow, downflow (counterflow), and horizontal.

Electric furnaces are not usually subdivided by the direction of airflow through them.

Items such as gas pipe connections, venting requirements, return air location, supply air location, minimum clearances, noise, accessibility, length of duct runs, and thermostat location should be considered when locating a gas furnace.

When considering the venting requirements for a gas-fired furnace, any and all local codes and ordinances must be followed. An unsafe furnace should not be operated because of the hazard to life and property.

When a gas furnace is to be installed in a living space, closet or any other tightly enclosed room, openings must be provided to ensure that adequate combustion air is available to the unit.

The central return is the most popular method of directing the return air to the furnace because of the added expense of the return air ducts.

Under no circumstances should the equipment closet be used as the return air plenum.

The air supply is connected to the air outlet of the furnace. The supply air distribution system is connected to the plenum and directs the air to the desired points in the living area.

The supply air openings should be located so that the air will be evenly distributed throughout the room, or where the air will be directed to the point of greatest heat loss or gain.

Under no circumstances should combustible material be located within the minimum clearances given in the manufacturer's literature.

Because of air noise and equipment vibration, the unit should not be located near bedrooms or the family room if at all possible.

When considering possible locations for the equipment, sufficient room must be allowed for servicing the unit.

The indoor unit should be located so that the length of the duct runs to each room will be as short as possible. The number of turns and elbows must also be kept to a minimum. These precautions help to reduce heat loss and heat gain through the air distribution system.

The satisfactory operation of an air conditioning system depends greatly on careful placement of the thermostat.

Upflow furnaces are best suited for closet, garage, or basement installations.

Downflow furnaces are best suited for closet, garage, or in second floor attics.

Horizontal flow furnaces are best suited for crawl space or attic installations.

There is much more flexibility in locating electric furnaces than with gas furnaces because there is no vent or gas pipe needed for the installation.

The Btuh rating of an electric furnace can be changed by changing the number of heating elements used in it.

Basically, there are two types of cooling systems: split-system and self-contained (packaged) units. The split-system evaporator and condensing unit are in different locations. The self-contained unit has the evaporator and condensing unit in one cabinet.

The following considerations should be taken into account when choosing the location for split-system units: support, access, air supply, comfort considerations, length of the refrigerant lines, and length of the electric lines.

Horizontal self-contained units are generally used on mobile homes, in crawl spaces, and for through-the-wall installations.

Chapter 6

Commercial Heat Load Calculation

In this chapter we will use the TU Electric Commercial Heat Load Calculation form, and information applicable to the TU Electric service area, but the general steps are the same for any area served. The electric utility company in your area probably has the same, or similar charts, tables, and information. Check with the utility company for assistance. For this example we will calculate the heat load for a cafe located in Hillsboro, Texas and we will estimate the cost of operation using a heat pump, electric furnace, and gas heating units.

INTRODUCTION

We will use the TU Electric Weather Form, the Infiltration Construction Definition Chart, an example, and a floor plan for a restaurant to which the example applies. There will also be instructions for estimating the cost of operation using various sources of heating. Check with the power company in your area for this information and forms. They may appear different but the procedure is about the same.

Use the following procedure:

Step 1: Refer to the TU Electric Form 219-C (Figure 6-1) and the TU Electric Weather Chart (Chart 6-1).

The cooling design conditions for the Hillsboro, Texas area are 100°F outside temperature and 70°F inside temperature. Place this information in the proper space at the top of the form (Figure 6-2). Subtract the two numbers to obtain the temperature difference (30°F).

Step 2: The heating design is usually calculated on a 75°F inside temperature. From the TU Electric Weather Chart, the outside winter design temperature is selected for the geographical location involved. For example, Hillsboro, Texas has a 25°F outside temperature design (Figure 6-3).

Figure 6-1. Commercial heating and cooling form. (Courtesy of TU Electric Co.)

HEATING AND COOLING ESTIMATE
COMMERCIAL

Customer ________________
Address ________________ Town ________________
Prepared by ________________ Date ________________

DESIGN CONDITIONS
COOLING D.B. °F
Outside Temp. ________
Inside Temp. ________
Temp.Diff. ________

HEATING D.B. °F
Inside Temp. ________
Outside Temp. ________
Temp.Diff. ________

INSULATION
Ceiling ________
Wall ________
Floor ________
Window ________

STRUCTURE FACES
☐ North ________
☐ East ________
☐ South ________
☐ West ________

OVERHANG OR SHADING

CONSTRUCTION	SC	COOLING FACTOR: No Shading	Inside Shading & Overhang 0′ 1′ 2′ 3′ 4′	Outside Shades	HEATING FACTOR	QUANTITY	COOLING Btuh	HEATING Btuh
1. GLASS--SOLARGAIN Windows Doors Skylights (Sq.Ft.) (*Note A)								
North		32	14 0 0 0 0	0				
Northeast		50	23 21 19 19 17	17				
East		50	23 20 19 17 14	23				
Southeast		80	36 26 19 8 0	21				
South		100	45 33 10 0 0	26				
Southwest		155	70 51 39 17 0	42				
West		190	86 82 80 67 57	57				
Northwest		128	58 54 51 51 49	37				
Horizontal		185						

CONSTRUCTION	U	Cooling 20	Cooling 22	Cooling 25	Cooling 30	Heating 65	Heating 60	Heating 55	Heating 50	QUANTITY	COOLING Btuh	HEATING Btuh
DESIGN D. B. TEMP. DIFFERENTIAL	U	20	22	25	30	65	60	55	50			
2. GLASS TRANSMISSION (Sq.Ft.)		U X TD				U X TD						
Standard--Single Glazing	1.13	22.6	24.9	28.2	33.9	73.4	67.8	62.2	56.5			
Insulating--Double Glazing	.78	15.6	17.2	19.5	23.4	50.7	46.8	42.9	39.0			
Storm or Insulating GlassW/T Bk.	.56	11.2	12.3	14.0	16.8	36.4	33.6	30.8	28.0			
3. DOORS--(Sq.Ft.)		U X TD				U x TD						
Solid Wood, Hollow Core or Metal	.55	11.0	12.1	13.8	16.5	35.8	33.0	30.3	27.5			
4a. FLOORSLAB (Linear Ft.Exp. Edge)		U X Zero				U X TD						
No Edge Insulation	.81					56.6	48.6	44.6	40.5			
R-4 Edge Insulation	.68					44.2	40.8	37.4	34.0			

4b. FLOORS: ENC. CRAWLSPACE(Sq. Ft.)		U X Zero				U X (TD X 20)						
No Insulation R-7 Insulation	.270 .093					12.2 4.2	10.8 3.7	9.4 3.2	8.1 2.8			
R-13 Insulation	.060					2.7	2.4	2.1	1.8			
5. WALLS: (Sq. Ft.) Net		U X TD				U X TD						
No Insulation--Masonry No Insulation---Wood Siding	.389 .320	7.8 6.4	8.6 7.0	9.7 8.0	11.7 9.6	25.3 20.8	23.3 19.2	21.4 17.6	19.5 16.0			
No Insulation---Brick Veneer R-4 Insulation	.240 .138	4.8 2.8	5.3 3.0	6.0 3.4	7.2 4.1	15.6 8.9	14.4 8.3	13.2 7.6	12.0 6.9			
R-7 Insulation R-11 Insulation	.109 .075	2.2 1.5	2.4 1.7	2.7 1.9	3.3 2.3	7.1 4.9	6.6 4.5	6.0 4.1	5.5 3.8			
R-13 Insulation	.065	1.3	1.4	1.6	2.0	4.2	3.9	3.6	3.3			
6a. CEILINGS-W/Attic (Sq. Ft.)		U X (TD x 40)				U X TD						
No Insulation R-4 Insulation	.598 .176	35.9 10.6	37.1 10.9	38.9 11.4	41.9 12.3	38.9 11.4	35.9 10.6	32.9 9.7	29.9 8.8			
R-7 Insulation R-11 Insulation	.114 .079	6.8 4.7	7.1 4.9	7.4 5.1	8.0 5.5	7.4 5.1	6.8 4.7	6.3 4.3	5.7 4.0			
R-13 Insulation R-19 Insulation	.068 .048	4.1 2.9	4.2 3.0	4.4 3.1	4.8 3.4	4.4 3.1	4.1 2.9	3.7 2.6	3.4 2.4			
R-22 Insulation R-26 Insulation	.042 .036	2.5 2.2	2.6 2.2	2.7 2.3	2.9 2.5	2.7 2.3	2.5 2.2	2.3 2.0	2.1 1.8			
6b. Ceilings--Attic (Sq. Ft.)		U X (TD X 45)				U X TD						
No Insulation R-4 Insulation	.470 .160	30.6 10.4	31.5 10.7	32.9 11.2	35.3 12.0	30.6 10.4	28.2 9.6	25.9 8.8	23.5 8.0			
R-7 Insulation R-11 Insulation	.109 .076	7.1 4.9	7.3 5.1	7.6 5.3	8.2 5.7	7.1 4.9	6.5 8.6	6.0 4.2	5.5 3.8			
R-13 Insulation R-19 Insulation	.064 .047	4.2 3.1	4.3 3.1	4.5 3.3	4.8 3.5	4.2 3.1	3.8 2.8	3.5 2.6	3.2 2.4			
R-22 Insulation	.041	2.7	2.7	2.9	3.0	2.7	2.5	2.3	2.0			
7. VENTILATION OR INFILTRATION (*Note C)	CFM	22	24	28	33	72	66	61	55			
8. SENSIBLE HEAT Btuh-Sub-total		Line 1 thru 7										

(*Continued*)

Figure 6-1 (*Concluded*)

CONSTRUCTION	COOLING FACTOR	HEATING FACTOR	QUANTITY	COOLING Btuh	HEATING Btuh
8. SENSIBLE HEAT Btuh-Sub-Total	Line 8 Reverse Side				
9. SENSIBLE HEAT Btu W/DUCT LOSS Sub-Total	Line 8 x Factor From Table 3				
10. VENTILATION OR INFILTRATION LATENT HEAT GAIN-EXTERIOR (Note C)	23				
11. TOTAL HEAT Btuh W/OUT INTERNAL LOAD	LINE 9 + 10				
12. INTERIOR SENSIBLE HEAT GAIN					
A. People-Ave. Number (Table 2)					
B. Lighting-kW	3413				
C. Motors- HP	2550				
D. Other Internal Load-kW	3413				
E. Gas Cooking (*Note D)	5880				
F. Electric Cooking (Note D)	2400				
13. INTERIOR SENSIBLE HEAT GAIN Sub-Total	Line A thru F				
14. INTERIOR SENSIBLE HEAT GAIN W/DUCT LOSS Sub-Total	Line 13 X Factor From Table 3				
15. LATENT HEAT GAIN--INTERIOR					
A. People-Ave. (*Table 2)					
B. Other interior Load					
C. Gas Cooking-Number of Burners	2520				
D. Electric Cooking (*Note D)	1000				
16. LATENT HEAT GAIN INTERIOR Sub-Total	Line 15 A thru D				
17. TOTAL INTERIOR HEAT GAIN	Line 14 + 16				
18. TOTAL LOAD Btuh	Cooling Line 11 + 17	Heating Line 9 thru 14			

INSTRUCTIONS FOR HEATING & COOLING ESTIMATES - FOR 219C

*Note A -
1. For Solor Gain use only the one direction producing the greatest Btuh.
2. Solor factors may be multiplied by .90 for plate glass, by .85 for double pane glass, by.50 for glass block.
3. No solor load is used if overhang exceeds 4' per floor for nominal windows.
4. Windows shaded all day by buildings or other permanent objects should be included under window transmission only.

*Note B - For light colored walls or roofs, COOLING factor may be multiplied by .75.

*Note C -
1. Infiltration formula: Volume = CFM
2. For extra heavy people loads, calculate ventilation CFM requirements from Table 1. Use this CFM quantity in place of infiltration CFM above, if it is greater.
3. If the quantity of ventilation air is stated by owner, architect or engineer, use the amount they specify.

*Note D - Use 50% of nameplate kW for quantity and multiply by sensible and latent factors.

TABLE 1 - Ventilation CM Per Person

No Somking	20
Some Smoking	40
Heavy Smoking	60

TABLE 2 - People - Btuh Per Person

	Sensible	Latent
Inactive	200	250
Moderate Activity	300	500
Active	400	800

TABLE 3 - Duct Loss Factors

Ductwork Location	Inches of Duct Insulation	Duct Factor Cooling	Heating
Attic - Vented	1	1.15	1.25
- Vented	2	1.10	1.15
- Unvented	1	1.20	1.20
- Unvented	2	1.15	1.10
Crawl Space - Unvented	1	1.05	1.10
Within Conditioned Area	1	1.00	1.00
Within Slab	1	1.20	1.25

TABLE 4 - Equipment Efficiency

Resistance Heat: 3,413 MBtu/kWh
Heat Pump: EER/SEER - Cooling
HSPF - Heating
Gas Furnace: AFUE x 10

Fuel Cost:
Electricity: Ave. kWh $__________
Gas: Ave. MCF $__________

1 MBtu + 1,000 Btu

Chart 6-1. TU Electric Weather Chart. (Courtesy of TU Electric Co.)

	Outside Temperature Design (°F)*		Location Multiplier	Location Multiplier Multiplier
	Summer	Winter	Heating	Cooling
DALLAS DIV. & CENTRAL REGION: Arlington, Dallas, Duncanville, Euless, Ft. Worth, Garland, Grand Prairie, Grapevine, Farmers Branch, Irving, Lancaster, Mesquite, Plano, Richardson	100	20	1.00	1.00
EASTER REGION: Athens, Corsicana, Crockett, Lufkin, Nacogdoches, Palestine, Terrell, Tyler	100	25	0.96	0.91
NORTHERN REGION: Decatur, Denison, Gainsville, McKinney, Mineral Wells, Paris, Sherman, Sulphur Springs	100	20	1.15	0.85
NORTHWEST REGION: Archer City, Breckenridge, Burkburnett, DeLeon, Eastland, Electra, Graham, Henrietta, Iowa Park, Wichita Falls	100	20	1.20	0.88
SOUTHERN REGION: Brownwood, Cleburne, Hillsboro, Killeen, Round Rock, Stephenville, Taylor, Temple, Waco, Waxahatchie	100	25	0.93	0.97
WESTERN REGION: Andrews, Big Spring, Colorado City, Crane, Lamesa, Midland, Monahans, Odessa, Snyder, Sweetwater	100	20	1.15	0.80

*Generalized from ASHRAE Fundamentals Handbook 1985, Chapter 24.

Figure 6-2. Entering cooling design conditions on heat load form. (Courtesy of TU Electric Co.)

DESIGN CONDITIONS	
COOLING D.B.	°F
Outside Temp.	100
Inside Temp.	70
Temp. Diff.	30

Figure 6-3. Entering heating design conditions on the heat load form. (Courtesy of TU Electric Co.)

DESIGN CONDITIONS	
HEATING D.B.	°F
Inside Temp.	75
Outside Temp.	25
Temp. Diff.	50

Step 3: Insert the inside and outside design temperatures on the form and subtract the two numbers to obtain the temperature differential. Insert these figures in the spaces provided on the chart.

Step 4: Refer to the set of floor plans for the restaurant to determine the door and window schedule and other design criteria (Figure 6-4).

In addition to the floor plan, reference should be made to other parts of the plans, such as the roof plan, foundation plan, cross sections and elevations. If some of the needed information cannot be found on the plans, it can be obtained by consulting with the architect, builder, or the building owner.

Much of the design information can be found on the floor plan (Figure 6-4), but more is needed. The following also applies:

1. The building faces north.

2. The amount of overhang is: East = 0'; North = 4'; West = 0'; South = 0'.

3. The duct is located in the conditioned area.

4. It will be a roof top unit.

Figure 6-4. Type F-Floor plan (30 × 80). (Courtesy of TU Electric Co.)

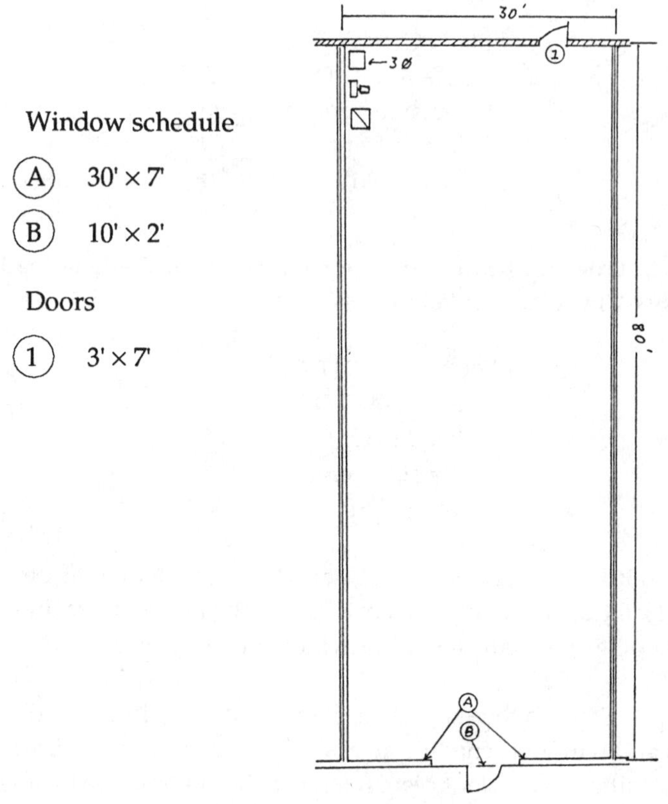

5. Ventilation: There will be about 40 people with some of them smoking.

6. The people will be inactive.

7. The restaurant will be open six days a week from 8:00 a.m. to 10:00 p.m.

8. No economizer will be used.
   Fuel costs are: Electricity kWh = \$0.07 Summer; \$0.06 Winter.
   Gas MCF = \$5.07 + Monthly Minimum. Minimum = \$6.50 Summer; \$10.50 Winter.

9. Equipment Efficiency: Heat pump, cooling = 9 EER. Heating = 7.4 HSPF.

10. Lighting:

24 fixtures × 4 lamps × 50 watts	=	4,800 watts
11 fixtures × 1 lamp × 100 watts	=	1,100 watts
1 fixture × 1 lamp × 100 watts	=	100 watts
Total	=	6,000 watts

Motors: 1 Refrigerator 1 HP
1 Freezer 1 HP

11. Cooking:

1 - Fryer	12	kW
1 - Griddle	12	kW
1 - Convection Oven	11	kW
1 - Hot Food Server	2	kW
1 - Coffee Maker	1	kW
Total kW	= 38	kW

Note: D - 38 kW ÷ 2 = 19 kW
Sensible 19 kW × 2,400 = 45,600
Latent 19 kW × 1,000 = 19,000

12. Glass Solar Gain. (SC) at 0.85.
North (30' × 7') + (2' × 10') = 230 sq. ft.
East = 0
South = 0
West = 0

13. Glass Transmission: 230 sq. ft.

14. Doors: 1 – 3' × 7' = 21 sq. ft.

15. Floor. Slab = 30' × 2 = 60 linear ft.
No additional footage because of inside walls.

16. Walls:

North: (30' × 15') – [(30' × 7') + (10' × 2')]	=	220	sq. ft.
East: - Next to conditioned space.	=	0	sq. ft.
South: (30' × 15') – (3' × 7')	=	429	sq. ft.
West: - Next to conditioned space.	=	0	sq. ft.
Total	=	649	sq. ft.

17. Ceiling: 30' × 80' – 2,400 sq. ft.
9' drop ceiling
Insulation on roof 15'.

18. Ventilation Infiltration:

40 people × 40 CFM/Person	= 1600 CFM Sensible
	= 1600 CFM Latent
People: 40 × 200 Sensible	= 8,000 Btuh
40 × 250 Latent	= 10,000 Btuh

19. Insulation: Ceiling is R-19
Walls is R-11
Glass is double glazed Slab floor

Step 5. Use the data which applies to the building construction to fill in the remainder of the data block at the top of the form (Figure 6-5).

Figure 6-5. Completing data block. (Courtesy of TU Electric Co.)

Insulation		Structure Faces	Overhang or Shading
Ceiling	R-19	___ North	4'
Wall	R-11	___ East	0
Floor	Slab	___ South	0
Window	Double Glazing	___ West	0

Step 6. Item 1. Glass Solar Gain: Locate the correct column under the cooling factor heating. In this example, only the North side of the building has a roof overhang. It is 4 feet. Circle the 4' under the "Inside Shading Overhang Column." The factors directly below the 4' figure are the number of Btus of radiant heat gain per sq. ft. of glass for each direction of exposure. In this example the figure is 0 because it is on the north side of the building (Figure 6-6).

Step 7. Item 1: Circle the solar gain factors under the 4' column which apply to each direction of glass exposure. In this example there are none.

Step 8. Item 1: Under the quantity column heading, insert the square feet of glass for the corresponding direction of exposure after first determining the glass area for each exposure. Refer to the floor plan and window schedule (Figure 6-4).

The glass area is found for each window by multiplying the window width by the window height in feet. The area for the doors is found in the same manner. The total area of glass exposure for each direction is the sum of the individual window areas and any glass door areas located facing that direction.

Fill in the square feet of glass in the quantity column opposite the direction of exposure, and under the appropriate shading category. Insert a value for SC (shading coefficient) to compensate for plateglass or multiple layers as shown on Form 219-C (Figure 6-6).

Step 9. Item 1: Multiply the cooling factor times the quantity of glass for each direction of exposure and insert these figures in the "Cooling Btuh" Column. Note A gives shading factors to modify the cooling Btuhs should other than standard single pane window glass be used. For instance, the Btuh for double glass would be the product of 0.85 times the cooling factor times the area of glass (Figure 6-6).

For all practical purposes, the sun can impose a maximum solar heat gain from only one vertical wall or exposure at any given time. Therefore, only the one largest vertical calculated Btuh solar heat gain will be included in the total cooling heat load. In our example, this will be "0." Identify the largest cooling Btuh figure by marking it in some manner, such as with an asterisk, or by crossing-out the other amounts.

A skylight was not used in this building, so the "Horizontal" area of glass category listed under "Glass-Solar Gain" is ignored. If horizontal glass (i.e., skylight) was used, then the total square feet (area) of the horizontal quantity should be multiplied by the solar gain cooling factor to obtain a cooling Btuh for the solar gain from the horizontal glass. This should be marked with an asterisk also, and be added along with the largest vertical area solar gain Btuh to obtain the total cooling Btuh load when calculating the subtotal for Item 9.

Step 10. Item 1: From the established design conditions determined in Steps 1 and 3, determine the temperature differentials for heating and cooling. In the last line of Item 1, circle the heating and cooling temperature differentials (Figure 6-6).

Step 11. Items 2 through 7: Circle all of the heating and cooling factors which apply and are to be used in Items 2 through 6. These values are located directly under the circled design dry bulb (D.B.) temperature differential for both heating and cooling and directly across from the type of construction involved. Item 7 has only one heating and one cooling factor for each design dry bulb temperature differential. See Step 12 for additional instructions on Item 7.

The following characteristics apply to the building. (The factors which apply are circled under the 30°F Cooling and the 50°F Heating Differential Columns for Items 2, 3, 4a, 5, 6a, and 7 on the form 219-C.)

Figure 6-6. Entering construction and heating and cooling details on heat load form. (Courtesy of TU Electric Co.)

CONSTRUCTION	SC	COOLING FACTOR: No Shading	Inside Shading & Overhang 0′	1′	2′	3′	4′	Outside Shades	HEATING FACTOR	QUANTITY	COOLING Btuh	HEATING Btuh
1. GLASS--SOLAR GAIN Windows Doors Skylights (Sq.Ft.) (*Note A)												
North Northeast	.85	32 50	14 23	0 21	0 19	0 19	0 (circled) 17	0 17		230	0	
East Southeast	.85	50 80	23 36	20 26	18 19	17 8	14 0	23 21		0		
South Southwest	.85	100 155	45 70	33 51	10 39	0 17	0 0	26 42		0		
West Northwest	.85	190 128	86 58	82 54	80 51	67 51	57 49	57 37		0		
Horizontal		185										
DESIGN D. B. TEMP. DIFFERENTIAL		U	20		22		25	30 (circled)	65 60 55 50 (circled)			
2. GLASS TRANSMISSION (Sq.Ft.)			U X TD						U X TD			
Standard--Single Glazing Insulating--Double Glazing		1.13 .78	22.6 15.6		24.9 17.2		28.2 19.5	33.9 23.4 (circled)	73.4 67.8 62.2 56.5 50.7 46.8 42.9 39.0 (circled)	230	2382	8970
Storm or Insulating GlassW/T Bk.		.56	11.2		12.3		14.0	16.8	36.4 33.6 30.8 28.0			
3. DOORS--(Sq.Ft.)			U X TD						U x TD			
Solid Wood, Hollow Core or Metal		.55	11.0		12.1		13.8	16.5 (circled)	35.8 33.0 30.3 27.5 (circled)	21	347	578
4a. FLOOR SLAB (Linear Ft. Exp. Edge)			U X Zero						U X TD			
No Edge Insulation R-4 Edge Insulation		.81 .68							56.6 48.6 44.6 40.5 (circled) 44.2 40.8 37.4 34.0	60		2430

4b. **FLOORS: ENC. CRAWL SPACE** (Sq. Ft.)(*Note B)		U X Zero				U X (TD X 20)						
No Insulation R-7 Insulation	.270 .093					12.2 4.2	10.8 3.7	9.4 3.2	8.1 2.8			
R-13 Insulation	.060					2.7	2.4	2.1	1.8			
5. **WALLS:** (Sq. Ft.) Net		U X TD				U X TD						
No Insulation--Masonry No Insulation---Wood Siding	.389 .320	7.8 6.4	8.6 7.0	9.7 8.0	11.7 9.6	25.3 20.8	23.3 19.2	21.4 17.6	19.5 16.0			
No Insulation---Brick Veneer R-4 Insulation	.240 .138	4.8 2.8	5.3 3.0	6.0 3.4	7.2 4.1	15.6 8.9	14.4 8.3	13.2 7.6	12.0 6.9			
R-7 Insulation R-11 Insulation	.109 .075	2.2 1.5	2.4 1.7	2.7 1.9	3.3 2.3	7.1 4.9	6.6 4.5	6.0 4.1	5.5 3.8	649	1493	2466
R-13 Insulation	.065	1.3	1.4	1.6	2.0	4.2	3.9	3.6	3.3			
6a. **CEILINGS-W/Attic** (Sq. Ft.)		U X (TD x 40)				U X TD						
No Insulation R-4 Insulation	.598 .176	35.9 10.6	37.1 10.9	38.9 11.4	41.9 12.3	38.9 11.4	35.9 10.6	32.9 9.7	29.9 8.8			
R-7 Insulation R-11 Insulation	.114 .079	6.8 4.7	7.1 4.9	7.4 5.1	8.0 5.5	7.4 5.1	6.8 4.7	6.3 4.3	5.7 4.0			
R-13 Insulation R-19 Insulation	.068 .048	4.1 2.9	4.2 3.0	4.4 3.1	4.8 3.4	4.4 3.1	4.1 2.9	3.7 2.6	3.4 2.4			
R-22 Insulation R-26 Insulation	.042 .036	2.5 2.2	2.6 2.2	2.7 2.3	2.9 2.5	2.7 2.3	2.5 2.2	2.3 2.0	2.1 1.8			
6b. **Ceilings--Attic** (Sq. Ft.)(*N.B)		U X (TD X 45)				U X TD						
No Insulation R-4 Insulation	.470 .160	30.6 10.4	31.5 10.7	32.9 11.2	35.3 12.0	30.6 10.4	28.2 9.6	25.9 8.8	23.5 8.0			
R-7 Insulation R-11 Insulation	.109 .076	7.1 4.9	7.3 5.1	7.6 5.3	8.2 5.7	7.1 4.9	6.5 8.6	6.0 4.2	5.5 3.8			
R-13 Insulation R-19 Insulation	.064 .047	4.2 3.1	4.3 3.1	4.5 3.3	4.8 3.5	4.2 3.1	3.8 2.8	3.5 2.6	3.2 2.4	2400	8400	5760
R-22 Insulation	.041	2.7	2.7	2.9	3.0	2.7	2.5	2.3	2.0			

1. Insulating glass with double glazing (Item 2).
2. Solid wood doors (Item 3).
3. Slab floor with no edge insulation (Item 4a).
4. R-11 wall insulation (Item 5).
5. Ceiling with no attic, R-19 insulation (Item 6b).

Step 12. Item 7: To determine the ventilation or infiltration air load, first circle the factors in Item 7 under the design dry bulb temperature difference entered above. In this example we would circle 33 under the "Cooling Factor" and 55 under the "Heating Factor" headings. Multiply each of these factors by 1600, the ventilation infiltration cfm indicated in the building requirements. Enter each of these figures in the column under the correct heading (Figure 6-7).

Figure 6-7. Determining ventilation or infiltration air load. (Courtesy of TU Electric Co.)

7. VENTILATION OR INFILTRATION (*Note C) CFM	22 24 28 (33)

72 66 61 (55)	1600	52800	88000

Step 13. Item 8: Determine the sensible heat load Btuh by adding all of the figures in each of the columns headed "Cooling Btuh" and "Heating Btuh." Enter these figures on this line under the proper heading (Figure 6-8).

Figure 6-8. Determining sensible heat load. (Courtesy of TU Electric Co.)

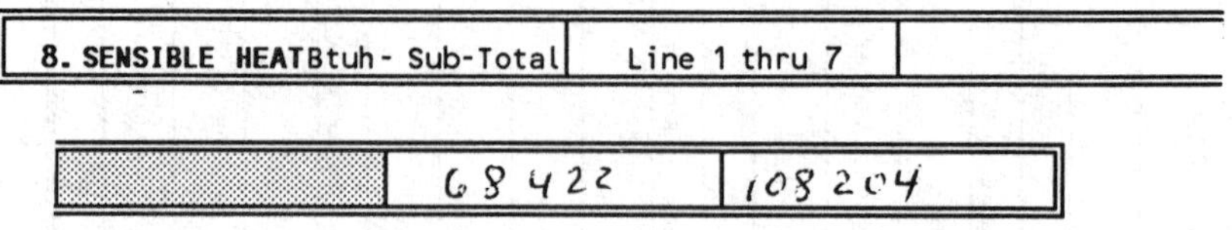

8. SENSIBLE HEAT Btuh - Sub-Total	Line 1 thru 7	

	68422	108204

Step 14. Item 9: Determine the sensible heat loss through the duct system in Btuh. This is done by multiplying line 8 times the factor from table 3 on the bottom of the heat load calculation form. This factor is found to be 1.00 because the ducts are located in the conditioned space

and have no heat loss or gain. Therefore, enter the same figure from line 8 on line 9 (Figure 6-9).

Step 15. Item 10: Determine the ventilation or infiltration latent heat gain-exterior. This is done by multiplying the cooling factor (23 on the form) times the latent cfm for the building (1600 cfm in our example). Enter this figure under the cooling Btuh heading on line 10 (Figure 6-9).

Step 16. Item 11: The total heat load without the internal heat load is calculated and entered on line 11. To calculate this figure add line 9 plus line 10 and enter this figure on line 11 of the form (Figure 6-9).

Step 17. Item 12A: From Table 2 on Form 219-C find the heat load factors for people. Circle the heat loads for the various activities of the people in the Table. This is 200 sensible heat and 250 latent heat for each of the people estimated to be present. Enter 200 under the "Cooling Factor" heading and the number of people estimated to be present is 40, from the information given previously. Enter 40 under the "Quantity" heading on the form (Figure 6-9). Multiply the sensible heat factor times the number of people that are estimated to be present. Enter this figure in the "Cooling Btuh" column.

Step 18. Item 12B: To determine the amount of kW required for the lights multiply 3413 times the number of light bulbs to be used. Each kW of electricity is 3413 Btuh, on the form. The number of lamps used for the lighting is found to be 6 from the previously given information about the building. Enter this figure in the "Quantity" column on the form. Multiply 3413 times 6. Enter this figure on line 12A under the heading "Cooling Btuh" column (Figure 6-9).

Step 19. Item 12C: To determine the amount of heat given off by the electric motors used, multiply the total motor horsepower by 2550. In our example there are two, one-horsepower electric motors that will be used. Enter 2 in the "Quantity" column. Multiply 2550 times 2 and enter this figure on line 12C under the heading "Cooling Btuh" (Figure 6-9).

Step 20. Item 12F: To determine the amount of heat produced by electric cooking multiply 2400 times one-half the number of the total kW to be used for cooking appliances. In our example this is 19. Enter 19 on line 12F under the "Quantity" heading. Multiply 2400 times 19 and enter this figure on line 12F under the heading "Cooling Btuh" (Figure 6-9).

Figure 6-9. Determining sensible heat loss through the duct system. (Courtesy of TU Electric Co.)

CONSTRUCTION	COOLING FACTOR	HEATING FACTOR	QUANTITY	COOLING Btuh	HEATING Btuh
8. SENSIBLE HEAT Btuh-Sub-Total	Line 8 Reverse Side			68422	108204
9. SENSIBLE HEAT Btu W/DUCT LOSS Sub-Total	Line 8 x Factor From Table 3			68422	108204
10. VENTILATION OR INFILTRATION LATENT HEAT GAIN-EXTERIOR (*Note C)	23		1600	36800	
11. TOTAL HEAT Btuh W/OUT INTERNAL LOAD	LINE 9 + 10			105222	
12. INTERIOR SENSIBLE HEAT GAIN					
A. People-Ave. Number (*Table2)	200		40	8000	
B. Lighting-kW	3413		6	20478	
C. Motors- HP	2550		2	5100	
D. Other Internal Load-kW	3413				
E. Gas Cooking (*Note D)	5880				
F. Electric Cooking (*Note D)	2400		19	45600	
13. INTERIOR SENSIBLE HEAT GAIN Sub-Total	Line A thru F			79178	
14. INTERIOR SENSIBLE HEAT GAIN W/DUCT LOSS Sub-Total	Line 13 X Factor From Table 3			79178	79178
15. LATENT HEAT GAIN--INTERIOR					
A. People-Ave. (*Table 2)			40	1000	
B. Other Interior Load					
C. Gas Cooking-Number of Burners	2520				
D. Electric Cooking (*Note D)	1000		19	19000	
16. LATENT HEAT GAIN INTERIOR Sub-Total	Line 15 A thru D			24000	
17. TOTAL INTERIOR HEAT GAIN	Line 14 + 16			108178	
18. TOTAL LOAD Btuh	Cooling Line 11 + 17	Heating Line9 thru14		213400	24026

INSTRUCTIONS FOR HEATING & COOLING ESTIMATES - FOR 219C

*Note A - 1. For Solor Gain use only the one direction producing the greatest Btuh.
2. Solor factors may be multiplied by .90 for plate glass, by .85 for double pane glass, by .50 for glass block.
3. No solor load is used if overhang exceeds 4′ per floor for nominal windows.
4. Windows shaded all day by buildings or other permanent objects should be included under window transmission only.

*Note B - For light colored walls or roofs, COOLING factor may be multiplied by .75.

*Note C - 1. Infiltration formula: Volume = CFM
2. For extra heavy people loads, calculate ventilation CFM requirements from Table 1. Use this CFM quantity in place of infiltration CFM above, if it is greater.
3. If the quantity of ventilation air is stated by owner, architect or engineer, use the amount they specify.

*Note D - Use 50% of nameplate kW for quantity and multiply by sensible and latent factors.

TABLE 1 - Ventilation CM Per Person

No Smoking	20
Some Smoking	40
Heavy Smoking	60

TABLE 2 - People - Btuh Per Person

	Sensible	Latent
Inactive	200	250
Moderate Activity	300	500
Active	400	800

TABLE 3 - Duct Loss Factors

Ductwork Location	Inches of Duct Insulation	Duct Factor Cooling	Duct Factor Heating
Attic - Vented	1	1.15	1.25
- Vented	2	1.10	1.15
- Unvented	1	1.20	1.20
- Unvented	2	1.15	1.10
Crawl Space - Unvented	1	1.05	1.10
Within Conditioned Area	1	1.00	1.00
Within Slab	1	1.20	1.25

TABLE 4 - Equipment Efficiency

Resistance Heat: 3,413 MBtuh/kWh
Heat Pump: EER/SEER - Cooling
HSPF - Heating
Gas Furnace: AFUE x 10

Fuel Cost:
Electricity: Ave. kWh $ ______
Gas: Ave. MCF $ ______

1 MBtu = 1,000 Btu

Step 21. Item 13: To determine the sensible heat gain sub total add lines 12A through 12F. Enter this figure on line 13 under the heading "Cooling Btuh."

Step 22. Item 14: The sub total of the interior sensible heat gain with duct loss is determined by multiplying line 13 times the duct heat loss factor from Table 3. This factor was determined to be 1.00 because the ducts are located inside the conditioned area. Thus, line 13 multiplied by a factor of 1.00 equals the value of line 13. Enter this figure on line 14 under both headings "Cooling Btuh" and "Heating Btuh" (Figure 6-9).

Step 23. Item 15A: The latent heat gain for the people inside the building is 250 Btuh per person, from Table 2. The estimated number of people that will be present is 40. Enter 250 in the column under the heading "Cooling Factor" and 40 in the "Quantity" column on line 15A. To determine the Latent heat in Btuh, multiply 250 times 40 and enter this figure on line 15A under the heading "Cooling Btuh" (Figure 6-9).

Step 24. Item 15D: To determine the latent heat caused by cooking inside the building multiply 1000, from the chart, times one-half the total kW used for the cooling appliances. In our example this is 19. Enter this figure on line 15D under the heading "Cooling Btuh" (Figure 6-9).

Step 25. Item 16: The sub-total of the interior latent heat gain is determined by adding lines 15A through 15D. Add these figures and enter the total on line 16 under the heading "Cooling Btuh" (Figure 6-9).

Step 26. Item 17: The total interior heat gain is determined by adding lines 14 and 16. Enter this figure on line 17 under the heading "Cooling Btuh" (Figure 6-9).

Step 27. Item 18: The total cooling heat load is determined by adding lines 11 and 17. Enter this figure on line 18 under the heading "Cooling Btuh." Determine the heating heat loss by subtracting line 14 from line 9. Enter this figure on line 18 under the heading "Heating Btuh" (Figure 6-9).

Thus, the total cooling load is 213,400 Btuh, or 18 tons. The total heating loss is 29,026 Btuh, or 8.5 kW. The equipment chosen should have at least this much capacity.

OPERATING COST ESTIMATE

The following chart (Chart 6-2) will help in estimating the cost of operating the equipment that is to be installed in this building. This chart will be used along with the 219-C Commercial Heat Load calculation form that we just completed.

Use the following steps to estimate the operating cost of this equipment:

Step 1. HEATING LOAD CONSTANT: Use the following formula to determine this factor.
HBLC = Design Heating MBtuh (line 9 on form 219-C) ÷ Design DB Temp. Diff. (indoor – outdoor) = 2.16.

108.204 ÷ 75 – 50 = 108.204 ÷ 50 = 2.16.

Step 2. COOLING BUILDING LOAD CONSTANT: Use the following formula to determine this factor.
CLBC = Design Cooling MBtuh (line 11 on form 219-C) ÷ Design DB Temp. Diff (outdoor – indoor) = 3.51.

105.222 ÷ 100 – 70 = 105.22 ÷ 30 = 3.51.

Step 3. HEATING EQUALIZATION TEMPERATURE: Use the following formula to determine this factor.
EQ $Temp_H$ = Indoor Operating Temp. – Interior Sensible Heat Gain MBtuh (line 14 on form 219-C). HBLC = 38.

75 – (79.178 ÷ 2.16) = 75 – 37 = 38.

Step 4. COOLING EQUALIZATION TEMPERATURE. Use the following formula to determine this factor.
EQ $Temp_C$ = Indoor Operating Temp. – Interior Heat Gain Total (line 17 on form 219-C). CLBC = 39.

70 – (108.178 ÷ 3.51) = 70 – 31 = 39.

Step 5. ANNUAL HEATING COST ESTIMATE. Use the following formula to determine this factor.
AHCE = HBLC × Heating Degree Hours (from Chart 6-3) × Weather Chart Multiplier × Fuel Cost. Equipment Efficiency (AFUE × 10 or HSPF or 3.413) = $22.00 Heating.

2.16 × 1324 × 0.93 × 0.06.- 7.4 = 22.

Step 6. ANNUAL COOLING COST ESTIMATION (7 Day Operation).

Chart 6-2. Cafe example—operating cost estimate. (Courtesy of TU Electric Co.)

Heating Building Load Constant:

HBLC = Design Heating MBtuh (Line 11)/Design D.B. Temperature Difference (Indoor – Outdoor) = <u>2.16</u>

$$\frac{108.204}{75-25} = \frac{108.204}{50} = 2.16$$

Cooling Building Load Constant:

CBLC = Design Cooling MBtuh (Line 11)/Design D.B. Temperature Difference (Outdoor – Indoor) = <u>3.51</u>

$$\frac{105.222}{100-70} = \frac{105.222}{30} = 3.51$$

Heating Equalization Temperature:

EQ $Temp_H$ = Indoor Operating Temperature – Interior Sensible Heat Gain MBtuh (Line 14)/HBLC = <u>38</u>

$75 - (79.178/2.16) = 75 - 37 = 38$

Cooling Equalization Temperature:

EQ $Temp_C$ = Indoor Operating Temperature – Interior Heat Gain Total Btuh (Line 17)/CBLC = <u>39</u>

$70 - (1008.178/3.51) = 70 - 31 = 39$

Annual Heating Cost Estimate (7-day operation):

$$\frac{\text{HBLC} \times \text{Heating Degree Hours (from Chart)} \times \text{Weather Chart Multiplier} \times \text{Fuel Cost}}{\text{Equipment Efficiency (AFUE} \times 10 \text{ or HSPF or 3.413)}} = 22 \text{ Heating.}$$

$2.16 \times 1324 \times 0.93 \times 0.06/7.4 = 22$

Annual Cooling Cost Estimate (7-day operation):

$$\frac{\text{CLBC} \times \text{Cooling Degree Hours (from Chart)} \times \text{Weather Chart Multiplier} \times \text{Fuel Cost}}{\text{Equipment Efficiency (EER/SEER)}} = 4165 \text{ Cooling.}$$

$3.51 \times 157274 \times 0.97 \times 0.07/9 = 4165$

Heating with electric furnace = $2.16 \times 1324 \times 0.93 \times 0.06/3413$ = \$47.00

Heating with gas furnace 65% AFUE

Monthly minimums = $(6 \times 10.50) + 6 \times 6.50) = 102$	102.00
Heating = $2.16 \times 1324 \times 0.93 \times 5.07/650 = 21$	22.00
Furnace electrical = $21 \times 1.07 = 22$	
Standing pilot = $2 \times 7 \times 5.07 = 71$ (2 gas furnaces)	\$71.00

Note: For 5-day operation, multiply operating cost × 5/7.
For gas furnace operation, multiply operating cost × 1.07 for gas furnace electrical use. Add 7 MCF gas for each standing pilot.
Annual Heating Cost Estimate:
Six-day operation
Cooling \$4165 + 7.4 HSPF heat pump \$22.00 = \$4187× 6/7 = \$3589.00

Chart 6-3. Adjusted annual heating and cooling degree hours for 7-day operation from TMY weather data for D/FW Airport. (Courtesy of TU Electric Co.)

	Adjusted Heating Degree hours					Adjusted Cooling Degree Hours (No Economizer)					
Eq Temp	1 shift (7a-6p)	1.5 shift (8a-10p)	2 shift (6a-2a)	3 shift (24 hr)	Annual Heating Deg Days	1 shift (7a-6p)	1.5 shift (8a-10p)	2 shift (6a-2a)	3 shift (24 hr)	Annual Cooling Deg Days	Eq Temp
75	38,474	48,904	80,609	103,762	4,323	19,618	25,232	27,884	28,781	1,199	75
74	36,286	46,136	76,315	98,398	4,100	21,391	27,508	30,687	31,878	1,328	74
73	34,187	43,472	72,178	93,219	3,884	23,256	29,893	33,654	35,152	1,465	73
72	32,164	40,899	68,173	88,204	3,675	25,198	32,371	36,757	38,603	1,608	72
71	30,222	38,422	64,306	83,353	3,473	27,225	34,950	40,002	42,223	1,759	71
70	28,352	36,035	60,564	78,653	3,277	29,324	37,617	43,373	45,993	1,916	70
69	26,582	33,772	56,978	74,133	3,098	31,523	40,408	46,899	49,944	2,081	69
68	24,897	31,606	53,530	69,781	2,908	33,804	43,295	50,560	54,060	2,253	68
67	23,285	29,543	50,228	65,601	2,733	36,158	46,282	54,365	58,365	2,431	67
66	21,759	27,578	47,072	61,608	2,567	38,593	49,363	58,309	62,807	2,617	66
65	20,293	25,697	44,045	57,768	2,407	41,087	52,525	62,376	67,416	2,809	65
64	18,897	23,898	41,134	54,063	2,253	43,646	55,763	66,553	72,151	3,006	64
63	17,572	22,188	38,359	50,523	2,105	46,269	59,083	70,853	77,039	3,210	63
62	16,314	20,568	35,710	47,137	1,964	48,955	62,485	75,270	82,068	3,420	62
61	15,129	19,042	33,205	43,937	1,831	51,707	65,974	79,818	87,265	3,636	61
60	14.003	17,597	30,816	40,874	1,703	54,511	69,535	84,470	92,586	3,858	60
59	12,918	16,209	28,518	37,917	1,580	57,354	73,147	89,204	98,003	4,083	59
58	11,897	14,897	26,330	35,095	1,462	60,254	76,829	94,039	103,543	4,314	58
57	10,938	13,669	24,267	32,421	1,351	63,210	80,586	98,985	109,215	4,551	57
56	10,026	12,508	22,309	29,880	1,245	66,208	84,404	104,027	115,008	4,792	56
55	9,166	11,408	20,460	27,481	1,145	69,245	88,278	109,169	120,931	5,039	55
54	8,365	10,391	18,728	25,227	1,051	72,353	92,226	114,414	126,983	5,291	54
53	7,609	9,431	17,089	23,082	963	75,493	96,225	119,745	133,134	5,547	53
52	6,898	8,521	15,541	21,046	877	78,673	100,270	125,158	139,384	5,808	52
51	6,237	7,670	14,085	19,130	797	81,898	104,368	130,653	145,742	6,073	51
50	5,619	6,881	12,715	17,321	722	85,162	108,522	136,228	152,198	6,342	50

49	5,042	6,141	11,423	15,626	651	88,465	112,722	141,722	158,756	6,615	49
48	4,501	5,443	10,201	14,020	584	91,799	116,958	147,580	165,395	6,891	48
47	3,994	4,796	9,062	12,516	521	95,163	121,241	153,362	172,126	7,172	47
46	3,530	4,203	8,000	11,105	463	98,567	125,573	159,215	178,942	7,456	46
45	3,116	3,681	7,040	9,805	409	102,016	129,968	165,160	185,857	7.744	45
44	2,749	3,223	6,818	8,619	359	105,508	134,422	171,196	192,876	8,036	44
43	2,423	2,817	5,414	7,551	315	109,036	138,924	177,314	200,001	8,333	43
42	2,135	2,464	4,739	6,612	276	112,599	143,472	183,517	207,243	8,635	42
41	1,871	2,144	4,123	5,765	240	116,184	148,050	189,773	214,569	8,940	41
40	1,622	1,845	3,567	5,007	209	119,782	152,649	196,083	221,975	9,249	40
39	1,393	1,577	3,065	4,321	180	123,398	157,274	202,442	229,446	9,560	39
38	1,177	1,324	2,601	3,690	154	127,026	161,912	208,834	236,966	9,874	38
37	975	1,086	2,169	3,106	129	130,668	166,567	215,257	244,530	10,189	37
36	790	872	1,786	2,582	108	134,323	171,241	221,723	252,247	10,506	36
35	631	691	1,455	2,124	88	138,003	175,944	228,237	259,824	10,826	35
34	504	540	1,167	1,723	72	141,710	180,675	234,790	267,552	11,148	34
33	396	415	922	1,383	58	145,437	185,431	241,381	275,337	11,472	33
32	313	323	730	1,116	46	149,186	190,216	248,021	283,187	11,799	32
31	259	256	586	910	38	152,952	195,023	254,702	291,092	12,129	31
30	199	198	467	737	31	156,729	199,838	261,408	299,027	12,459	30
29	155	147	366	589	25	160,513	204,661	268,129	306,985	12,791	29
28	117	103	278	462	19	164,301	209,488	274,862	314,960	13,123	28
27	82	66	201	350	15	168,093	214,324	281,606	322,951	13,456	27
26	57	42	144	266	11	171,895	219,178	288,367	330,967	13,790	26
25	42	25	102	200	8	175,703	224,021	295,142	338,998	14,125	25
20	5	3	8	23	1	194,787	248,342	329,122	379,296	15,804	20
15	0	0	0	0	0	213,900	272,680	363,179	419,731	17,489	15
10	0	0	0	0	0	233,017	297,019	397,243	460,187	19,174	10

Use the following formula to determine this factor.
ACCE = CBLC × Cooling Degree Hours (from Chart 6-3) × Weather chart Multiplier × Fuel Cost ÷ Equipment Efficiency (EER ÷ SEER) = 4165.

$3.51 \times 157274 \times 0.97 \times 0.07 \div 9 = 4165.$

Step 7. HEATING WITH AN ELECTRIC FURNACE. When an electric furnace is considered, use the following formula to determine this factor. HWEF = HBLC × Heating Degree Hours (from Chart 6-3) × Weather Chart Multiplier × Fuel Cost ÷ Equipment Efficiency (AFUE × 10 or HSPF or 3.413) = $47.00.

$2.16 \times 1324 \times 0.93 \times 0.06 \div 3.413 = 47$

Step 8. HEATING WITH A GAS FURNACE. When a gas furnace is being considered, use the following steps to determine this factor. The furnace is considered to have an AFUE of 65.

Monthly Minimums = $(6 \times 10.50) + (6 \times 6.50) = 102$
Heating (Use the formula in Step 5) =
$2.16 \times 1324 \times 0.93 \times 5.07 - 650 = 21$
Furnace Electrical Cost = 21 × 1.07* = 22
Standing Pilot = $2 \times 7 \times 5.07 = 71$ (2 gas furnaces required)

Step 8. COOLING SIX DAY OPERATION. To determine the cost for six day operation, use the following formula.
CSDO = ACCE + HSPF of heat pump × 6/7* = 3589.
$(4165 + 22) \times 6/7 = 3589.00$

For gas furnace operation, multiply the operating cost times 1.07 for the electricity that the gas furnace will use. Add 7 MCF gas for each standing pilot.

*Note: For five day operation, multiply the operating cost by 5/7.

Chapter 7
Commercial Equipment Sizing

The sizing and selection of commercial air conditioning equipment must be undertaken with special care. These steps are just as important to the overall satisfactory operation of this type of equipment as estimating the heat load of the building. When the equipment to be installed is too small, the desired indoor conditions will probably not be achieved, especially when the peak load is experienced. When the equipment is sized too large there will be lack of humidity control accompanied by excessive operating costs.

INTRODUCTION

In order to properly size and select commercial air conditioning equipment a complete heat load estimate must be completed. This process was explained in Chapter 6. The equipment size is taken from the estimate form which was completed in Chapter 6. The equipment is then selected on the basis of manufacturer's charts, tables, and specifications. From this information we can select a commercial air conditioning system that will fulfill our needs.

It must be remembered that if the equipment size is not properly chosen, unsatisfactory results may cause many problems in the near future for the air conditioning company as well as the building owner and occupants. A unit which is too small may cost less to purchase and install but it will run continuously and produce poor temperature and humidity control inside the space. The continuous operation will wear out the equipment much sooner than is desired. When the unit is too large, the cost of installation will be more than is necessary, there will be poor humidity control, the cost of operation will be excessive, and the customer will usually be dissatisfied with the complete installation.

HEATING EQUIPMENT SIZING

Selecting a unit that has as closely as possible the same capacity as the load will result in a system that will operate efficiently and as satisfactorily as possible. When a system is selected with the desired balance between the unit capacity and the heat load the system will perform as expected and desired.

In the present market there is an infinite number of combinations between components available. Thus, the proper unit capacity can be selected that will very closely match the heat load of the building. However, there will seldom be a unit that will exactly match the building load. It then becomes a point of making an educated compromise between these two components. It must be realized that there are some considerations in making this decision that cannot be compromised. It is also important to learn how equipment capacity is determined, the performance characteristics of the different models, sizes, and combinations, and how all these components operate.

It is our good fortune that the equipment manufacturers have made most of these decisions for us during the designing and manufacturing stages. The factory engineers have completely analyzed most possible problems and have placed all this information in tables and charts for our use in selecting the equipment. We will use our example of the restaurant to select the equipment.

GENERAL INFORMATION

The selection of self-contained (PTAC) units is relatively simple because the system components have already been chosen by the manufacturer. Because of this all we must learn to do is match the system to the building heat load and the utilities that are available. This step involves choosing a unit that has the correct amount of cfm, capacity, and the proper electrical characteristics.

Data Required for Heating Unit Selection

There is certain information that must be known before attempting to select any piece of equipment. The following information is required to properly select a heating unit for a commercial building:

1. The inside and outside design temperatures
2. The total heat loss in Btuh

3. The type of heating unit to be used
4. The required cfm
5. The external static pressure of the unit

The Inside and Outside Design Temperatures

The inside and outside dry bulb (DB) temperatures are all that must be known for this step. Usually the inside design temperatures are between 70 and 75°F.

The Total Heat Loss in Btuh

The total heat loss required is determined by properly completing the heat load form. When the heat loss from the duct work installed in unconditioned areas is needed, it can be determined in two different ways: (1) use the method presented on the heat load form, and (2) multiply the calculated heat loss by 0.95; thus the actual amount of heat delivered to the building is 95% of the output capacity of the heating unit selected.

Types of Heating Equipment

The type of heating equipment selected will depend on several application variations, such as the type of energy available, or desired, and the type of air distribution system desired or required.

The type of energy used—gas, oil, or electricity—will depend to a great deal on the availability and economy of each source. The convenience and dependability of each heat source should also be considered.

The type of air distribution desired will depend on whether the building is divided into several rooms, or spaces, or if it is just one large room; where the occupants are most likely to be most of the time; where the greatest heat load will be centered; also if one side of the building will require cooling and another heating at the same time, or other such requirements.

The type of energy to be used and the direction of the airflow must be known before the equipment manufacturers' tables can be used. When the heating unit is to be part of a year-round type system, the type and size of cooling equipment may influence the type of heating unit chosen.

The Required CFM

To properly distribute the heated air in the conditioned space requires that the cfm be properly calculated before designing the duct system. There are various methods of designing duct systems for commercial applications that are published by various organizations, such as the Natural Warm Air Heating and Air-Conditioning Association and ASHRAE, and of course the local utility companies. The method used will depend on personal preference.

External Static Pressure Caused by the Ductwork

The external static pressure at the required cfm is a vital part of the complete duct system design calculations. It should therefore be included in the duct system specifications.

HEATING EQUIPMENT SELECTION

When the total heat loss has been calculated, there are only two steps left in selecting a heating unit for a commercial application. They are (1) selecting the heating unit from the manufacturers' tables, and (2) determining the necessary cfm adjustment for the heating cycle.

In our selection we will use the building specifications that were presented in Chapter 6 (Figure 7-1).

Selection: We determined that the total heat loss for the restaurant was 29,026 Btuh, or 8.5 kW. This figure includes the heat loss of the structure and the duct heat loss. Thus the heating unit we select must have an output capacity of at least 29,026 Btuh.

1. Some manufacturers' capacity data tables list output capacities, whereas others list input capacities. When the output capacity is listed, this figure may be used for selecting the heating unit. If the input capacity is listed, the output capacity can be determined by multiplying the input capacity by the percent efficiency rating of the heating unit.

2. Next, we select a heating unit that has a capacity equal to or greater than 29,026 Btuh. Figure 7-2 shows a manufacturer's specifications chart for a roof-mounted self-contained heating and cooling unit.

It must be noted that in this particular installation, the heating requirement is very small. This is often the case with applications such as restaurants and other buildings having a low heat loss factor. We must take into account that the cooling load is much greater than the heating load. Therefore, we will select a heating unit that will have sufficient cooling capacity. Notice that the unit, model 580C036, having the closest

Figure 7-1. Cafe example floor plan. (Courtesy of TU Electric Co.)

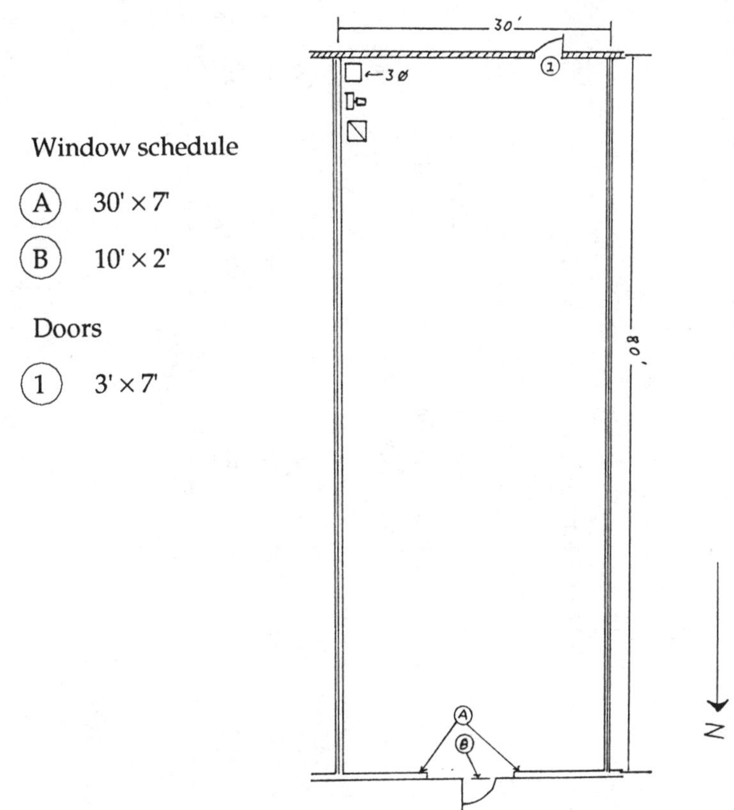

heating value to the amount that we need only has a cooling capacity of 35,800 Btuh nominal capacity. This will not be any where near the 213,400 Btuh, or 18 nominal tons, cooling capacity that the building needs. To get a unit having this capacity we will need model number 579D216 which has a heating capacity of 154,000 Btuh in the low heating mode and 270,000 Btuh in the high heating mode and 212,000 Btuh cooling capacity. The unit would operate in the low heating mode most, if not all, of the time. This would provide sufficient heat in almost any outdoor temperature encountered in this locality. It is better to have too much heat than not enough cooling. The required cooling capacity of 213,400 Btuh is just a little over the rating of the unit which is 212,000 Btuh. However, the cost of the larger equipment and the cost of operation would not normally be justified just for 1,400 Btuh. In most installations this small amount would not be noticeable. Many manufacturers do not manufacture equipment in all

Figure 7-2. Specifications Models 579D 12.5 tons. (Courtesy of BDP Co.)

Model	Nom. Tons	ARI Cooling Performance		MBH Heat Input		Nom. CFM	Electrical Phase Available			Dimensions (in.)		
		Btuh	EER	First Stage	Second Stage		208/230	460	575	W	H	D
579D150	12.5	142,000	9.0	154	231	5000	3	3	-	83-1/2	45	86-1/2
579D180	15	178.000	8.5	154	270	6000	3	3	-	83-1/2	45	86-1/2
579D216	18	212,000	8.6	154	270	7200	3	3	-	83-1/2	45	86-1/2

sizes in the smaller commercial equipment. Therefore, sometimes it is necessary to oversize the unit for the load just to have sufficient capacity.

3. The unit has an air-handling capacity of 7200 cfm.

4. The unit is 83-1/2 in. wide, 45 in. high, and 86-1/2 in. deep. The roof curbing must be installed that will fit these measurements. The roof of the building must be checked to make certain that the structure will support the weight of the unit which can be obtained from the shipping invoice, or from the manufacturer of the unit.

COOLING EQUIPMENT SIZING

The proper balance of equipment and load is very important. If the unit is too small it will run continuously with inadequate temperature control, and high operating costs. If the unit is too large, it will not run enough to remove sufficient humidity and there will also be dead air spots throughout the building. The operating costs will be excessively high, and the cost of the equipment and installation will be more than if the smaller unit were chosen. Also, the balance between the system components, such as the condensing unit and the evaporator, and the air delivery and the duct system, is important if comfort is to be maintained at maximum efficiency.

GENERAL INFORMATION

The selection of cooling equipment, whether it is a remote split system or a packaged unit, has been made relatively simple because the equipment manufacturers have already determined what components will produce a given capacity. Therefore, only the balancing of the unit capacity to the building heat load and the determination of such items as the cfm delivered by the unit and the required electrical characteristics are left to the field engineer.

DATA REQUIRED FOR COOLING UNIT SELECTION

There is certain information that must be determined before attempting to select the cooling equipment. The following is a list of the most important items. Others may be required for the particular application:

1. The summer design outdoor dry bulb (DB) and wet bulb (WB) temperatures
2. The conditioned space design DB and WB temperatures
3. The required ventilation air in cfm
4. The total cooling load in Btuh
5. The cfm air delivery required by the building
6. The condensing medium (air or water) and the temperature of the medium entering the condenser
7. The duct external static pressure

The Design Outdoor Air Temperatures

The design outdoor air conditions are necessary for two reasons: (1) to complete the cooling (heat gain) load estimate, and (2) to establish the conditions of the air that enters the evaporator coil when positive ventilation is provided.

The design outdoor air temperatures used should be those commonly used in the geographical area where the building is located. The design outdoor air temperatures for most large, U.S. cities can be found in tables (Table 3-2). If the area under consideration is not listed in the table, use the closest city within 100 miles. When no listing is shown, within a suitable distance, the accepted outdoor design temperatures can be obtained from the heating, ventilating, and air conditioning guide, a local engineer, or a local weather bureau.

The Conditioned Space Design DB and WB Temperatures

The conditioned space design temperatures are also necessary for two reasons: (1) the heat load calculation, and (2) the evaporator entering air temperature calculation. The air entering the evaporator is a mixture of outside air and the air returned from the room to the evaporator. Thus, when the temperatures of the return air (indoor design) and the ventilation air (outdoor design), together with the percentages of ventilation air, are known, the temperature of the air entering the evaporator can be easily calculated.

When considering bid-specification jobs, the indoor air conditions will be specified by the engineer. When other types of jobs are considered, the design indoor air conditions are established which are acceptable to the owner and which are consistent with local experience and recommendations by the industry.

The Required Ventilation Air in cfm

The cfm of ventilation air required must be known before attempting to determine the temperature of the air entering the evaporator or attempting to determine the sensible heat capacity of the air conditioning system. The ventilation air requirement is determined from the cooling load calculation and should be obtained from the load calculation form.

The Total Cooling Load in Btuh

The total cooling load (heat gain) is the basis for the complete air conditioning unit selection. The tentative and final equipment selection is made on the basis of this calculation.

The heat gain load is found on the heat gain calculation form. There are several short forms available for making this estimate. However, should special conditions be present, or if the construction features are considerably different from those described on the calculation form, a more detailed load estimating method should be used.

The cfm Air Delivery Required

When the long-form heat gain calculation form is used, the cfm required can be calculated from the total sensible heat load and the rise in the DB temperature of the air supply. On some load calculation forms no cfm requirement is specified. In such cases, the unit is chosen on the Btuh capacity alone. It is then assumed that the rated air delivery is 400 cfm per ton of refrigeration. This figure of 400 cfm is used in making duct static pressure calculations.

The Condensing Medium (Air or Water) and the Temperature of the Medium Entering the Condenser

When water-cooled condensers are used with well water or city water, the entering water temperature must be determined to calculate properly the gallons per minute (gpm) of water that must flow through the condenser to achieve the design condensing temperatures. When a cooling tower is used, the discharge water temperature at the tower must be known to calculate the gpm flow and the final condensing temperature.

The city or well water temperature can be measured at the source of supply. The cooling tower condenser water temperatures are obtained from the tower manufacturer's literature (Table 7-1).

The capacity of an air-cooled condenser is determined on the difference between the condensing temperatures and the design outdoor air DB temperatures. Most equipment manufacturers publish these data in tables (Table 7-2). The entering air temperatures for air-cooled condensers are the various design air temperatures.

Table 7-1. Cooling Tower Rating Chart. (Courtesy of The Marley Company)

No.	Capacity Tons-Refrig. 78° WB	Overall Dimensions Wide	Deep	High	Pump Head In Feet	H.P.	Motor Voltage	Cold Water Outlet	Ship. Wt.	Max. Oper. Wt.	FOB Louisville Ea.	Part No.	Basin Cover Ea.	Part No.	Air Inlet Screen Ea.
91000	5	32½"	69"	60 5/16"	4.5	¼	115/230/60/1	2" FIPT	460 Lbs.	960 Lbs.	$1066.00			91055	$ 38.25
91001	7	32½"	69"	60 5/16"	4.5	⅓	115/230/60/1	2" FIPT	460 Lbs.	960 Lbs.	1079.00			91055	38.25
91002	10	32½"	69"	61⅛"	4.5	⅓	115/230/60/1	2" FIPT	480 Lbs.	1060 Lbs.	1158.00			91056	40.75
91003	15	32½"	69"	70 3/16"	5.3	½	115/230/60/1	2" FIPT	520 Lbs.	1100 Lbs.	1342.00				
91004	15	32½"	69"	70 3/16"	5.3	½	200/60/3	2" FIPT	520 Lbs.	1100 Lbs.	1342.00	91050	12.00	91057	52.75
91005	15	32½"	69"	70 3/16"	5.3	½	230/460/60/3	2" FIPT	520 Lbs.	1100 Lbs.	1342.00				
91006	20	32½"	72¾"	80½"	6.5	½	115/230/60/1	4" IPT	580 Lbs.	1020 Lbs.	1500.00				
91007	20	32½"	72¾"	80½"	6.5	½	200/60/3	4" IPT	580 Lbs.	1020 Lbs.	1500.00			91058	58.00
91008	20	32½"	72¾"	80½"	6.5	½	230/460/60/3	4" IPT	580 Lbs.	1020 Lbs.	1500.00				
91009	25	46½"	75⅜"	80½"	6.5	½	115/230/60/1	4" IPT	750 Lbs.	1410 Lbs.	1645.00				
91010	25	46½"	75⅜"	80½"	6.5	½	200/60/3	4" IPT	750 Lbs.	1410 Lbs.	1645.00				
91011	25	46½"	75⅜"	80½"	6.5	½	230/460/60/3	4" IPT	750 Lbs.	1410 Lbs.	1645.00				
91012	30	46½"	75⅜"	80½"	6.5	1	115/230/60/1	4" IPT	750 Lbs.	1410 Lbs.	1776.00				
91013	30	46½"	75⅜"	80½"	6.5	1	200/60/3	4" IPT	750 Lbs.	1410 Lbs.	1776.00	91051	19.75	91059	72.25
91014	30	46½"	75⅜"	80½"	6.5	1	230/460/60/3	4" IPT	750 Lbs.	1410 Lbs.	1776.00				
91016	40	46½"	75⅜"	80½"	8.5	2	200/60/3	4" IPT	750 Lbs.	1410 Lbs.	2053.00				
91017	40	46½"	75⅜"	80½"	8.5	2	230/460/60/3	4" IPT	750 Lbs.	1410 Lbs.	2053.00				
91018	55	64"	83 1/16"	91⅜"	9.0	2	200/60/3	4" IPT	1170 Lbs.	2210 Lbs.	2539.00				
91019	55	64"	83 1/16"	91⅜"	9.0	2	230/460/60/3	4" IPT	1170 Lbs.	2210 Lbs.	2539.00				
91020	65	64"	83 1/16"	91⅜"	8.0	3	200/60/3	4" IPT	1190 Lbs.	2230 Lbs.	2829.00	91052	27.75	91060	108.00
91021	65	64"	83 1/16"	91⅜"	8.0	3	230/460/60/3	4" IPT	1190 Lbs.	2230 Lbs.	2829.00				
91022	75	76"	89 1/16"	91⅜"	8.0	3	200/60/3	6" IPT	1400 Lbs.	2770 Lbs.	3158.00	91053	30.25	91061	117.00
91023	75	76"	89 1/16"	91⅜"	8.0	3	230/460/60/3	6" IPT	1400 Lbs.	2770 Lbs.	3158.00	91053	30.25	91061	117.00
91024	90	88"	92¾"	91⅜"	8.0	5	200/60/3	6" IPT	1610 Lbs.	3250 Lbs.	3671.00	91054	33.00	91062	138.00
91025	90	88"	92¾"	91⅜"	8.0	5	230/460/60/3	6" IPT	1610 Lbs.	3250 Lbs.	3671.00	91054	33.00	91062	138.00

Table 7-2. Specifications and Information. (Courtesy of ARCOAIRE)

		ALP013		ALP016		ALP021	
COOLING PERFORMANCE	COOLING BTUH ①	136,800		188,400		256,800	
	TOTAL BTUH	③ 148,300		② 185,000		② 246,000	
	SENSIBLE BTUH	③ 120,900		② 141,900		② 183,500	
ELECTRICAL DATA	NOMINAL VOLTAGE 3PH/60HZ	208/230	460	208/230	460	208/230	460
	ALLOWABLE VOLTAGE RANGE	187-253	414-506	187/253	414-506	187-253	414-506
	MIN. WIRE SIZE AMPACITY	53.6	27.7	79	39.3	107.9	53.7
	MAX. RECOMMENDED FUSE SIZE	70	35	100	50	150	70
COMPRESSOR DATA	TYPE (QUANTITY)	HERMETIC-2 SPEED (1)					
	NOMINAL HP (NO. CYLINDERS)	10 (3)	10 (3)	15 (6)	15 (6)	20 (6)	20 (6)
	OIL CHARGE—PINTS	6	6	15	15	15	15
	RLA ⑥	30	15	46	23	68	34
	LRA	192	96	248	124	362	181
CONDENSER COIL	TYPE	ALUMINUM FINS ON COPPER TUBES WITH INTEGRAL SUBCOOL					
	FACE AREA—SQ. FT.	30.3	30.3	30.3	30.3	30.3	30.3
	NO. OF CIRCUITS	1	1	1	1	1	1
	FINS PER INCH (ROWS DEEP)	16 (2)	16 (2)	12 (2)	12 (2)	16 (2)	16 (2)
CONDENSER FANS AND MOTORS	TYPE	PROPELLER					
	DIAMETER INCHES (QUANTITY)	24″ (2)		24″ (2)		24″ (2)	
	TOTAL CFM	10,150	10,150	10,550	10,550	10,150	10,150
	MOTOR TYPE	DIRECT DRIVE					
	MOTOR H.P. (QUANTITY)	.5 (2)	.5 (2)	.5 (2)	.5 (2)	.5 (2)	.5 (2)
	RPM	1075	1075	1075	1075	1075	1075
	RLA—EACH MOTOR	3.5	1.8	3.5	1.8	3.5	1.8
PHYSICAL DATA	OPERATING CHARGE—LBS R22 ④	11 LB. 0 OZ.		12 LB. 0 OZ.		12 LB. 0 OZ.	
	PUMP DOWN CAPACITY 90% FULL @90°F ⑤	25.0	25.0	26.0	26.0	26.0	26.0
	UNIT WEIGHT—LBS	600	600	690	690	720	720

① NOMINAL CAPACITY BASED ON 95°F AMBIENT AND 45°F SUCTION TEMP.
② BTUH WHEN USED WITH MATCHING AIR HANDLER AT 400 CFM/TON BASED ON 80°F DB AND 67°F WB ENTERING EVAPORATOR COIL AND 95°F AIR ENTERING CONDENSER.
③ BTUH WHEN USED WITH DB150 AIR HANDLER AT 5200 CFM.
④ UNITS FROM FACTORY ARE FURNISHED WITH A HOLDING CHARGE ONLY. SEE INSTRUCTIONS FOR ADDITIONAL CHARGE REQUIRED FOR VARIOUS LINE SIZES AND LENGTHS.
⑤ TOTAL PUMP DOWN CAPACITY WILL BE UNIT PUMP DOWN CAPACITY.
⑥ RLA OF 2 SPEED COMPRESSOR RUNNING ON LOW SPEED WILL BE 50% OF RLA STATED.

The Duct External Static Pressure

When an air conditioning unit is to be connected to a duct system, the external static pressure must be known to determine the fan speed and fan motor horsepower required. The maximum external static pressure for the equipment is determined by the manufacturer and is indicated on the equipment name-plate. The external static pressure for the duct system is calculated when the ductwork is designed and should be included in the duct system specifications. In any case it should never exceed the manufacturer's specifications.

SPLIT-SYSTEM EQUIPMENT SELECTION PROCEDURE

Basically, the same selection procedure is used for both residential and commercial split-system equipment applications. Before the actual equipment selection can be done, the heat gain calculations must be completed. The equipment can then be chosen according to the appropriate following application and installation considerations:

1. The temperatures of the air entering the evaporator and condenser are determined. The design outdoor air DB temperature is the condenser entering air temperature. In applications where there is no ventilation air brought in through the equipment, the design conditioned space DB and WB temperatures are used as the evaporator entering air temperatures. When the ventilation air is introduced through the evaporator, the entering air temperatures may be calculated or obtained from a table (Table 7-3).

2. Next a split system with sufficient capacity is selected from the manufacturer's tables. When sizing commercial units, the total capacity should be equal to or greater than the total heat gain load, and the sensible capacity should be enough to take care of the sensible heat load.

3. The air delivery by the unit against the calculated external static pressure must be determined. When the cooling cfm requirement is considerably more than the heating requirement, the heating unit may be equipped with a two-speed fan motor.

For an example of the selection of a split-system air conditioning unit, let's use the restaurant for which we calculated the heat load in the

Table 7-3. Resulting Temperatures of a Mixture of Return Air and Outdoor Air (to be used only for return air having an 80°F DB temperature and 50% humidity.)

Percentage of outside air	*Dry Bulb Temperature of the Air Mixture*			*Wet Bulb Temperature of the Air Mixture*							*Percentage of Outside Air*
	Outside Dry Bulb Temp.			*Outside Wet Bulb Temperature*							
	95°	100°	105°	74°	75°	76°	77°	78°	79°	80°	
2%	80.3	80.4	80.5	67.2	67.2	67.2	67.2	67.3	67.3	67.3	2%
4	80.6	80.8	81.0	67.3	67.3	67.4	67.4	67.5	67.5	67.6	4
6	80.9	81.2	81.5	67.5	67.5	67.6	67.7	67.7	67.8	67.9	6
8	81.2	81.6	82.0	67.6	67.7	67.8	67.9	68.0	68.1	68.2	8
10	81.5	82.0	82.5	67.8	67.9	68.0	68.1	68.2	68.4	68.5	10
12%	81.8	82.4	83.0	67.9	68.0	68.2	68.3	68.5	68.6	68.8	12%
14	82.1	82.8	83.5	68.1	68.2	68.4	68.5	68.7	68.9	69.1	14
16	82.4	83.2	84.0	68.2	68.4	68.6	68.8	69.0	69.2	69.4	16
18	82.7	83.6	84.5	68.3	68.5	68.8	69.0	69.2	69.4	69.6	18
20	83.0	84.0	85.0	68.5	68.7	69.0	69.2	69.4	69.7	69.9	20
22%	83.3	84.4	85.5	68.6	68.9	69.1	69.4	69.7	69.9	70.2	22%
24	83.6	84.8	86.0	68.8	69.1	69.3	69.6	69.9	70.2	70.5	24
26	83.9	85.2	86.5	68.9	69.2	69.5	69.8	70.1	70.4	70.8	26
28	84.2	85.0	87.0	69.1	69.4	69.7	70.0	70.4	70.7	71.0	28
30	84.5	86.0	87.5	69.2	69.6	69.9	70.2	70.6	70.9	71.3	30

previous chapter. Let us assume that a year-around air conditioning unit is to be installed. The heat gain calculation was completed in chapter 6. We will use the same construction details that were used then.

Selection:

In Chapter 6 it was determined that the total heat gain for our building was 213,400 Btuh. This includes the sensible heat gain of the structure, the duct heat gain, and the latent heat gain. Thus the cooling unit selected must have a total Btuh capacity of 213,400 Btuh.

The heating unit must have an output (bonnet) capacity rating of 29,026 Btuh.

1. Some manufacturers' data lists the output capacities at different outdoor ambient temperatures, whereas others just list them at one temperature.

2. Next, we select a condensing unit that has the capacity equal to or slightly greater than 213,400 Btuh. Using Table 7-4, locate the proper outdoor design temperature for our building (100°F) across the top of the chart, then read down the MBH column until the proper suction temperature is reached (40°F).

We find that this unit has a cooling capacity of 224,400 Btuh under these conditions. The unit will use 25.2 kW of electricity per hour with an EER of 8.9.

3. We must match the coil to the condensing unit in Btuh capacity. Checking the manufacturer's specifications chart (Table 7-4) we find under the 40°F suction heading that when the return air is at a wet bulb temperature of 64°F, the coil will have a capacity of 232,000 Btuh when the blower is moving 8,000 cfm. These types of coils may be mounted either in the vertical discharge position or the horizontal discharge position (Figure 7-3).

VERTICAL PACKAGED (PTAC) UNIT SELECTION PROCEDURE

These units are sometimes referred to as PTAC units (packaged terminal air conditioning units). The manufacturers' recommended selection procedure will vary with different makes and models of self-contained air conditioning units; however, all of the procedures are basically

Table 7-4. Capacity Data. (Courtesy of ARCOAIRE)

ALP013C

SUCTION TEMP. °F	Ambient Air Temperature																				
	90°			95°			100°F			105°F			110°F			115°F			120°F		
	MBH	KW	EER	MBH	KW	EER	MBH	KW	EER	MBH	KW	EER	MBH	KW	EER	MBH	KW	EER	MBH	KW	EER
35.0	115.2	11.7	9.8	110.4	12.0	9.2	104.4	12.2	8.6	99.6	12.4	8.0	93.6	12.6	7.5	87.6	12.7	6.8	80.4	12.9	6.3
40.0	128.4	12.3	10.5	123.6	12.5	9.8	117.6	12.8	9.2	111.6	13.0	8.5	104.4	13.3	7.9	97.2	13.5	7.2	90.0	13.7	6.6
45.0	142.8	12.8	11.2	136.8	13.1	10.4	129.6	13.4	9.6	123.6	13.7	9.0	115.2	14.0	8.3	108.0	14.2	7.6	100.8	14.5	6.9
50.0	156.0	13.3	11.7	150.0	13.6	11.0	142.8	14.0	10.2	135.6	14.4	9.4	128.4	14.7	8.7	120.0	15.0	8.0	112.8	15.3	7.4

ALP016C

SUCTION TEMP. °F	Ambient Air Temperature																				
	90°			95°			100°F			105°F			110°F			115°F			120°F		
	MBH	KW	EER	MBH	KW	EER	MBH	KW	EER	MBH	KW	EER	MBH	KW	EER	MBH	KW	EER	MBH	KW	EER
35.0	163.2	16.8	9.7	154.8	17.1	9.1	146.4	17.3	8.5	138.0	17.5	7.9	129.6	17.7	7.3	120.0	17.8	6.7	110.4	17.9	6.1
40.0	180.0	17.7	10.2	171.6	18.1	9.5	163.2	18.4	8.9	154.8	18.7	8.3	145.2	19.0	7.6	134.4	19.2	7.0	124.8	19.4	6.4
45.0	198.0	18.6	10.6	188.4	19.1	9.9	180.0	19.5	9.3	171.6	19.9	8.6	160.8	20.2	7.9	150.0	20.5	7.3	139.2	20.8	6.7
50.0	216.0	19.5	11.0	206.4	20.0	10.3	196.8	20.6	9.6	187.2	21.0	8.9	176.4	21.4	8.2	164.4	21.7	7.6	153.6	22.1	6.9

ALP021C

SUCTION TEMP. °F	Ambient Air Temperature																				
	90°			95°			100°F			105°F			110°F			115°F			120°F		
	MBH	KW	EER	MBH	KW	EER	MBH	KW	EER	MBH	KW	EER	MBH	KW	EER	MBH	KW	EER	MBH	KW	EER
35.0	222.0	23.0	9.7	206.4	23.3	9.1	202.8	23.7	8.6	193.2	23.5	8.1	181.2	24.2	7.5	168.0	24.3	6.9	156.0	24.5	6.4
40.0	244.8	24.3	10.0	234.0	24.8	9.5	224.4	25.2	8.9	213.6	25.6	8.4	201.6	25.8	7.8	187.2	26.1	7.2	175.2	26.4	6.6
45.0	267.6	25.7	10.4	256.8	26.2	9.8	246.0	26.7	9.2	235.2	27.2	8.7	222.0	27.5	8.4	206.4	27.9	7.4	194.4	28.2	6.9
50.0	290.4	27.1	10.7	279.6	27.7	10.0	268.8	28.3	9.5	256.8	28.8	8.9	242.4	29.2	8.3	228.0	29.6	7.7	214.8	30.0	7.1

(1) KW and EER are for entire condensing unit.

Continuing Engineering research results in steady improvement. Therefore, these specifications are subject to change without notice. Always consult equipment nameplate for exact electrical requirements.

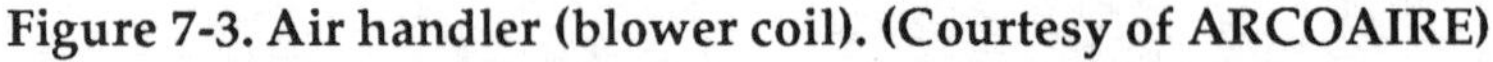

Figure 7-3. Air handler (blower coil). (Courtesy of ARCOAIRE)

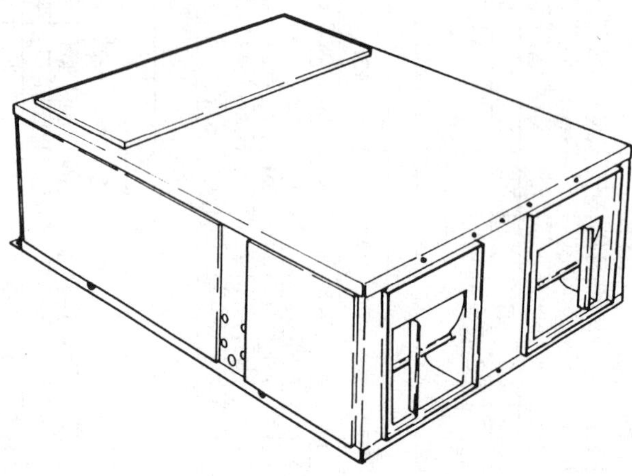

the same. We will use the following general procedure in the example that follows.

Before beginning the selection process, the heat gain estimate form must be completed and the necessary items discussed under the heading "Data Required for Cooling Unit Selection," must be gathered. Once these tasks are completed, the packaged unit may be selected.

Selection:

1. The first step is to make a tentative selection based on the heat gain calculation in Chapter 6. This selection is made from manufacturer's nominal unit capacity tables (Table 7-5).

2. Compare the rated cfm delivery for the unit selected, when provided in the table, to the air delivery requirement for the building. If the rated cfm is within 20% of the required cfm, the selection should be made on the basis of the required cfm. If there is a difference of more than 20% in the cfm, a special built-up system may be considered.

3. Calculate the entering dry bulb and wet bulb temperatures, or get them from Table 7-3. Regardless of which method is used, the indoor and outdoor design temperatures and the ventilation requirements must be known.

Table 7-5. Water-cooled unit data.

UF Series Condensed Specifications

Model	UF036W	UF060W	UF090W	UF120W	UF180W	UF240W	UF300W	UF360W
Dimensions (in.): Height	78	78	78	78	78	79½	79½	79½
Width	30	42	54	65	89	76¼	96½	96½
Depth	20	20	20	28	28	38½	38½	38½
Ship Wt. (lbs.)	370	510	672	1020	1267	1445	1865	2025
Filters— †Disposable ‡Permanent, cleanable	†(1)20×25×1 †(1)16×25×1	†(1)20×25×1 †(1)16×25×1	†(3)16×25×1	‡(3)19×27×1	‡(8)15×20×1	‡(4)20×25×1 (2)20×20×1	‡(6)25×30×1	‡(6)25×30×1
Performance: Cooling, std rating (Btuh)	36,000	60,000	86,000	120,000	176,000*	240,000*	300,000*	360,000*
Power input (watts)	3,950	6,700	9,700	13,100				

*Units above 120,000 Btuh not ARI rated

UR Series Condensed Specifications

Model	UR030	UR036	UR048	UR060
Dimensions (in.): Height	26 11/32	26 11/32	30	30
Width	42 15/16	42 15/16	52⅜	52⅜
Depth	45 3/16	45 3/16	52⅜	52⅜
Ship Wt. (lbs.)	320	340	450	510
Filters (recommended), dimensions (in.) not included in unit	(2)16×20×1	(2)16×20×1	(2)20×20×1	(2)20×20×1
Performance: Cooling, standard rating (Btuh)	30,000	38,000	46,000	61,000
Power input (watts)	4,300	5,100	5,400	8,300

Electrical data:	UR030		UR048		
Phase/Cycle	1/60		1/60	3/60	3/60
Rated voltage	208-230		230	230	460
	UR036		UR060		
Phase/Cycle	1/60	3/60	1/60	3/60	3/60
Rated voltage	230	208-230	230	208-240	460

4. When a water-cooled unit is considered, the condensing temperature should be determined. A tentative condensing temperature of approximately 30°F higher than the entering water temperature is normally taken.

5. Determine the total unit capacity and the heat capacity of the unit at the conditions given for the unit selected. This information is usually listed in the manufacturers' equipment tables.
   a. When selecting water-cooled units, the evaporator entering air dry bulb and wet bulb temperatures and the calculated condensing temperatures are necessary.
   b. When considering air-cooled units, the outdoor design dry bulb and wet bulb temperatures of the air entering the evaporator are necessary. Use the smallest condenser recommended at this point of the selection procedure. When the capacity of the unit selected falls between the rating listed in the table, it will be necessary to interpolate to obtain the exact capacities of the unit. The unit capacities listed in these tables are for standard air delivery only. When quantities other than standard are to be used, correction factors must be applied to the listed capacities (Table 7-6).

Table 7-6. Capacity Correction Factors*

CFM Compared to Rated Quantity	–40%	–30%	–20%	–10%	Std.	+10%	+20%
Cooling Capacity Multiplier	88	93	96	98	1.00	1.02	1.03
Sensible Capacity Multiplier	80	85	90	95	1.00	1.05	1.10
Heating Capacity Multiplier	78	83	89	94	1.00	1.06	1.12

*To be applied to cooling and heating capacities

6. The total capacity of the unit should be equal to or greater than the total cooling load unless the design temperatures can be compromised. When additional capacity is required, a new selection must be made to handle the load.

a. When considering a water-cooled condenser, the condensing temperature should be lowered 5 to 10°F and the new unit capacities should be considered before selecting a larger unit. However, if the load remains larger than the unit capacity, the next larger unit should be selected.
b. When considering an air-cooled condenser and the unit capacity will not meet the load requirements, a larger-capacity condenser should be selected. Should the unit capacity still be too small, the next-larger size unit must be selected.

7. After establishing the actual unit capacity, compare the sensible heat capacity of the unit to the heat gain load of the building. If the unit sensible heat capacity is not equal to or greater than the cooling load, a larger unit must be selected.

8. When considering a water-cooled condenser, determine the gpm water flow required and the water pressure drop through the condenser. Use the final condensing temperature and the total cooling load in tons of refrigeration to determine the gpm required from a condenser water requirement table, when provided for the unit being considered. After the condenser gpm requirement is determined, the pressure drop can be read from the table prepared by the manufacturer, if this information is given.

9. Determine the fan speed and the fan motor horsepower required to deliver the required volume of air against the calculated external static pressure caused by the air distribution system. The air volume is read directly from the manufacturer's charts and performance tables. Should the required fan motor horsepower exceed the recommended limits, a larger fan motor must be used.

Vertical self-contained units are usually available in only fairly large capacity increments. Many situations are encountered when the actual heat gain load falls between two of these capacities. For example, the calculated heat gain load indicated a 12.5-ton load. The available units are either 10 or 15 tons of capacity. With the possibility of future expansion, the building owner may authorize a single 15-ton unit. However, if he is economy minded, he may decide to tolerate a few days which are warmer than normal and authorize the smaller 10-ton unit. The contractor can, however, offer a 7.5-ton unit and a 5-ton unit to match the load exactly. In this case the customer must be willing to pay the higher cost of a two-unit installation.

It is always wise, whatever the case may be, to compare the actual sensible and latent heat capacities of the unit with the calculated heat gain of the building. It is often desirable, for comfort reasons, that the unit have the capacity to handle the complete sensible heat load.

When the sensible heat percentage (the ratio of sensible heat gain to the latent heat gain) is normally high or low, a more detailed check should be made. This check can be made by consulting the equipment manufacturer's capacity tables.

VERTICAL PACKAGED (PTAC) EQUIPMENT SELECTION PROCEDURE

These units are sometimes referred to as PTAC units (packaged terminal air conditioning units). Let us select a vertical self-contained water-cooled condensing unit for use in an office building in Dallas, Texas. The condensing water is to be supplied from a cooling tower. The unit will be used with a free discharge air plenum. There are no specified air delivery requirements. The necessary data are:

1. The outdoor design temperatures are 100°F DB and 78°F WB.
2. The indoor design temperatures are 78°F DB and 65°F WB.
3. The ventilation air required is 300 cfm.
4. The total cooling load is 76 tons (84,000 Btuh).
5. The condensing medium is water at 85°F.
6. The air delivery required is 400 cfm/ton or 2800 cfm.
7. External static pressure is zero.

Selection:

1. From the manufacturers' rating chart, select unit model UF090W (Table 7-5).

2. When supplied, determine the rated cfm delivery for the model chosen. This will normally be 400 cfm/ton. In our case the amount of air delivered would be 2800 cfm.

3. Calculate the DB and WB temperatures of the air entering the evaporator:
   a. Ventilation requirement ÷ Standard cfm delivery = 300 ÷ 2800 = 10.7% or 10% outside air.
   b. Entering air DB(°F) = (0.10 × 100) + (0.90 × 78°F) = 82.2°F DB.

c. The WB temperature at 80.2°F WB with the given mixture is found on the psychrometric chart to be 66.4°F WB.

4. Calculate the condensing temperature: 85°F + 30 = 115°F condensing temperature.

5. Because there is no specified latent heat load, the sensible heat percentage is not considered in our example.

6. Determine the gpm water flow required. When the manufacturer's table does not include this information, it is customary to figure 3 gpm per ton of refrigeration. Thus; 7.16 × 3 = 21.49 gpm.

7. Determine the air delivery in cfm of the unit. When this information is not included in the manufacturer's table, it is customary to figure 400 cfm per ton of refrigeration. Thus 7.16 × 400 = 2864 cfm. The air can now be properly distributed to the needed areas.

HORIZONTAL PACKAGED (PTAC) UNIT SELECTION PROCEDURE

These units are sometimes referred to as PTAC units (packaged terminal air conditioning units). When considering horizontal packaged self-contained units for commercial installations, essentially the same procedure is used as when selecting vertical air-cooled self-contained units.

Before beginning the selection process, the heat gain estimate must be completed and the necessary items which were discussed under the heading "Data Required for Cooling Unit Selection," must be gathered. Once these tasks are complete, the packaged unit may be selected.

Selection:

1. From the manufacturer's capacity table, select a unit based on the total calculated heat gain.

2. Determine the standard cfm rating of the unit selected, if included, and compare it to the cfm required for the building.

3. Determine the WB and DB temperatures of the air entering the evaporator. Use the formula:
   Entering air DB = (% outside air × outdoor DB)
   + (% inside air × indoor DB).

4. Determine the total unit capacity and sensible heat capacity at the various conditions for the unit selected. This information is obtained from the manufacturer's capacity table, when provided, using the design outdoor DB and the return air DB temperatures.

5. When the total unit capacity is not sufficiently close to the calculated heat gain, another unit should be selected.

The capacity of air-cooled units cannot be adjusted as readily as water-cooled units. Nor do air-cooled packaged units offer a choice of evaporator or condenser coil for varying the system capacity. This problem is alleviated, however, because air-cooled packaged units are readily available in a large number of capacity ranges. When the calculated heat gain cannot be met exactly, the alternate choice of a larger or smaller unit, or a multiple-unit type of installation will depend on the owner's preference. The considerations for making this decision are basically the same as those listed for vertical self-contained units.

SUMMARY

The proper sizing and selection of air conditioning equipment is just as important as any other step in the estimating process.

The proper sizing and selection of air conditioning equipment can be accomplished only after an accurate heat loss and heat gain analysis is complete.

The selection of equipment is a problem of choosing the unit that is the closest to satisfying the demands of capacity, performance, and cost.

The information that must be determined before attempting to select a warm air heating unit includes such items as: (1) the inside and outside design temperatures, (2) the total heat loss in Btuh of the building, (3) the type of heating equipment to be used, (4) the required cfm, and (5) the external static pressure caused by the ductwork.

When considering the design temperatures, only the dry bulb temperatures are required.

The total heat loss figure is determined by completing a heat loss form.

The type of heating unit selected will depend on several application variations, such as the type of energy available, or desired, and what direction of air flow is desired.

The direction of air flow and the type of energy chosen must be determined before the manufacturer's tables can be used.

To distribute the heated air properly in the conditioned space will require that the cfm be properly calculated before the duct system is designed.

The external static pressure at the required cfm is a vital part of the complete duct system design calculations. It should, therefore, be included in the duct system specifications.

When the total heat loss has been calculated, there are only two steps left in selecting a heating unit for a heating-only system: (1) selecting the heating unit from the manufacturer's tables, and (2) determining the necessary cfm adjustment for heating.

The balancing of system components, such as the condensing unit and the evaporator, and the air delivery and the duct system, is important if comfort and maximum efficiency are to be maintained.

The selection of the cooling equipment, whether it be remote, split system or a packaged unit, has been made relatively simple because the equipment manufacturers have already determined what components will produce a given unit capacity.

The following information must be known before the proper selection of cooling equipment for commercial installations can be completed: (1) the summer design outdoor dry bulb and wet bulb temperatures, (2) the conditioned space design DB and WB temperatures, (3) the required ventilation air in cfm, (4) the cooling load in Btuh, (5) the cfm delivery required, (6) the condensing medium (air or water), and the temperature of the medium entering the condenser, and (7) the duct external static pressure.

The design outdoor air temperatures used should be those commonly used in the geographical area where the building is located.

The conditioned space design temperatures are also necessary for two reasons: (1) the load calculations, and (2) the evaporator entering-air-temperature calculation.

The required cfm of ventilation air must be known before attempting to determine the temperature of the air entering the evaporator or attempting to determine the sensible heat capacity of the air conditioning system.

The total cooling load (heat gain) is the basis for the complete air conditioning unit selection.

When the long-form heat gain calculation form is used, the cfm required can be calculated from the total sensible heat load and the rise in the DB temperature of the air supply. On some load calculation forms no cfm requirement is specified. In such cases, the unit is chosen on the Btuh capacity alone.

When water-cooled condensers are used with well water or city water, the entering water temperature must be determined to calculate properly the gpm of water that must flow through the condenser to achieve the design condensing temperatures.

The capacity of an air-cooled condenser is determined on the difference between the condensing temperatures and the design outdoor air DB temperatures.

When an air conditioning unit is to be connected to a duct system, the external static pressure must be known to determine the fan speed and the fan horsepower required.

After the heat gain calculations have been completed, the split system equipment can then be chosen according to the appropriate following application and installation considerations: (1) the temperature of the air entering the evaporator and condenser are determined, (2) a split system with sufficient capacity is selected from the manufacturer's tables, and (3) the air delivery by the unit must be calculated against the external static pressure.

Chapter 8

Commercial Equipment Location

Up to this point we have discussed psychrometrics, heat loss, heat gain, and equipment sizing and selection. In this chapter we will discuss some principal considerations regarding the location and installation of commercial air conditioning equipment.

INTRODUCTION

A well-planned location for the air conditioning and heating unit is one of the most important steps in a good installation. The following are some of the considerations that are involved in planning the location of the various units.

All units require some of the same considerations with regard to location and installation. Therefore, we will discuss at this time such items as gas pipe connections, venting requirements, return air location, supply air location, minimum clearances, noise, accessibility, length of duct runs, and thermostat location.

Gas Pipe Connections

The availability of the gas pipe to the heating unit is important because of the labor involved in making the connections. When the gas pipe is already in the equipment room, a certain amount of labor and materials will be saved. However, should the gas pipe be located on the other side of the building, the cost of labor and materials will be greatly increased, thus increasing the cost of the installation. The gas supply to the building must have the capacity to handle the extra load that the new equipment will place on it. Be sure to consider any restrictions or requirements that might be imposed because of local and/or national building codes.

Venting Requirements

When considering the venting requirements for gas-fired heating equipment, any and all local and national codes and ordinances must be followed. An unsafe heating unit should not be installed or operated due to the possible hazard to life and property. The heating unit should be located so that the vent pipe will provide as straight a path as possible for the flue gases to escape to the atmosphere. Never decrease the size of the vent piping from that recommended by the manufacturer. Vent pipe is very expensive; therefore, a short, straight run is not only easier but also more economical to install.

When the heating unit is to be installed in an equipment room, openings must be provided to make certain that adequate ventilation and combustion air is available to the unit. Be sure that the local and national codes are followed when locating and sizing these openings. It is more desirable from a safety viewpoint and for more economical operation if the combustion air is taken from outside the building space. See Figure 8-1.

Figure 8-1. Suggested method of providing combustion air supply.

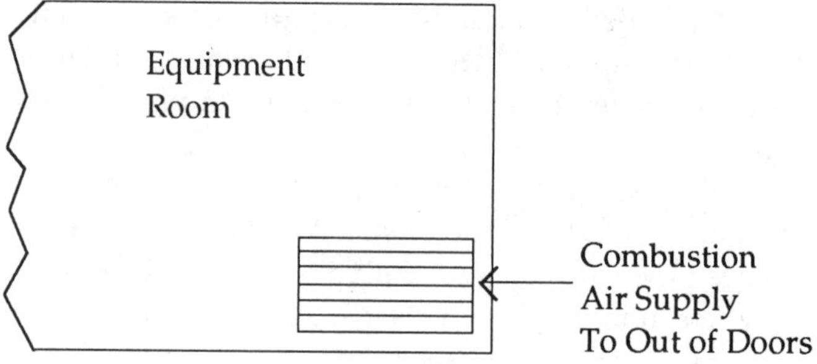

However, if it is not feasible to duct the combustion air from outside, it may be taken from inside the building. Adequate covering must be placed on the openings to prevent insects from entering through the duct or opening when it is located outside. Be sure to provide extra dimensions to the combustion air intake because of the added restriction caused by the grill or covering.

Return Air Location

There are several methods used for directing the return air to the heating unit. The most popular are: central return (see Figure 8-2), a return in each space (see Figure 8-3), and a combination of the two. It is obvious that a return air duct to each space would require more material and labor, thus increasing the cost of the installation. However, it is sometimes more desirable than the central return because of the added comfort and reduced noise.

Figure 8-2. Central return.

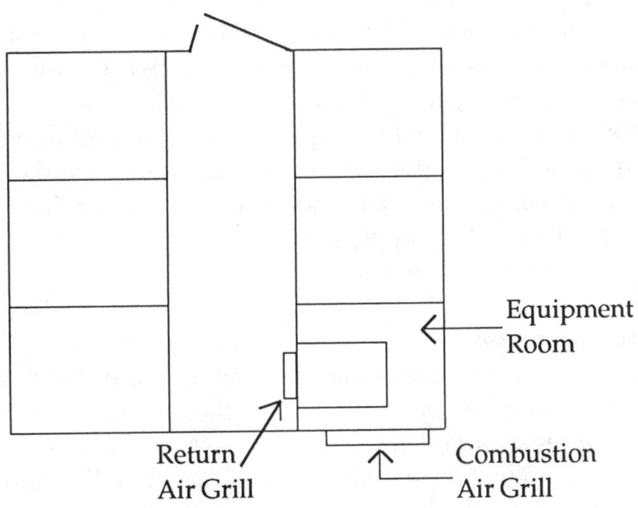

Figure 8-3. Return air outlet in each room.

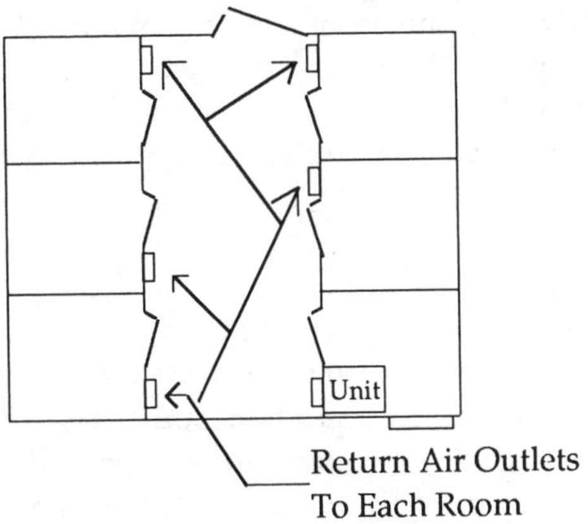

Under no circumstances should the equipment room be used as the return air plenum. The return air duct and plenum should completely isolate the return air from the combustion air in the equipment room. This precaution is to prevent the products of combustion from entering the occupied space.

Supply Air Location

The supply air plenum, if one is used, is located on the air outlet of the heating unit. The air distribution system that directs the air to the desired spaces is connected to the plenum. Usually, an extended plenum is used on commercial applications. See Figure 8-4.

Naturally, when more material and labor are required, the price of the installation will increase accordingly; therefore, it is desirable that the duct system be kept as simple as possible and still do the job properly.

The supply air openings should be located so that the air will be evenly distributed throughout the space, or where the air will be directed toward the point of greatest heat loss or gain. Generally, the air will be directed toward an outside wall or window. See Figure 8-5.

The location of the supply air opening will determine, to a great extent, the length of duct required.

Minimum Clearances

Under no circumstances should combustible material be located within the minimum clearances given by the equipment manufacturer. These clearances are set by the American Gas Association (AGA) for each model and they must be carefully followed. The unit dimensions should be correlated with the equipment room dimensions to ensure that sufficient room is available. If not, another location must be found or the room size increased to accommodate the unit safely. In most instances this will not be a problem with commercial equipment, but it is something that should be checked before the bid is presented. Otherwise, there could be considerable cost for the installation. If this cost becomes the responsibility of the contractor it could be the difference between making a profit and losing money on the job.

Noise

Because of air noise and equipment vibration, the unit should not be located near meeting rooms or other spaces where noise would be a distracting factor. When such a location cannot be avoided, acoustical materials can be placed in the return air plenum and the air handling section of the unit to reduce the air noise. Vibration can be reduced, or eliminated, by placing the unit on a solid foundation and using vibration-elimination materials.

Accessibility

When locating the unit, sufficient room must be allowed for servicing the unit. Some commercial air conditioning units have very large

Figure 8-4. Extended plenum system.

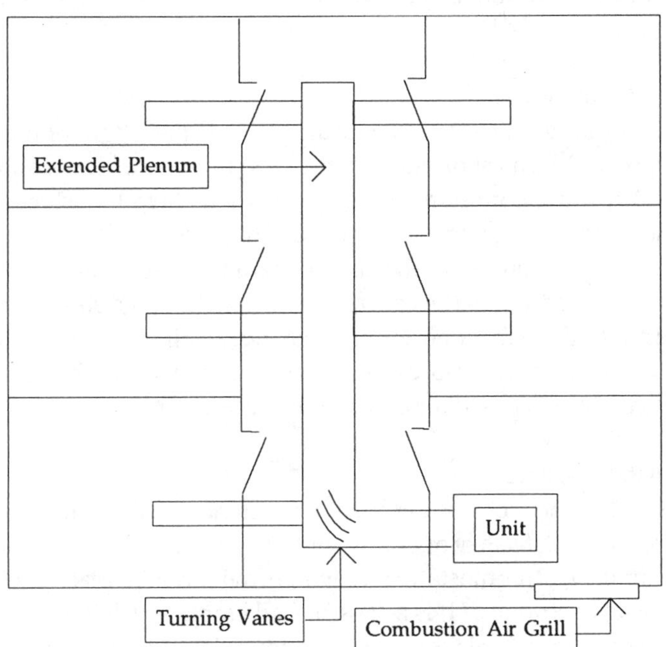

Figure 8-5. Supply air outlet location.

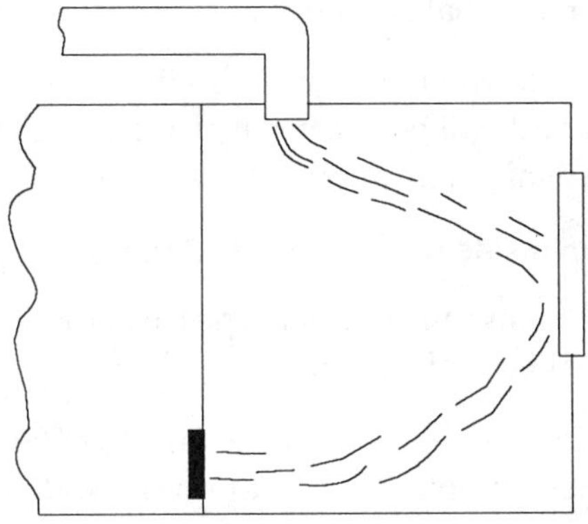

components, such as the fan shaft, which may need to be removed at some future date. Room must be left to make this possible. If room is not left there will be a much greater expense for repairs when the time comes to make them. There must be sufficient room for removing and replacing the filters, lubricating the bearings, lighting the pilot burner if needed, and other such routine operations. If sufficient room is not left these service operations may not be completed with the efficiency desired by the equipment owner.

Length of Duct Runs

The heating unit should be located so that the length of the ducts to each space will be as short as possible. The number of turns and elbows must be kept to a minimum. These precautions help to reduce heat loss and heat gain through the air distribution system, and allow a simpler designed installation. Also, the cost of making these fittings is very high which must be added to the cost of the installation. The shorter duct runs also help to reduce the amount of resistance to air flow, allowing the air delivery to be balanced more easily. Some states require that the air flow be balanced by a licensed air distribution technician.

Thermostat Location

The placement of the thermostat in a commercial installation is quite complex. The satisfactory operation of the unit depends greatly on careful placement of the thermostat. However, it must not be placed where it is accessible to everyone. If it is accessible, all employees will want to adjust the thermostat to suit their own comfort level. Naturally, this would cause much confusion and poor equipment operation. The following are points that must be considered when locating the thermostat:

1. It must be in a room that is conditioned by the unit it is controlling.
2. It must be exposed to normal free air circulation. Do not locate the thermostat where it will be affected by lamps, appliances, sunlight, or drafts from opening doors or outside windows.
3. It must be on an inside wall about 4 to 5 ft. from the floor.
4. It must not be behind furniture, drapes, or other objects that would affect the normal flow of free air.
5. It must not be located where it can be affected by vibrations, such as on a wall of the equipment room, or on the wall in which a door is located.

INDOOR UNIT LOCATION

When considering the placement of an indoor unit, such as an add-on or installing the unit in an existing building and there is no room designed to include the equipment under consideration, the following points must be addressed.

It must not interfere with the normal work activities of the space in which it is installed. If it will, the customer must be consulted before making a decision.

There must be sufficient structural strength to support the weight of the unit. If the unit is to be suspended from the ceiling there must be a place where it can safely be attached to the building. See Figure 8-6.

Check for the closest drain for the condensate. When one is not close a condensate pump will be needed. The distance and the size of the unit will determine the size of condensate pump unit needed.

When a gas-fired unit is to be installed check for the closest gas supply pipe. Make certain that it is large enough to meet the added demands that will be placed on it by the equipment. It may be necessary to install the pipe from the main gas supply to the unit.

When the heating is to be supplied by steam or hot water the supply lines must be located. Determine if they are large enough to meet the added requirements for the unit being installed.

When a duct system is to be installed there must be enough room to place the ducts without interfering with the normal work activities of the space. When they may, or will, interfere the building owner or occupant must be consulted before deciding where to install the ducts. When air distribution ducts are not required, make certain that the discharge air will not interrupt the normal work activities in the space. In most cases, if the air is directed on the occupants there will be complaints.

When installing the supply air ducts, place the air outlets where air

Figure 8-6. Commercial unit suspended from ceiling beams.

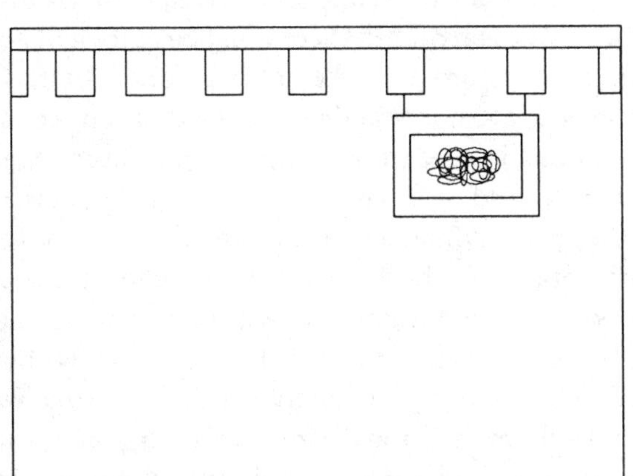

will not blow directly on the occupants unless it is desired. In some industrial applications it is desirable for the air to blow directly on the process or the equipment used during the process. In other applications air blowing directly on the process could completely upset the outcome of the product. This should be discussed with the occupant before locating the air outlets.

The electrical supply must be located. It is best to use the closest one possible to reduce the costs of wiring. Also, the wiring must be kept as short as possible to eliminate the possibility of low voltage or excessive installation costs. Be sure that the correct voltage is available.

Make certain that there is sufficient space around the unit for normal and proper service operations. Check to see if unit service operations will interfere with the normal work activities in the space. This is especially important for filter changing and bearing lubrication which should be done on a regular basis.

SELF CONTAINED INDOOR UNIT (Horizontal and Upflow)

The unit should be located so that it will not interfere with the normal work activities in the space. When this cannot be avoided, the occupants or the owner should be consulted so that a suitable location can be found. Usually this will not be a problem unless the unit is being installed in an existing building or space, or it is an addition to the existing system.

When the unit is to be suspended from the ceiling or some other part of the building the structure must be strong enough to support the unit safely. When the present structure will not support the unit some modifications should be completed to support the weight of the unit.

When ductwork is required there must be sufficient space to install it so that it will not interfere with the normal work activities in the space. There must also be a decision whether or not a return duct air duct system is needed. When one is needed there must be sufficient space for it. The supply air outlets should be located so that they will not interfere with the occupants of the space. When the unit is being installed for some commercial or industrial operation the air outlets must be located so that they will not interfere with the operation, but will still provide the desired air conditioning functions. When the unit is to be used without ductwork make certain that the air flow will not interrupt the normal work activities of the space and still provide the desired air conditioning functions.

When heat is to be supplied by steam or hot water, the closest supply pipes must be located. The pipes must be large enough to meet the additional requirements of the unit. If not, another source must be located. The steam or hot water lines must be insulated to reduce the amount of heat lost in the piping.

When heating is to be supplied with a gas-fired unit, the nearest gas supply that will meet the demands can be used. The distance that the gas supply is from the unit must be considered so that the proper size pipe can be installed.

Also, the vent must be considered. It should be as straight as possible and extend to the proper point above the roof. The vent is a very important factor in the installation because the materials are very expensive. When more vent pipe is required than estimated, a large part of the profits will be lost. See Figure 8-7.

When a water cooled unit is being considered there must be a place for the cooling tower, the water lines, and the pump. The cooling tower should be placed as close as possible to the unit to reduce the amount of piping required. When the tower is to be placed on top of the building the structure must be strong enough to support the weight of the tower, the pump, and the water in the pan. Check to make certain that the water lines connecting the unit to the tower can be installed. If the structural strength is not present, the tower will need to be placed where the building will properly support it or it may be installed on the ground. See Figure 8-8.

Most ordinances require that any unit installed on a roof must be placed on curbing to help prevent water leaks in the roof.

Figure 8-7. Self-contained unit vent installation.

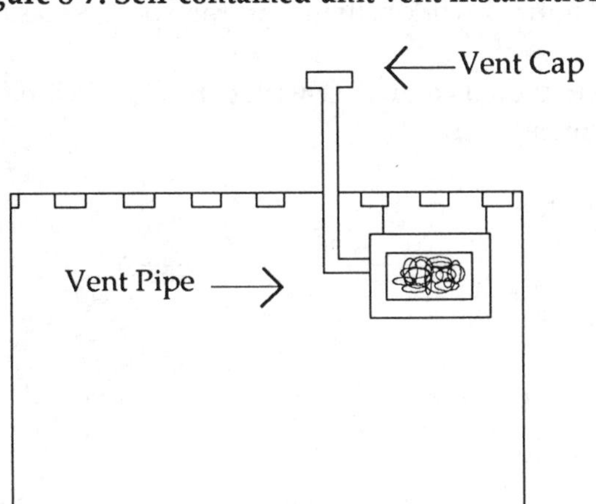

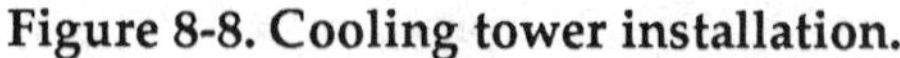

Figure 8-8. Cooling tower installation.

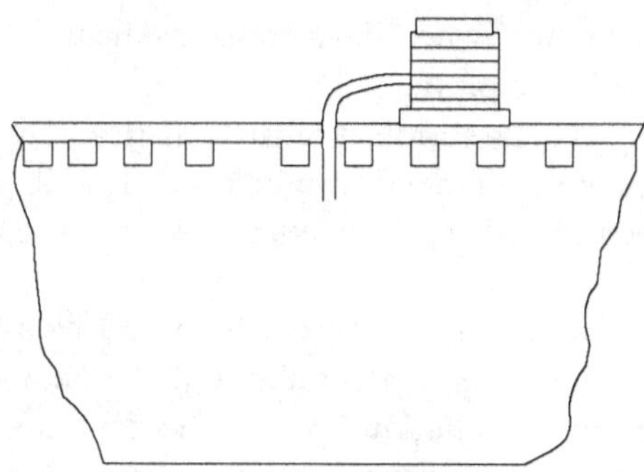

When an air cooled unit is being considered, the condenser is usually placed on the roof. It should be located as close to the unit as possible. Check to make certain that the refrigerant lines can be run through the roof. There should be no obstructions to the flow of air either to or from the condenser. It should be set directly on a strong member of the building or the curbing extended across two strong members to share the weight. See Figure 8-9.

Locate the nearest electrical supply that will provide the amount of power needed for the unit. The distance should be noted so that the proper size wire can be installed to prevent low voltage situations.

When more than one unit is to be installed in a single space make certain that their discharge airs do not work against each other. See Figure 8-10.

Make certain that the unit can be properly serviced without interfering with the normal work activities in the space, especially for routine

Figure 8-9. Air-cooled condenser with curbing (pitch pan) crossing 2 or more strong members.

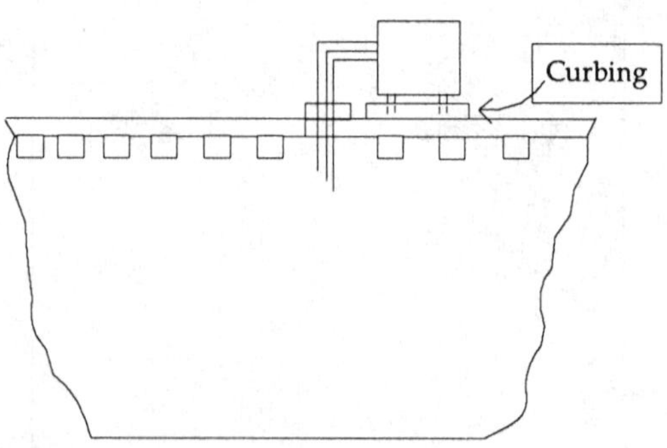

Figure 8-10. Correct installation of two indoor units.

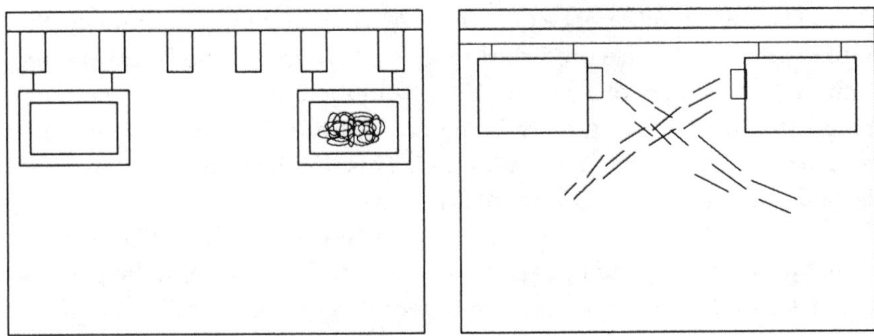

procedures such as filter changing, and bearing lubrication, which should be done on a regular schedule. There should be consideration given to the possibility that some time some of the larger components inside the unit will need to be removed and replaced. Make certain that sufficient space is available for this.

The nearest drain to where the condensate can be piped must be located. If there is not one available a condensate pump will be required. The proper size must be used so that the condensate drain can operate properly.

OUTDOOR UNIT (Condensing Unit)

When the condensing unit is to be located on the roof, the building must have the strength to support the weight. Sometimes these units are quite heavy. When the needed support is not at the nearest point to the indoor unit, a decision must be made to either relocate the condensing unit, relocate the indoor unit, install the additional refrigerant piping, or provide the needed support. Most codes require that the outdoor section of the unit be placed on curbing (pitch pans) to help prevent water leaks in the roof. See Figure 8-9.

When noise will be a factor, place the unit on vibration eliminating devices. There are several types on the market that work quite well. These should be added during the time of the installation because it will be very difficult, if not impossible, to add them after the unit has been installed.

Locate the nearest electrical supply. Determine if it has the capacity to provide the required electricity without voltage drop. It should be as close as possible to the unit to reduce the installation cost and the possible drop in voltage at the unit.

The unit should not be placed close to any building components that will restrict the service procedures required. There should be no obstruction to the free-flow of air either to or from the unit. It should be remembered that in the future there is the possibility that a major component will need to be removed and replaced. This will be quite difficult, if not impossible, when sufficient room is not left.

There must be access for installing the refrigerant lines to the indoor unit. These lines should be kept as short as possible to reduce the possibility of leaks and the amount of refrigerant required to fully charge the system.

The condensing unit should be located as close as possible to the indoor unit. However, the structural strength of the building will have a bearing on locating the unit.

When the building is higher than normal it must be determined if it is too tall for normal crane service. Some crane companies charge extra for lifting above a certain height. It may be required that a special crane be used. If so, the cost will increase dramatically.

The supply air ducts must be installed so that the air outlets are located to accommodate the work stations, an industrial process, or occupants of the space. For comfort applications, they will be placed to blow the air towards the point of greatest heat loss or heat gain for the space.

The return air, when used, should not interfere with the normal work activities of the space. When possible there should be a return for each space so that better air distribution can be possible. Make certain that there is sufficient space to install the return duct without the return air mixing with the combustion air to the unit.

SELF CONTAINED OUTDOOR UNIT

These units are quite heavy and must be supported properly to prevent damage to the building. The foundation that these units sit on must cross at least two supporting beams in the building. They should be placed so that the shortest duct possible can be used. Both the supply air duct and the return air duct must pass through the building roof. Some manufacturers have designed units that have the duct connections beneath the unit and others have them on the same side of the unit. These installations require that extra precautions be taken to prevent water leaks in the roof. See Figure 8-11.

Figure 8-11. Self-contained unit installed on roof.

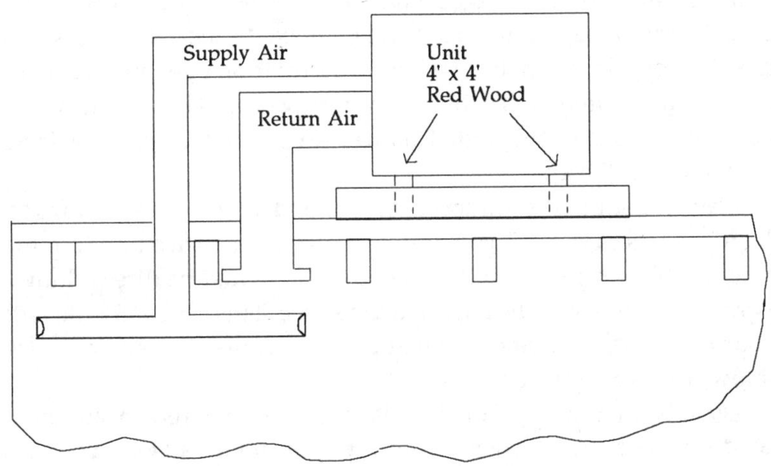

These units should be placed on curbing (pitch pans) to help in preventing water leaks through the roof. Curbing and sometimes special precautions are needed where the ducts penetrate the roof. Usually this requires that a certified roofing company be used to maintain the roofing bond. The ducts should not require that any building members be removed. When removal is required, there should be sufficient framing installed to prevent future problems with the roof and building structure.

Locate the nearest electrical supply. Generally, it is preferred that these units be provided with their own electrical supply from the main electrical panel, unless there is sufficient power in a secondary panel. The wire must be properly sized to prevent voltage drops between the panel and the unit.

Determine the shortest possible route for the condensate drain line. These drains should always discharge into a sewer drain. Most codes will not permit them to discharge into the storm drains located on the building roof. Many times a very long drain line is required. The condensate should not be allowed to drain onto the roof. To do so will sometimes cause the roofing bond to be canceled.

When the building is taller than normal check to make certain that the crane company can place the unit on the roof. If not, a special crane will be required which will usually cost extra.

When gas heating is to be used, the nearest gas supply must be located. The piping must be large enough to supply the added requirements for the unit. When this is not possible, piping will need to be installed from a supply main that will meet these demands.

Make certain that there will be sufficient room for servicing the unit. Remember that it is quite possible that some of the major components will need to be removed and replaced. This will be difficult if sufficient space is not allowed. Space should be left for normal maintenance such as the replacement of filters and lubrication of bearings. Also, room is needed for the installation of the unit. This will save time and perhaps increase the profits from the job.

Place the supply air outlets to accommodate the work stations or the industrial process requiring conditioned air. In an office building the air should not blow directly on the occupants. It should be directed towards the greatest heat loss or heat gain in the space. This is usually the outside wall or windows. In an industrial application it may be required that the air blow directly on the process.

Usually in this type of installation, a central return air outlet is located on the ceiling. When a comfort application is being considered, each individual space has either a return air grill or some type of passage to allow the air to move from the space to a hall or some other common space and then into the unit. For industrial applications the return air grill is usually placed on the ceiling and pulls the air equally from a larger space in which the supply air is discharged. See Figure 8-12.

COMMERCIAL AIR DISTRIBUTION SYSTEMS

Some of the more popular air distribution systems in use are the zoning, and variable air volume (VAV). We will briefly discuss them as follows. For a detailed description of these systems and how they are designed, consult one of the many manuals that are available.

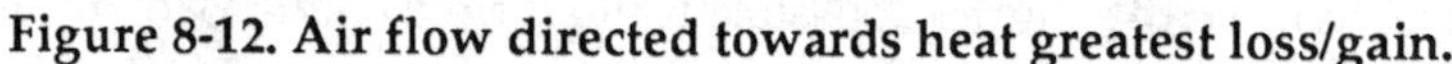

Figure 8-12. Air flow directed towards heat greatest loss/gain.

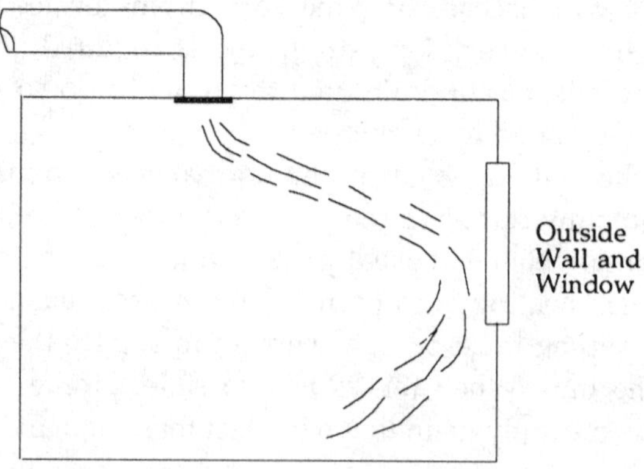

ZONING

(Also known as constant volume, single duct system.) This method is popular when a single system must maintain the temperature of one or more spaces at a desired temperature. An example of this is when a building has both north and south exposures. At the time the north zone could be needing heat, the south zone would be needing cooling. One way this is accomplished is by using a hot and cold deck with face and by-pass dampers. See Figure 8-13.

The hot deck is supplied with a heating medium, usually hot water or steam and the cold deck is supplied with a cooling medium, usually a chilled water system. The face and by-pass dampers are controlled by the thermostat located in each space. Each thermostat controls a set of dampers. Both the hot and cold decks are connected to each zone so that the

Figure 8-13. Air-handling unit for a single-zone.

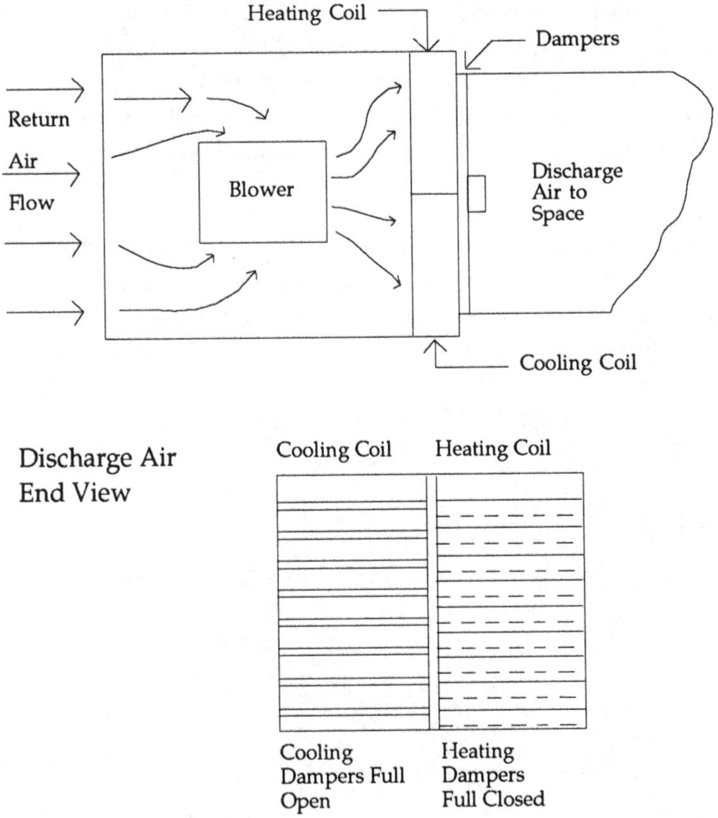

thermostat can control the zone temperature by varying the temperature of the air being delivered to it.

A modification of this is when a by-pass box is used. The system operates as a constant volume system in the equipment but as a variable air volume system in the conditioned space. This is possible through the use of a by-pass box installed in the duct to the space. The thermostat operates a set of dampers in response to the temperature inside the space. When less air is required, the damper system allows a part of the air to be dumped into the return air system, usually above the ceiling. The return air will be much closer to the temperature of the space and will require less conditioning, lowering the operating costs. See Figure 8-14.

VARIABLE AIR VOLUME

This type of system controls the space temperature by varying the amount of air delivered to the space. The VAV box, or device, is placed in the duct to the space being controlled. During operation, the air is held at an almost constant temperature. Some of the VAV boxes have reheat coils mounted in them so the air can be heated before delivery to the conditioned space. The most popular variations of the VAV system are: reheat, fan-powered systems, variable diffuser systems, and induction VAV box. The following is a brief description of these methods.

Reheat

The reheat system has a heating coil placed inside the VAV box. When the air needs to be warmed a heating medium such as steam or hot water is fed into the coil. A major advantage of this type of air distribution system is humidity control during low load periods. See Figure 8-15.

Figure 8-14. Use of by-pass VAV box.

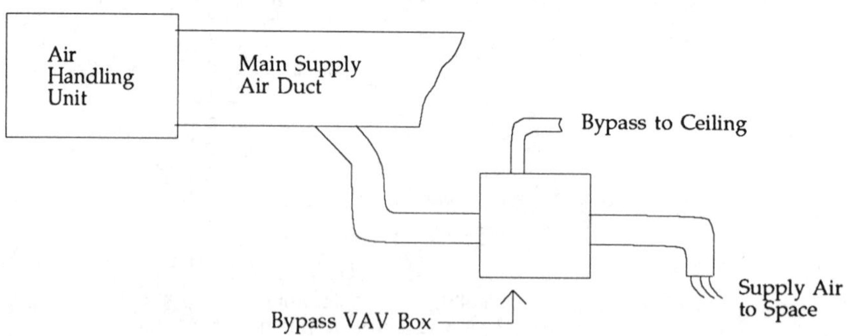

Figure 8-15. Typical reheat VAV system.

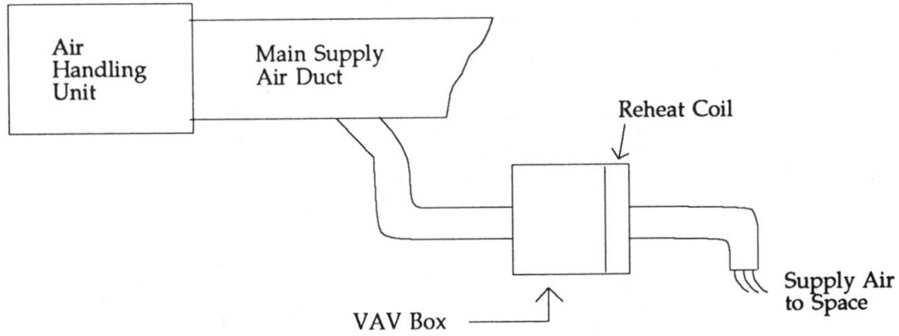

Fan-Powered Systems

There are two models of these systems: the series flow and parallel flow. These two systems are usually selected because they will allow more air to be circulated through the conditioned space when a low load condition occurs and still retain all the advantages of a VAV system during normal or peak load times.

Series Flow

When the series type is used, the fan is placed in the primary air stream. The fan runs continuously when used in this manner. Between the heating and cooling functions there is a point in which only return air is circulated into the space. In most commercial installations the space above the ceiling is used as the return air system. During this process the heat from the lighting system and other loads is used to reheat the air. Also, when the space is unoccupied, the main air-handling unit may not be in operation and the individual VAV boxes circulate and control the air inside the space. See Figure 8-16.

Parallel Flow

When the parallel system is used, the fan is placed outside the primary air system. The fan can be operated intermittently when used in this manner. Between the heating and cooling functions there is a point in which only return air is circulated into the space. In most commercial installations the space above the ceiling is used as the return air system. During this time the heat from the lighting system and other loads is used to reheat the air. Also, when the space in unoccupied, the main air-handling unit may not be in operation and each individual VAV box circulates and controls the air inside the space.

Figure 8-16. Typical series-flow VAV system.

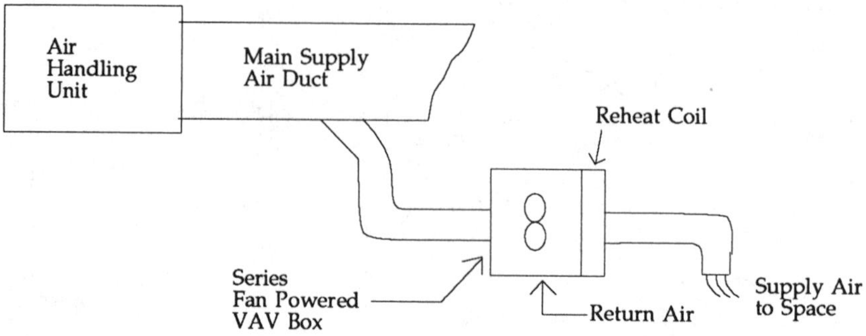

Air
Handling
Unit
Main Supply
Air Duct
Reheat Coil
Series
Fan Powered
VAV Box
Return Air
Supply Air
to Space

INDUCTION SYSTEM

The induction system is usually used as a terminal unit. One of its uses is to recirculate the return air (ceiling air) and reduce the cooling capacity by reducing the amount of primary air required from the main unit. See Figure 8-17.

Recirculation of the return air replaces the reheat coil in the VAV box. It uses the heat from the lighting system and other heat loads on the unit to maintain the required temperature. This system sometimes requires that heat be added during cold periods and when the space is not occupied. This is usually done with the main unit which is more expensive to operate than the single reheat coils in the other VAV boxes.

Figure 8-17. Typical induction type VAV system.

Air
Handling
Unit
Main Supply
Air Duct
Induced Air
Induction
Type VAV Box
Supply Air
to Space

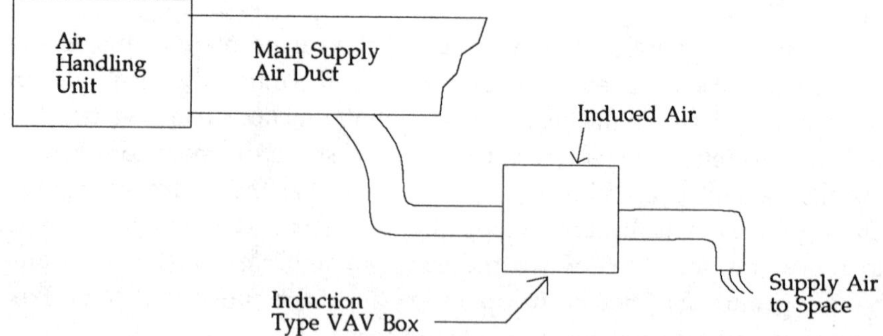

VARIABLE DIFFUSER

These types of VAV systems reduce the size of the opening in the air outlet of the diffuser. When this is done the amount of air induced into the diffuser from the room increases an almost like amount. This tends to keep the discharge air volume from the diffuser constant. This action mixes the cooler return air with the warmer air in the room.

SUMMARY

A well-planned location for the air conditioning and heating unit is one of the most important steps in a good installation.

The gas supply to the building must have the capacity to handle the extra load that the new equipment will place on it.

Be sure to consider any restrictions or requirements that might be necessary because of local and/or national building codes.

The heating unit should be located so that the vent pipe will provide as straight a path as possible for the flue gases to escape to the atmosphere.

When the heating unit is installed in an equipment room, an opening must be provided to make certain that adequate ventilation and combustion air is available to the unit.

Under no circumstances should the equipment room be used as the return air plenum.

The supply air openings should be located so that the air will be evenly distributed throughout the space, or where the air will be directed toward the point of greatest heat loss or gain.

Under no circumstances should combustible material be located within the minimum clearance given by the equipment manufacturer.

Because of the noise and equipment vibration, the unit should not be located near meeting rooms, or other places where noise would be a distracting factor.

When locating the unit, sufficient room must be allowed for servicing the unit.

The heating unit should be located so that the length of ducts to each space will be as short as possible. The number of turns and elbows must be kept to a minimum.

The indoor unit must not interfere with the normal work activities of the space in which it is installed.

There must be sufficient structural strength to support the weight of the unit.

When a drain is not close, a condensate pump will be needed.

Determine if the steam or hot water lines are large enough to meet the added requirements.

It is best to use the closest electrical supply that will support the equipment requirements and help prevent low voltage to the unit.

When the unit must be suspended from the ceiling or some other part of the building the structure must be strong enough to support the unit safely.

When a water cooled unit is being considered there must be a place for the cooling tower, the water lines, and the pump.

There should be no obstructions to the flow of air either to or from an air cooled condenser.

When more than one unit is to be installed in a single space make certain that their discharge airs do not work against each other.

When the condensing unit is to be placed on the roof, the building must have the strength to support the weight.

Most codes require the outdoor section of the roof to be placed on curbing (pitch pans) to help prevent water leaks in the roof.

When noise will be a factor, place the unit on vibration eliminating devices.

There must be access for installing the lines that connect to the units.

The indoor and outdoor units should be located as close to each other as possible.

Self-contained units (PTACs) are quite heavy and require strong support to prevent damage to the building.

Water running on the roof will sometimes cause the roofing bond to be canceled.

Some of the more popular commercial air distribution systems in use are the zoning and variable air volume (VAV).

Zoning (constant air volume) is popular when a single system must maintain the desired temperature in one or more spaces.

Zoning is accomplished through the use of hot and cold decks with face and by-pass dampers.

The variable air volume (VAV) system controls the space temperature by varying the amount of air delivered to the space.

There are two models of fan-powered systems. They are the series flow and parallel flow.

The variable diffuser system is another type of VAV system that reduces the size of the opening in the air outlet of the diffuser.

Chapter 9

Refrigerant Lines

When units are installed with remote condensing units or other components that are not included in a single cabinet, it is necessary to connect these components together with field-engineered refrigerant lines. When a standard catalog combination of units is being considered, the piping size calculations will usually have already been made by the equipment manufacturer's engineers. The pipe diameter and maximum lengths will be listed in the installation literature. When non-catalog combinations, such as one 10-ton condensing unit used with two 5-ton evaporators is being considered, the line sizes should be calculated by the methods that are discussed throughout the remainder of this chapter.

INTRODUCTION

The procedures used for sizing refrigerant piping is a series of compromises, at best. A good piping system will have a maximum capacity, be economical, provide proper oil return, provide minimum power consumption, require a minimum of refrigerant, have a low noise level, provide proper refrigerant control, and allow flexibility in system performance from 0 to 100% capacity of the unit without lubrication problems. Obviously, it is impossible to obtain all of these goals because some of them are in direct conflict with others. To make an intelligent decision as to what type of compromise is the most desirable, it is essential that the piping designer have a thorough understanding of the basic effects on the system performance of the piping design in the different points of the refrigeration system.

PRESSURE DROP

The pressure drop of the refrigerant as it flows through the piping, in general, tends to decrease system capacity and increase the amount of electric power required by the compressor. Therefore, excessive pressure drops should be avoided. The amount of pressure drop that is allowable will vary depending on the particular segment of the system that is involved. Each part of the system must be considered separately. There are probably more charts and tables available which cover refrigerant line capacity at a given pressure, temperature, and pressure drop than any other single subject in the field of refrigeration and air conditioning.

The piping designer must realize that there are several factors which govern the sizing of refrigerant lines; pressure drop is not the only criterion to be used is designing a refrigerant system. It is often required that refrigerant velocity, rather than pressure drop, be the determining factor in system design. Also, the critical nature of oil return can produce many system problems. A reliable pressure drop is far more preferable than oversized lines which might hold refrigerant far in excess of that required by the system. An overcharge of refrigerant can result in serious problems of liquid refrigerant control and the flywheel effect of large quantities of liquid refrigerant in the low-pressure side of the system which can result in erratic operation of the refrigerant flow control device.

The size of the refrigerant line connection on the service valve supplied with the compressor, or the size of the connection on the evaporator, condenser, or some other system accessory does not determine the correct size of line to be used. Equipment manufacturers select a valve size or connection fitting on the basis of its application to an average system and other factors, such as the application, length of connecting lines, type of control system, variation in the load, and a multitude of other factors. It is entirely possible for the required refrigerant line size to be either smaller or larger than the fittings on the various system components. In such cases, reducing fittings must be used.

OIL RETURN

Since oil must pass through the compressor cylinders to provide proper lubrication, a small amount of oil is always circulating through the system with the refrigerant. The oils used in the refrigeration systems are soluble in liquid refrigerant, and at normal room temperatures they will mix completely. Oil and refrigerant vapor, however, do not mix readily, and the oil can be properly circulated through the system only if the velocity of the refrigerant vapor is great enough to carry the oil along with it. To assure proper oil circulation, adequate refrigerant velocities must be maintained not only in the suction and discharge lines, but in the evaporator circuits as well.

Several factors combine to make oil return most critical at low evaporating temperatures. As the suction pressure drops and the refrigerant vapor become less dense, it becomes more difficult for the refrigerant to carry the oil along. At the same time, as the suction pressure drops, the compression ratio increases and the compressor capacity is decreased. Therefore, the weight of the refrigerant circulated decreases. Refrigeration oil alone becomes as thick as molasses at temperatures below 0°F. As long as it is mixed with a sufficient amount of liquid refrigerant, however, it flows freely. As the percentage of oil in the mixture increases, the viscosity increases.

At low-evaporating conditions, several factors start to converge and can create a critical condition. The density of the refrigerant vapor decreases, the velocity decreases, and as a result more oil starts to accumulate in the evaporator. As the oil and refrigerant mixture become thicker, the oil may start logging in the evaporator rather than returning to the compressor. This results in wide variations in the compressor crankcase oil level in poorly designed systems.

Oil logging in the evaporator can be minimized with adequate refrigerant velocities and properly designed evaporators, even at extremely low evaporating temperatures. Normally, oil separators are necessary for operation at evaporating temperatures below –50°F, to minimize the amount of oil in circulation.

EQUIVALENT LENGTH OF PIPE

Each valve, fitting, and bend in a refrigerant line adds friction pressure drop because of its interruption or restriction of smooth flow. Because of the detail and complexity of computing the pressure drop of each individual fitting, normal practice is to establish an equivalent length of straight pipe. Pressure drop and line sizing tables and charts are normally set up on a basis of a pressure drop per 100 ft. of straight pipe, so the use of equivalent lengths allows the data to be used directly (Table 9-1).

For accurate calculations or pressure drop, the equivalent length for each fitting should be calculated. As a practical matter, an experienced piping designer may be capable of making an accurate overall percentage allowance unless the piping system is extremely complicated. For long runs of piping of 100 ft. or greater, an allowance of 20 to 30% of the actual length may be adequate. For short runs of piping, an allowance as high as 50 to 75% or more of the lineal length may be necessary. Judgement and experience are necessary in making a good estimate, and estimates should

Table 9-1. Equivalent lengths of straight pipe for valves and fittings in feet (m).

(OD, In.) Line Size (mm.)	Globe Valve	Angle Valve	90° Elbow	45° Elbow	Tee Line	Tee Branch
1/2(12.7)	9(2.74)	5(1.52)	0.9(0.27)	0.4(0.12)	0.6(0.18)	2.0(0.61)
5/8(15.87)	12 (3.66)	6(1.83)	1.0(0.3048)	0.5(0.15)	0.8(0.24)	2.5(0.76)
7/8(22.23)	15(4.57)	8(2.44)	1.5(0.45)	0.7(0.21)	1.0(0.3048)	3.5(1.07)
1-1/8(28.57)	22(6.70)	12(3.66)	1.8(0.55)	0.9(0.27)	1.5(0.46)	4.5(1.37)
1-3/8(34.93)	28(8.53)	15(4.57)	2.4(0.73)	1.2(0.36)	1.8(0.55)	6.0(1.83)
1-5/8(41.27)	35 (10.67)	17 (5.18)	2.8(0.85)	1.4(0.43)	2.0(0.61)	7.0(2.13)
2-1/8(53.97)	45(13.72)	22(6.70)	3.9(1.19)	1.8(0.55)	3.0(0.91)	10.0(3.05)
2-5/8(66.67)	51(15.54)	26(7.92)	4.6(1.40)	2.2(0.67)	3.5(1.07)	12.0(3.66)
3-1/8(79.37)	65(19.81)	34(10.36)	5.5(1.68)	2.7(0.82)	4.5(1.37)	15.0(4.57)
3-5/8(92.07)	80 (24.38)	40(12.19)	6.5(1.98)	3.0(0.91)	5.0(1.52)	17.0(5.18)

be checked frequently with actual calculations to ensure near accuracy.

For items such as solenoid valves and pressure-regulating valves, where the pressure drop through the valve is relatively large, data are normally available from the manufacturer's catalog so that items of this nature can be considered independently of lineal length calculations.

PRESSURE DROP TABLES

There are pressure drop tables available which allow the combined pressure drop for all refrigerants (Figures 9-1 and 9-2).

These figures show the pressure drops for refrigerants R-22 and R-502. Pressure drops in the discharge line, suction line, and the liquid line can be determined from these charts for condensing temperatures ranging from 80 to 120°F.

To use the chart, start in the upper right-hand corner with the design capacity. Drop vertically downward on the line representing the operating condition desired. Then move horizontally to the left. A vertical line dropped down from the intersection point with each size of copper tubing to the design condensing temperature line allows the pressure drop in pounds per square inch (psi) per 100 ft. of tubing to be read directly from the chart. The diagonal pressure drop lines at the bottom of the chart represent the change in pressure drop due to a change in condensing temperature.

For example, in Figure 9-2 for R-502, the dashed line represents a pressure drop determination for a suction line in a system having a design capacity of 7.5 tons, or 90,000 Btuh operating with an evaporating temperature of 40°F, air conditioning application. The 2-5/8-in. suction line illustrated has a pressure drop of 0.23 psi per 100 ft. at 100°F condensing temperature. The same line with the same capacity would have a pressure drop of 0.19 psi per 100 ft. at 80°F condensing temperature, and 0.30 psi per 100 ft. at 120°F condensing temperature.

In the same manner, the corresponding pressure drop for any line size and any set of operating conditions within the range of the chart can be determined.

SIZING OF HOT GAS DISCHARGE LINES

On large commercial systems, it is sometimes required to install hot gas discharge lines from the compressor to the condenser. On units that

Figure 9-1. Refrigerant-22 pressure drop in lines corresponding to a 65°F evaporator outlet. (Copyright 1968 by the DuPont Company, reprinted by permission)

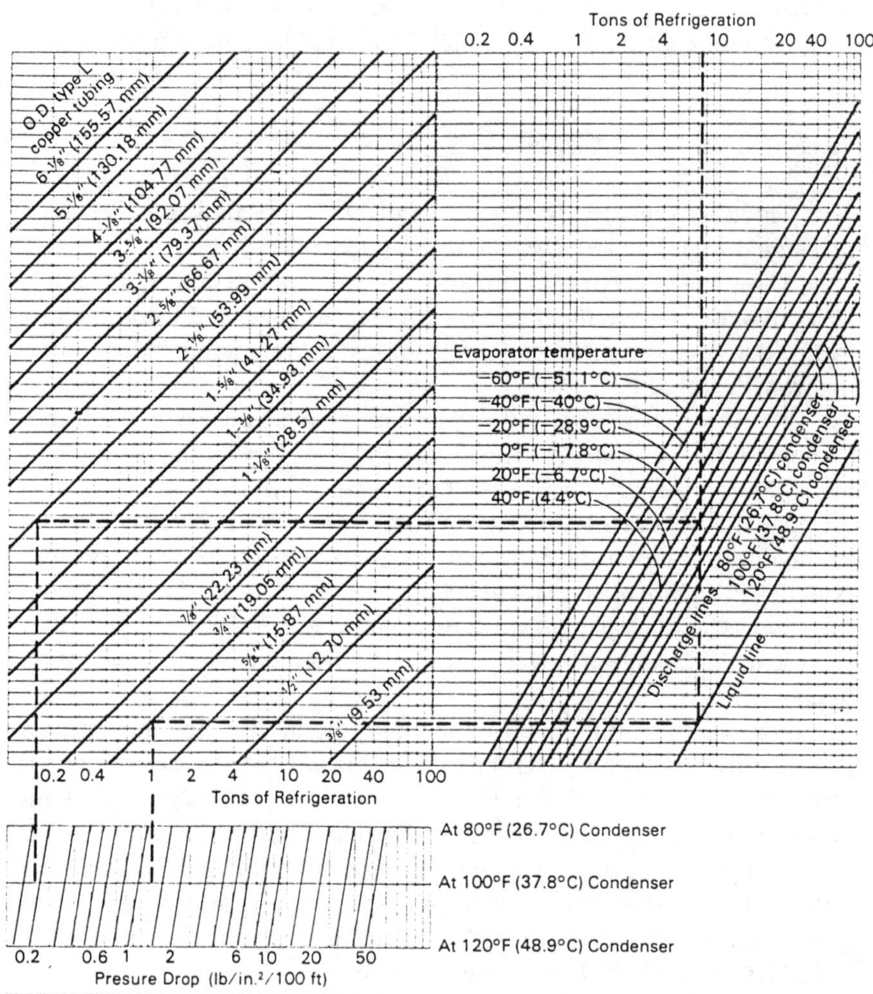

Note:

Pressure drops do not allow for pulsating flow.
If flow is pulsating, use next larger pipe size.
Liquid line determined at 0°F (−17.8°C) evaporator and 80°F (26.7°C) condenser.
Discharge lines at 0° (−17.8°C) evaporator.
Other conditions do not change results appreciably.
Vapor at evaporator outlet is assumed to be at 65°F (18.3°C).

Example:

7.5 tons at 40°F (4.44°C) evaporator and 100°F (37.8°C) condensing temperatures.
2⅛" (53.99 mm) suction line with a pressure drop of 0.2 psi/100 ft.
¾" (19.05 mm) liquid line with a pressure drop of 1.3 psi/100 ft.

Figure 9-2. Refrigerant-502 pressure drop in lines corresponding to a 65°F evaporator outlet. (Copyright 1968 by the DuPont Company, reprinted by permission)

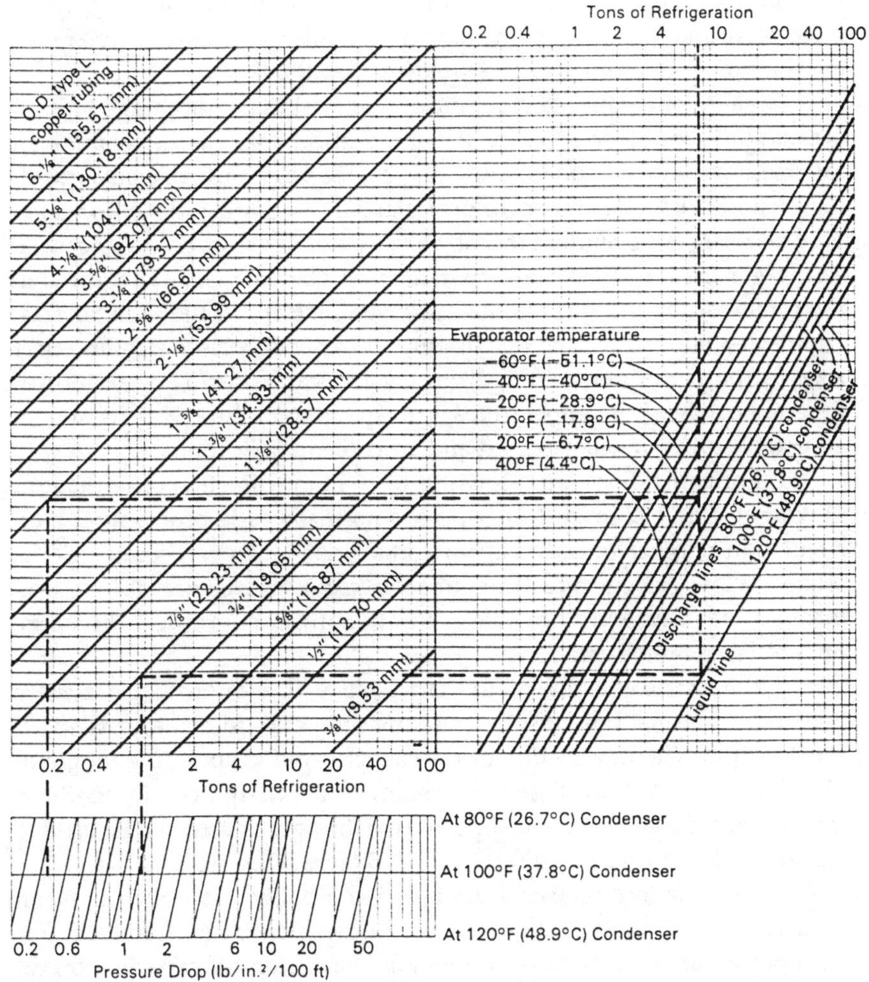

Note:

Pressure drops do not allow for pulsating flow.
If flow is pulsating, use next larger pipe size.
Liquid line determined at 0°F (−17.8°C) evaporator and 80°F (26.7°C) condenser.
Discharge lines at 0° (−17.8°C) evaporator.
Other conditions do not change results appreciably.
Vapor at evaporator outlet is assumed to be at 65°F (18.3°C).

Example:

7.5 tons at 40°F (4.44°C) evaporator and 100°F (37.8°C) condensing temperatures.

2⅛" (53.99 mm) suction line with a pressure drop of 0.2 psi/100 ft.

⅞" (22.2 mm) liquid line with a pressure drop of 1.2 psi/100 ft.

have the compressor and condenser combined in a single housing (condensing unit) this procedure is usually done at the factory during the manufacture of the unit.

Actually, a reasonable pressure drop in the discharge line is often desirable to dampen compressor discharge pulsations, and thereby reduce noise and vibration. Some discharge mufflers actually derive much of their efficiency from the pressure drop through them.

Because of the high temperatures that exist in the discharge line, oil circulation through both horizontal and vertical lines can be maintained satisfactorily with reasonably low refrigerant velocities. Since oil traveling up a riser usually creeps up the inner surface of the pipe, oil travel in vertical risers is dependent on the velocity of the gas at the tubing wall. The larger the pipe diameter, the greater will be the velocity required at the center of the pipe to maintain a given velocity at the wall surface. Figures 9-3 and 9-4 list the maximum recommended discharge line riser sizes for proper oil return for varying system capacities. The variation at different condensing temperatures is not great, so the line sizes are shown acceptable on both water-cooled and air-cooled applications.

If horizontal lines are run with a pitch in the direction of refrigerant flow of at least 1/2 in. in 10 ft., there is normally little problem with oil circulation at lower velocities in horizontal lines. However, because of the relatively low velocities required in a vertical discharge line, it is recommended that whenever possible both horizontal and vertical discharge lines be sized on the same basis.

To illustrate the use of the chart we will assume that a system operating with R-12 at 40°F evaporating temperature has a capacity of 100,000 Btuh. The intersection of the capacity and evaporating temperature lines at point X on Figure 9-3 indicates the design condition. Since this is below the 2-1/8 in. -OD line, the maximum size that can be used to ensure oil return up a vertical riser is a 1-5/8-in. -OD line.

Oil circulation in discharge lines is normally a problem only on systems where large variations in system capacity are encountered. For example, an air conditioning system may have steps of capacity control allowing it to operate during periods of light load at capacities possibly as low as 20 to 33% of the design capacity. The same situation may exist on commercial refrigeration systems, where compressors connected in parallel are cycled for capacity control. In such cases, vertical discharge lines must be sized to maintain velocities above the minimum velocity necessary to circulate oil properly at the minimum load condition.

For example, consider an air conditioning system using R-22 having a maximum design capacity of 300,000 Btuh with steps of capacity reduc-

Figure 9-3. Recommended maximum vertical compressor discharge line sizes to provide proper oil return using R-12 refrigerant .

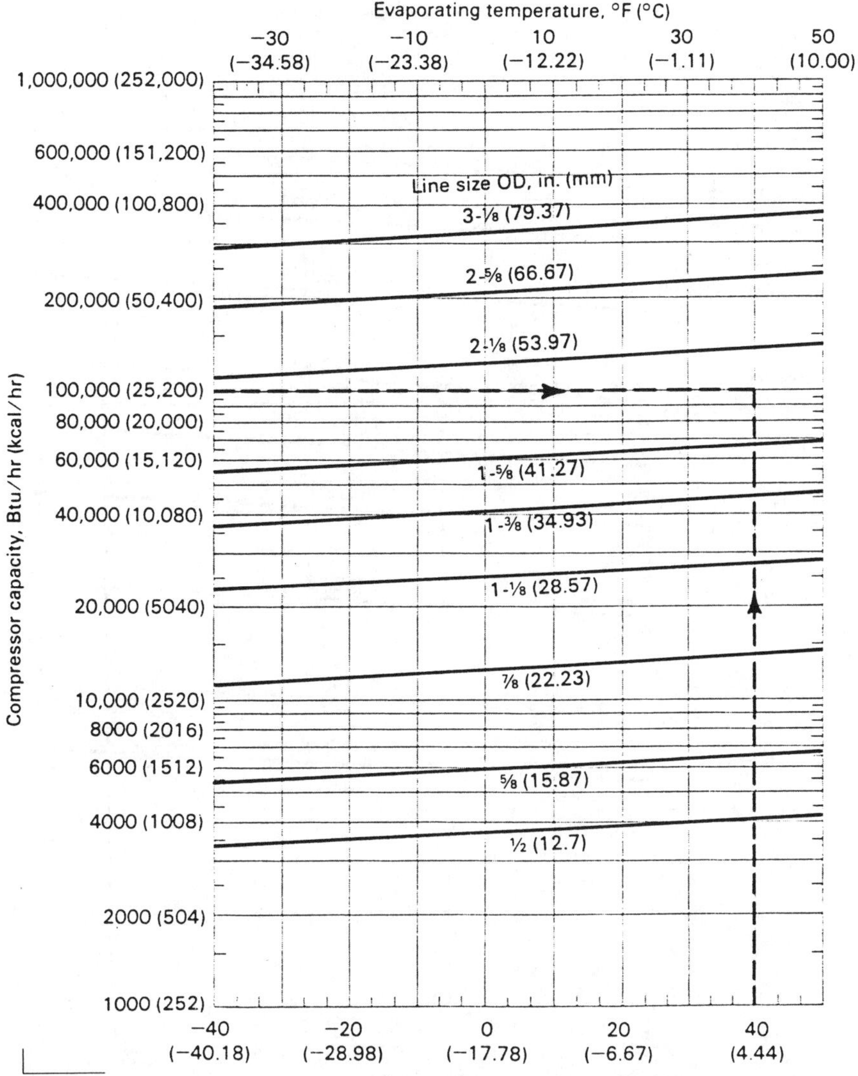

Figure 9-4. Recommended maximum vertical compressor discharge line sizes to provide proper oil return using R-22 and R-502 refrigerants.

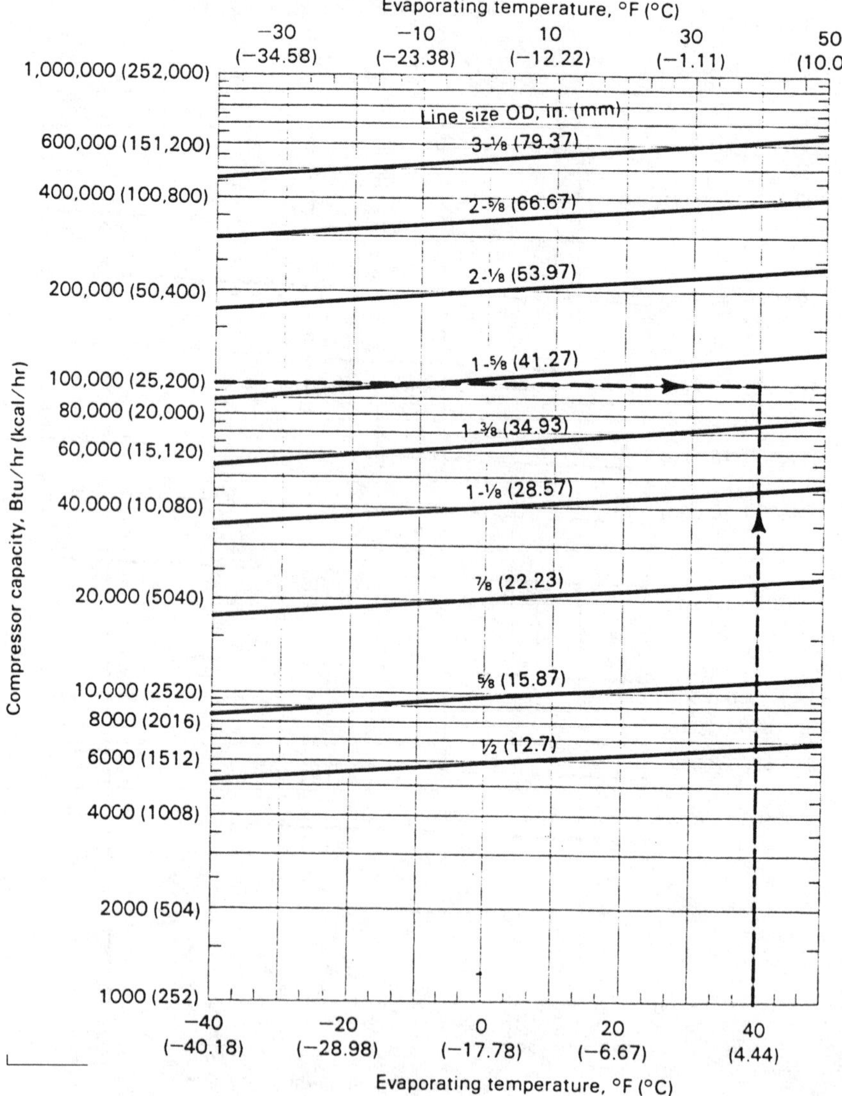

tion of up to 66%. Although the 300,000 Btuh condition could return oil up to a 2-5/8 in. -OD riser at light load conditions, the system would have only 100,000 Btuh capacity, so a 1-5/8 in. -OD rise must be used. In checking the pressure drop chart in Figure 9-4 at maximum load conditions, a 2-5/8 in. -OD pipe will have a pressure drop of approximately 4 psi per 100 ft. at a condensing temperature of 120°F. If the total equivalent length of pipe exceeds 150 ft., in order to keep the total pressure drop within reasonable limits, the horizontal line should be the next larger size, or 2-1/8 in.-OD, which would result in a pressure drop of only slightly over 1 psi per 100 ft.

Because of the flexibility in line sizing that the allowable pressure drop makes possible, discharge lines can almost always be sized satisfactorily without the necessity of double risers. If modifications are made to a system which result in the existing discharge line being oversized at light load conditions, the addition of an oil separator to minimize oil circulation will normally solve the problem.

One other limiting factor in discharge line sizing is that excessive velocity can cause noise problems. Velocities of 3000 ft. per minute (fpm) or more may result in high noise levels, and it is recommended that maximum refrigerant velocities be kept well below this level. Figures 9-5 and 9-6 give equivalent discharge line gas velocities for varying capacities and line sizes over the normal refrigeration and air conditioning range.

SIZING LIQUID LINES

Since liquid refrigerant and oil mix completely, velocity is not essential for oil circulation through the liquid line. The primary concern in liquid line sizing is to make certain that a solid column of liquid refrigerant reaches the expansion valve, or flow control device. If the pressure of the liquid refrigerant falls below its saturation temperature, a portion of it will flash into vapor to cool the remaining liquid to the new saturation temperature. This can occur in a liquid line if the pressure drops sufficiently because of friction in the liquid line or the vertical lift of the refrigerant.

Flash gas in the liquid line has a detrimental effect on system performance in several ways:

1. It increases the pressure drop due to friction.
2. It reduces the capacity of the flow control devices.
3. It can erode the expansion valve pin and seat.

Figure 9-5. Recommended discharge line velocities for various Btu/hr (Kcal/hr) capacities using R-22 and R-502 refrigerants.

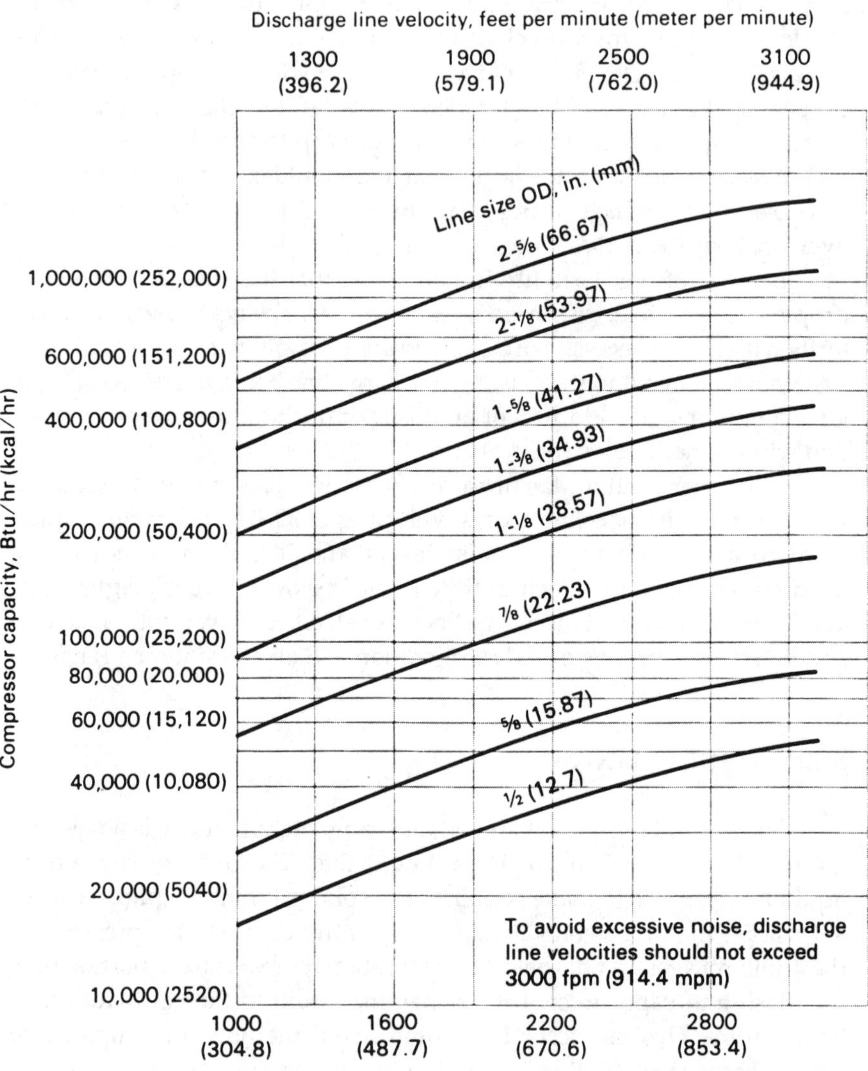

4. It can cause excessive noise.
5. It can cause erratic feeding of the liquid refrigerant to the evaporator.

For proper system performance, it is necessary that liquid refrigerant reach the flow control device at a subcooled temperature slightly

Figure 9-6. Recommended discharge line velocities for various Btu/hr (Kcal/hr) capacities using R-12 refrigerant.

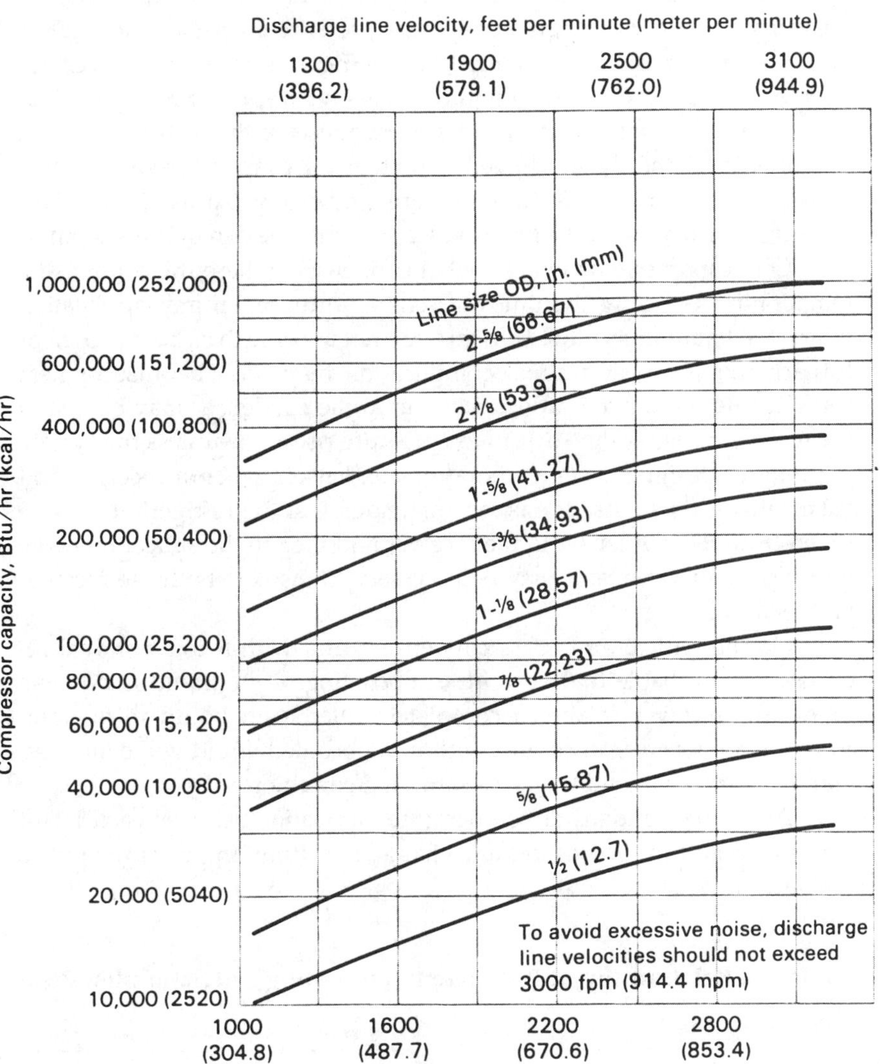

below the saturation temperature. On most systems, the liquid refrigerant is sufficiently subcooled as it leaves the condenser to provide for normal system pressure drops. The amount of subcooling necessary, however, depends on the individual system design.

On air-cooled and most water-cooled applications, the temperature of the liquid refrigerant is normally higher than the surrounding ambient temperature, so no heat is transferred into the liquid, and the only concern is the pressure drop in the liquid line. Besides the friction loss caused by the refrigerant flow through the piping, a pressure drop equivalent to the liquid head is involved in forcing liquid refrigerant to flow up a vertical riser. A vertical head of 2 ft. of liquid refrigerant is approximately equivalent to 1 psi. For example, if a condenser or a receiver in the basement of a building is to supply liquid refrigerant to an evaporator three floors above, or approximately 30 ft., a pressure drop of approximately 15 psi for the liquid head alone must be provided for in the design of the system.

On evaporative or water-cooled condensers where the condensing temperature is below the ambient temperature, or on any application where the liquid lines must pass through hot areas such as boiler room or furnace rooms, an additional complication may arise because of heat transfer into the liquid. Any subcooling in the condenser may be lost in the receiver or liquid line due to temperature rise alone unless the system is properly designed. On evaporative condensers when a receiver and subcooling coils are used, it is recommended that the refrigerant flow be piped from the condenser to the receiver and then to the subcooling coil. In critical applications, it may be necessary to insulate both the receiver and the liquid line.

On the typical air-cooled condensing unit with a conventional receiver, it is probable that very little subcooling of the liquid is possible unless the receiver is almost completely filled with liquid refrigerant. Vapor in the receiver in contact with the subcooled liquid will condense, and this effect will tend toward a saturated condition.

At normal condensing temperatures, a relation between each 1°F of subcooling and the corresponding change in saturation pressure applies (Table 9-2).

Table 9-2. Relationship of Refrigerant Subcooling and Saturation Pressures.

Refrigerant Type	Subcooling	Equivalent Chan ge in Saturation Pressure
R-12	1°F (0.56°C)	1.75 psi (12.06 KN/m^2
R-22	1°F (0.56°C)	2.75 psi (18.96 KN/m^2
R-502	1°F (0.56°C)	2.85 psi (19.65 KN/m^2

To illustrate, 5°F subcooling will allow a pressure drop with R-22 of 13.75 psi and R-502 of 14.25 psi without flashing in the liquid line. For the previous example of a condensing unit in a basement requiring a vertical lift of 30 ft., or approximately 15 psi, the necessary subcooling for the liquid head alone would be 5.5°F for R-22 and 5.25°F for R-502.

The necessary subcooling may be provided by the condenser used, but for systems with abnormally high vertical risers a suction-to-liquid heat exchanger may be required. Where long refrigerant lines are involved, and the temperature of the suction gas at the condensing unit is approaching room temperatures, a heat exchanger located near the condenser may not have sufficient temperature differential to cool the liquid adequately. Individual heat exchangers at each evaporator may be necessary to gain the desired operation.

In extreme cases, where a great deal of subcooling is required, there are several alternatives; a special heat exchanger with a separate subcooling expansion valve can provide maximum cooling with no penalty on system performance. It is also possible to reduce the capacity of the condenser so that a higher operating condensing temperature will make greater subcooling possible. Liquid refrigerant pumps also may be used to overcome large pressure drops.

Liquid line pressure drop causes no direct penalty in electrical power consumption, and the decrease in system capacity due to friction losses in the liquid line is negligible. Because of this, the only real restriction on the amount of liquid line pressure drop is the amount of subcooling available. Most references on pipe sizing recommend a conservative approach with friction pressure drops in the range of 3 to 5 psi. Where adequate subcooling is available, however, many applications have successfully used much higher design pressure drops. The total friction includes line losses through such accessories as solenoid valves, filter-driers, and hand valves.

To minimize the refrigerant charge, liquid lines should be kept as small as practical, and excessively low pressure drops should be avoided. On most systems, a reasonable design criterion is to size liquid lines on the basis of a pressure drop equivalent to 2°F subcooling.

A limitation on liquid line velocity is possible damage to the piping from pressure surges or liquid hammer caused by the rapid closing of a liquid line solenoid valve, and velocities above 300 fpm should be avoided when they are used. If liquid line solenoid valves are not used, higher refrigerant velocities can be used. Figure 9-7 gives liquid line velocities corresponding to various pressure drops and line sizes.

Figure 9-7. Recommended liquid line velocities for various pressure drops using R-12, R-22, and R-502.

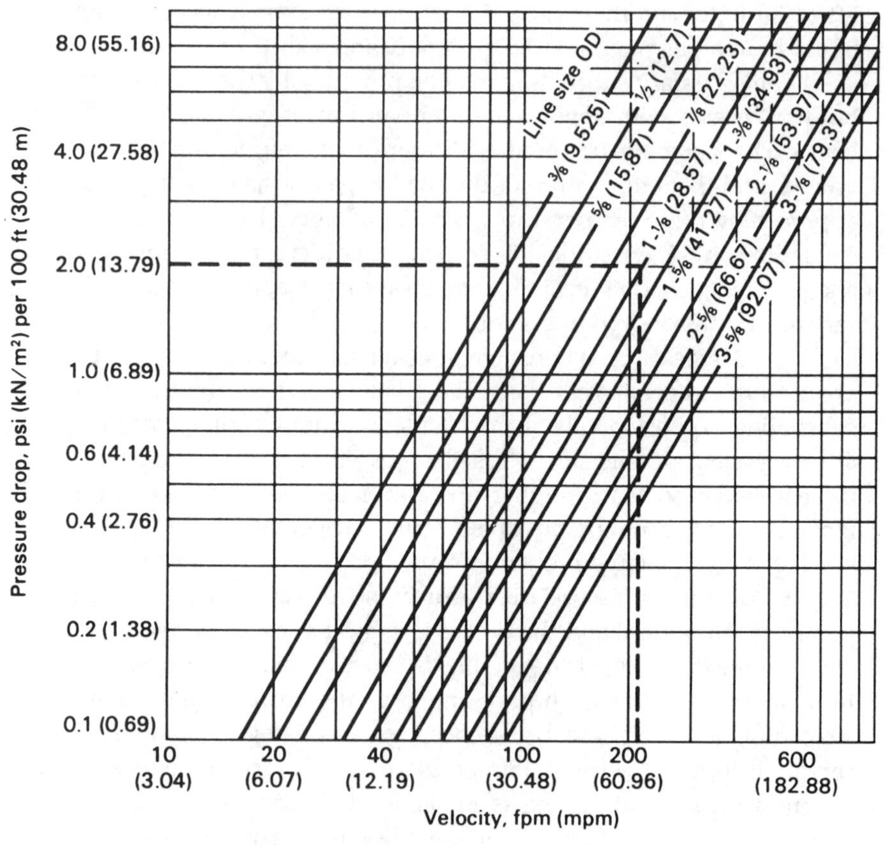

Example:

A pressure drop of 2 psi (13.79 kN/m^2) per 100 feet (30.48 m) in a 1-⅛" (28.57 mm) OD liquid line indicates a velocity of approximately 210 fpm (64 mpm).

SIZING SUCTION LINES

Suction line sizing is more important than that of the other lines from a design and system standpoint. Any pressure drop occurring due to a frictional resistance to flow results in a decrease in the refrigerant pressures at the compressor suction valve, compared with the pressure at the evaporator outlet. As the suction pressure is decreased, each pound of refrigerant that returns to the compressor occupies a greater volume, and the weight of the refrigerant being pumped by the compressor operating at a –40°F evaporating temperature will lose almost 6% of its rated capacity for each psi of suction line pressure drop.

The normally accepted design practice is to use a design criterion of a suction line pressure drop equal to a 2°F change in saturation temperature. Equivalent pressure drops for various operating conditions are developed and placed in tables (Table 9-3).

Table 9-3. Refrigerant pressure drop equivalents for a 2°F (1.11°C) change in the saturation temperature at various evaporating temperatures.

Evaporating Temperature F (C)	Pressure Drop, Psi (KN/m²)		
	R-12	R-22	R-502
45 (7.2)	2.0 (13.79)	3.0 (20.68)	3.3 (22.75)
20 (–6.7)	1.35 (9.31)	2.2 (15.17)	2.4 (16.55)
0 (–17.8)	1.0 (6.89)	1.65 (11.38)	1.85 (12.76)
–20 (–28.9)	0.75 (5.17)	1.15 (7.93)	1.35 (9.31)
–40 (–40)	0.5 (3.45)	0.8 (5.51)	1.0 (6.89)

The maintenance of adequate velocities to return the lubricating oil to the compressor properly is also of great importance when sizing suction lines. Studies have shown that oil is most viscous in a system after the suction gas has warmed up a few degrees higher than the evaporating temperature so that the oil is no longer saturated with refrigerant. This condition occurs in the suction line after the refrigerant vapor has left the evaporator. The movement of oil through the suction line is dependent on both the mass and the velocity of the suction vapor. As the mass or density decreases, higher refrigerant velocities are required to force the oil along.

Nominal minimum velocities of 700 fpm in horizontal lines and 1500 fpm in vertical lines have been recommended and used successfully for many years as suction line sizing design standards. The use of one nominal refrigerant velocity provided a simple and convenient means of checking the velocities. However, tests have shown that in vertical risers the oil tends to crawl up the inner surface of the tubing, and the larger the tubing, the greater the velocity required in the center of the tubing to maintain tube surface velocities that will carry the oil. The exact velocity required in vertical suction lines is dependent on both the evaporating temperature and the size of the line. Under varying conditions, the specific velocity required might be either greater or less than 1500 fpm.

For better accuracy in line sizing, revised maximum recommended vertical suction line sizes based on the minimum gas velocities have been calculated and are plotted in chart form for easy use (Figure 9-8).

These revised recommendations supersede previous vertical suction riser recommendations. No change has been made in the 700 fpm minimum velocity recommendation for horizontal lines (Figure 9-9).

To illustrate, assume again that a system operating with R-22 at a 40°F evaporating temperature has a capacity of 100,000 Btuh. In Figure 9-8 the intersection of the evaporating temperature and capacity lines indicates that a 2-1/8 in.-OD line will be required for proper oil return in the vertical risers.

Even though the system might have a much greater design capacity, the suction line sizing must be based on the minimum capacity anticipated in operation under light load conditions after allowing for the maximum reduction in capacity from the capacity control, if used.

Since the dual goals of low pressure drop and high velocities are in direct conflict with each other, obviously some compromises must be made in both areas. As a general approach, in suction line design, velocities must be kept as high as possible by sizing lines on the basis of the maximum pressure drop that can be tolerated. In no case, however, should the gas velocity be allowed to fall below the minimum levels necessary to return the oil to the compressor. It is recommended that a tentative selection of suction line sizes be made on the basis of a total pressure drop equivalent to a 2°F change in the saturated evaporating temperature. The final consideration always must be to maintain velocities adequate to return the lubricating oil to the compressor, even if this results in a higher pressure drop than is normally desirable.

DOUBLE RISERS

On systems equipped with capacity control compressors, or where tandem or multiple compressors are used with one or more compressors cycled off for capacity control, single-suction line risers may result in either unacceptably high or low gas velocities. A line that is sized prop-

Figure 9-8. Recommended maximum suction line sizes to provide proper oil return using vertical risers and R-22 and R-502 refrigerants.

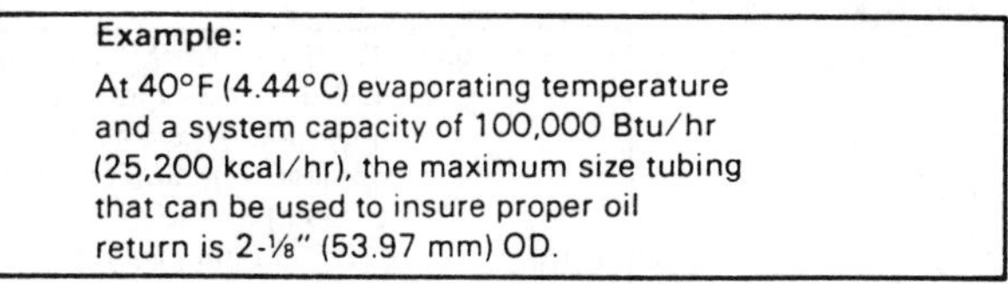

Example:

At 40°F (4.44°C) evaporating temperature and a system capacity of 100,000 Btu/hr (25,200 kcal/hr), the maximum size tubing that can be used to insure proper oil return is 2-⅛" (53.97 mm) OD.

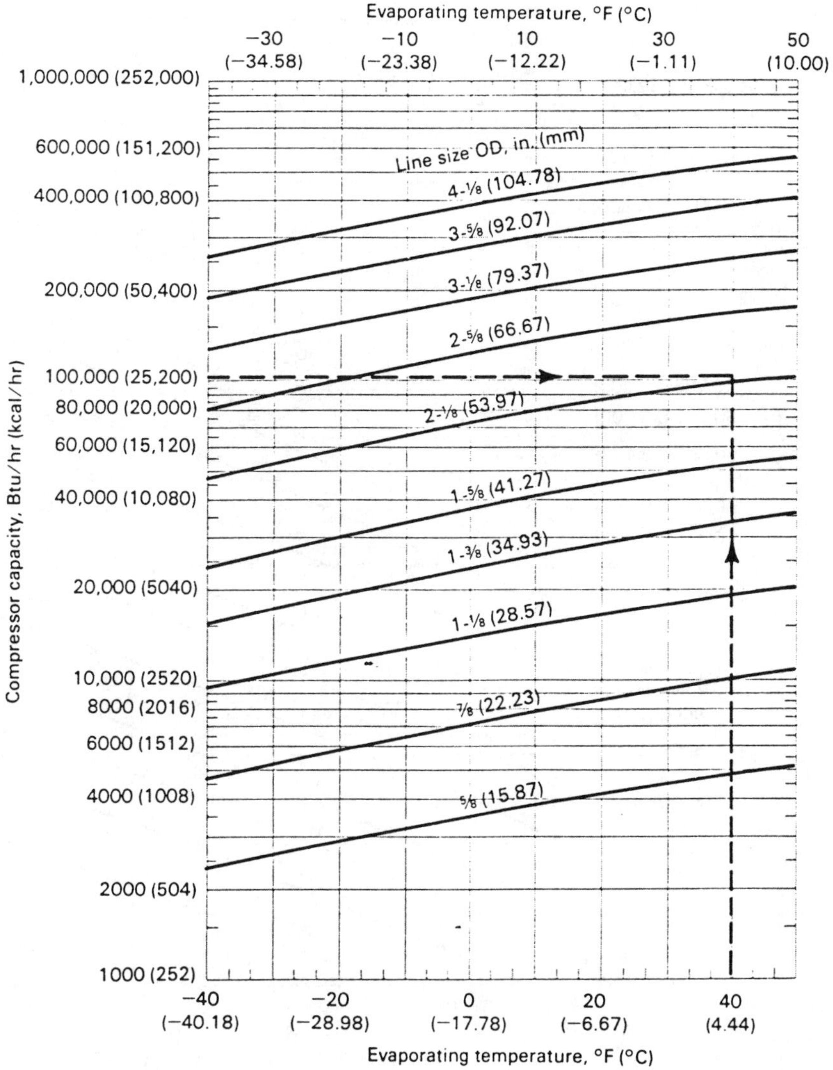

Evaporating temperature, °F (°C)

−30 (−34.58) −10 (−23.38) 10 (−12.22) 30 (−1.11) 50 (10.00)

Compressor capacity, Btu/hr (kcal/hr)

1,000,000 (252,000)
600,000 (151,200)
400,000 (100,800)
200,000 (50,400)
100,000 (25,200)
80,000 (20,000)
60,000 (15,120)
40,000 (10,080)
20,000 (5040)
10,000 (2520)
8000 (2016)
6000 (1512)
4000 (1008)
2000 (504)
1000 (252)

Line size OD, in. (mm)
4-⅛ (104.78)
3-⅝ (92.07)
3-⅛ (79.37)
2-⅝ (66.67)
2-⅛ (53.97)
1-⅝ (41.27)
1-⅜ (34.93)
1-⅛ (28.57)
⅞ (22.23)
⅝ (15.87)

−40 (−40.18) −20 (−28.98) 0 (−17.78) 20 (−6.67) 40 (4.44)

Evaporating temperature, °F (°C)

Figure 9-9. Recommended maximum horizontal suction line sizes to provide proper oil return using R-12 refrigerant.

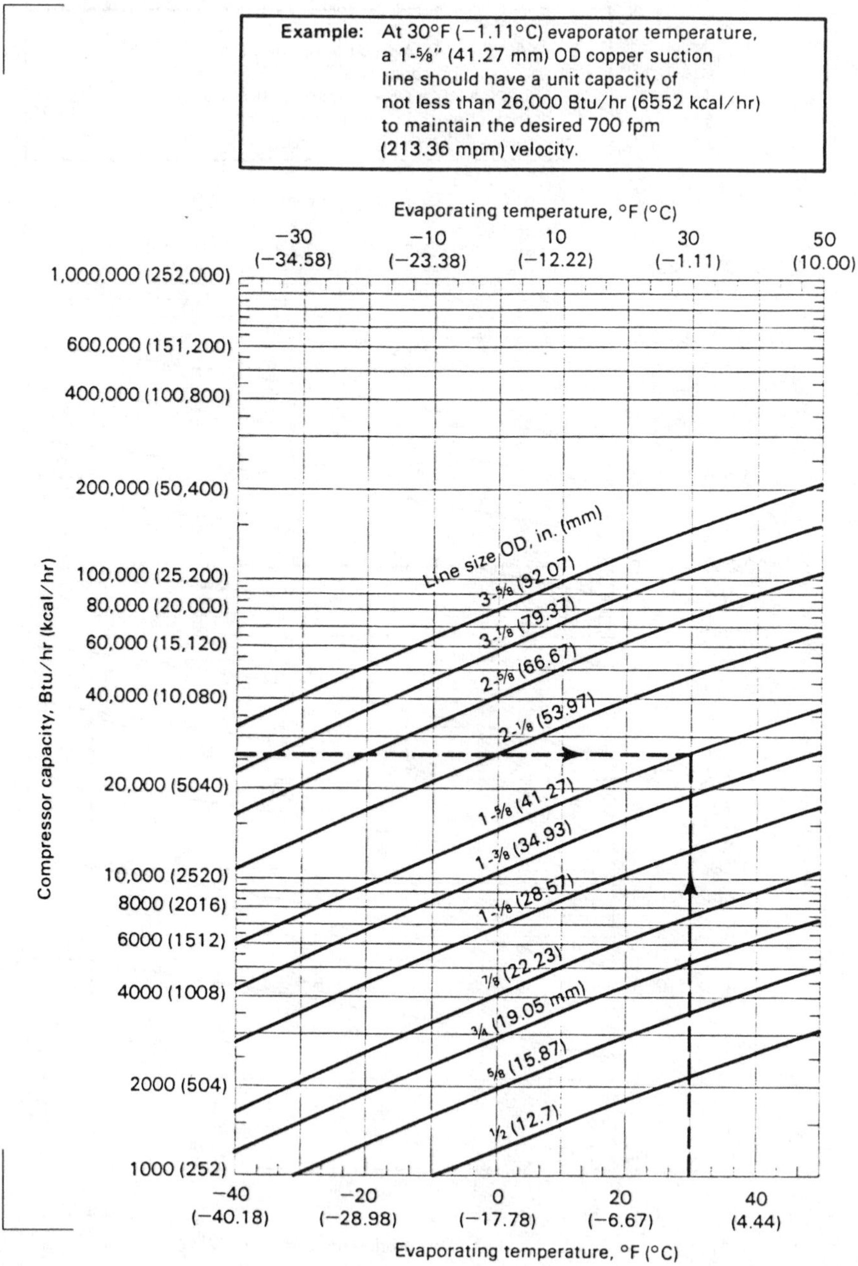

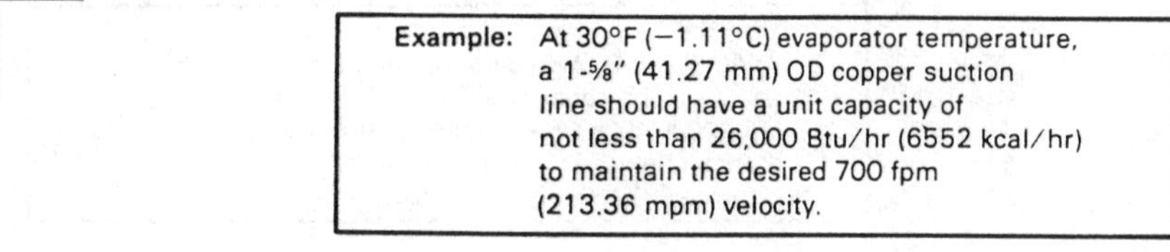

Example: At 30°F (−1.11°C) evaporator temperature, a 1-⅝" (41.27 mm) OD copper suction line should have a unit capacity of not less than 26,000 Btu/hr (6552 kcal/hr) to maintain the desired 700 fpm (213.36 mpm) velocity.

erly for light load conditions may have too high a pressure drop at maximum load. If the line is sized on the basis of full load conditions, the velocities may not be adequate at light load conditions to move the oil through the tubing. On cooling applications where somewhat higher pressure drops at maximum load conditions can be tolerated without any major penalty in overall system performance, it is usually preferable to accept the additional pressure drop imposed by a single vertical riser.

The two lines of a double riser should be sized so that the total cross-sectional area is equivalent to the cross-sectional area of a single riser that would have both satisfactory gas velocity and an acceptable pressure drop at maximum load conditions (Figure 9-10).

The two lines normally are of different sizes, with the larger line being trapped as shown. The smaller line must be sized to provide adequate velocities and acceptable pressure drop when the entire minimum load is carried in the smaller riser.

In operation, at maximum load conditions, gas and entrained oil will be flowing through both risers. At minimum load conditions, the gas velocity will not be high enough to carry the oil up both risers. The entrained oil will drop out of the refrigerant gas flow, and accumulate in the P trap, forming a liquid seal. This will force all the flow up the smaller

Figure 9-10. Refrigerant double riser suction line.

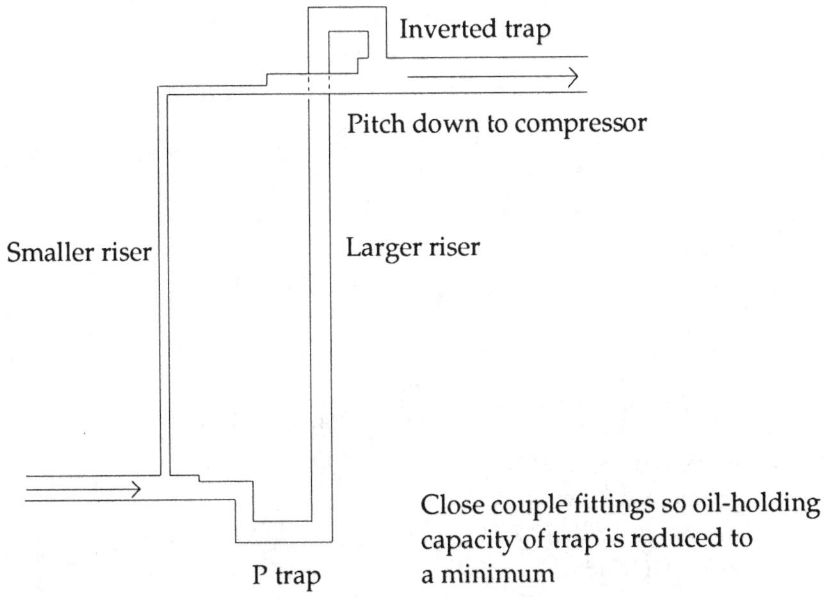

riser, thereby raising the velocity and assuring oil circulation through the system.

PIPING DESIGN FOR HORIZONTAL AND VERTICAL LINES

Horizontal suction and discharge lines should be pitched downward in the direction of flow to aid in oil drainage, with a downward pitch of at least 1/2 in. in 10 ft. Refrigerant lines should always be as short as possible and should run as directly as possible.

Piping should be located so that access to system components is not hindered, and so that any components that could possibly require future maintenance are easily accessible. If piping must be run through boiler rooms or other areas where they will be exposed to abnormally high temperatures, it may be necessary to insulate both the suction and liquid lines to prevent excessive heat transfer into the lines.

Every vertical riser greater than 3 to 4 ft. in height should have a P trap at the base to facilitate oil return up the riser (Figure 9-11).

Figure 9-11. Refrigerant suction line riser.

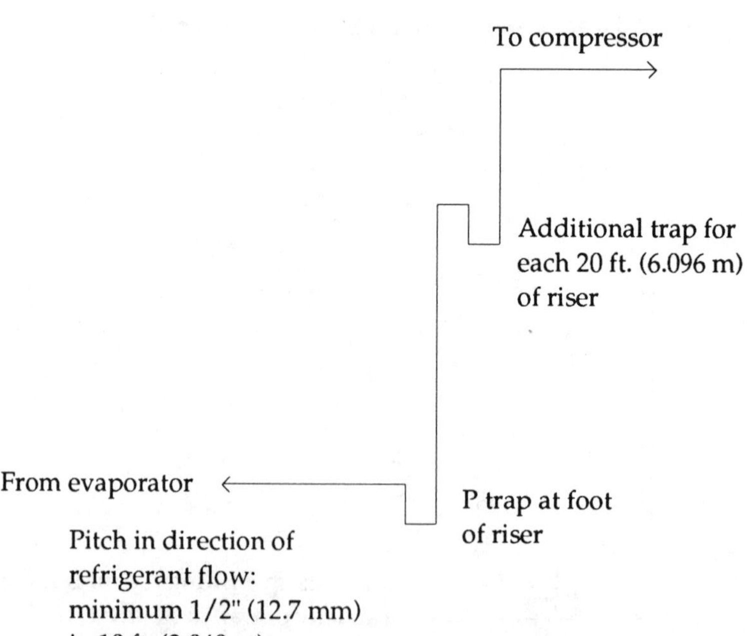

To avoid the accumulation of large quantities of oil, the trap should be of minimum depth and the horizontal section should be as short as possible.

Prefabricated wrought copper traps are available, or a trap can be made by using two street ells and one regular ell. Traps at the foot of hot gas risers are normally not required because of the easier movement of oil at higher temperatures. However, it is recommended that the discharge line from the compressor be looped to the floor prior to being run vertically upward to prevent the drainage of oil back to the compressor head during shutdown periods (Figure 9-12).

For long vertical risers in both suction and discharge lines, additional traps are recommended for each full length of pipe (approximately 20 ft.) to ensure proper oil movement.

In general, trapped sections of the suction line should be avoided except where necessary for oil return. Oil or liquid refrigerant accumulating in the suction line during the off cycle can return to the compressor at high velocity as a liquid slug on startup. This slug can break the compressor valves or cause other damage.

Figure 9-12. Compressor discharge line riser.

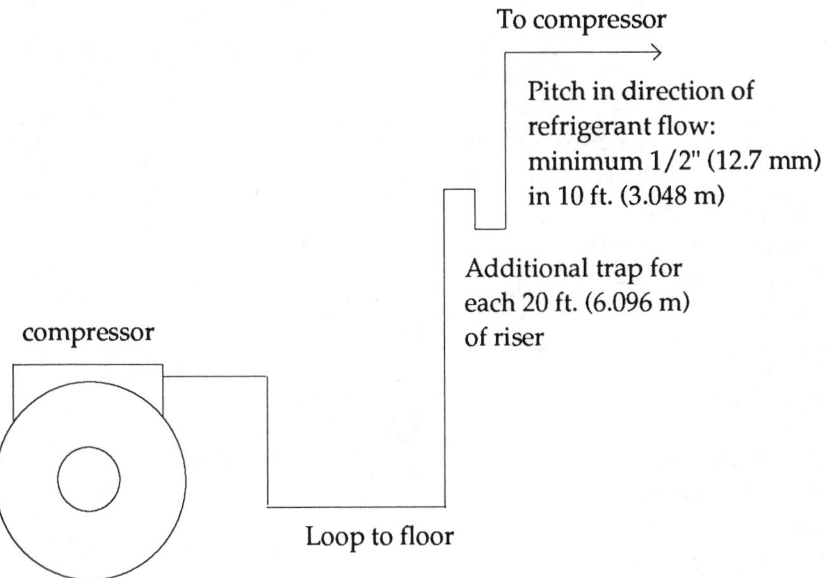

SUCTION LINE PIPING DESIGN AT THE EVAPORATOR

If a pumpdown control system is not used, each evaporator must be trapped to prevent liquid refrigerant from draining back to the compressor by gravity during the off cycle. Where multiple evaporators are connected to a common suction line, the connections to the common suction line must be made with inverted traps to prevent the drainage from one evaporator from affecting the expansion valve bulb control of another evaporator.

Where a suction riser is taken directly upward from an evaporator, a short horizontal section of tubing and a trap should be provided ahead of the riser so that a suitable mounting for the thermal expansion valve bulb is available. The trap serves as a drain area and helps to prevent the accumulation of liquid under the bulb, which could cause erratic expansion valve operation. If the suction line leaving the evaporator is free-draining, or if a reasonable length of horizontal piping precedes the vertical riser, no trap is required unless necessary for oil return (Figure 9-13).

Figure 9-13. Typical multiple evaporators suction line piping schematic.

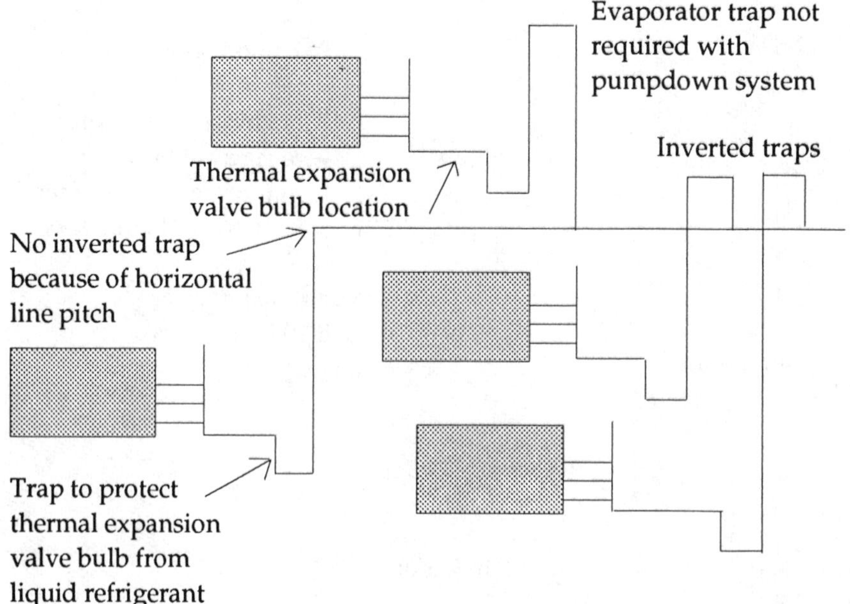

SUMMARY

A good refrigerant piping system will have a maximum capacity, be economical, provide proper oil return, provide minimum power consumption, require a minimum amount of refrigerant charge, have a low noise level, provide proper refrigerant control, and allow perfect flexibility in system performance from 0 to 100% of unit capacity without lubrication problems.

Pressure drop, in general, tends to decrease system capacity and increase the amount of electric power required by the compressor.

It is often required that refrigerant velocity, rather than pressure drop, be the determining factor in system design.

The critical nature of oil return can produce many system difficulties.

An overcharge of refrigerant can result in serious problems of liquid refrigerant control, and the flywheel effect of large quantities of liquid refrigerant in the low-pressure side of the system can result in erratic operation of the refrigerant flow control device.

The size of the refrigerant line connection on a service valve supplied with a compressor, or the size of the connection on an evaporator, condenser, or some other system accessory, does not determine the correct size of the refrigerant line to be used.

Oil and refrigerant vapor do not mix readily, and the oil can be properly circulated through the system only if the velocity of the refrigerant vapor is great enough to carry the oil along with it.

To assure proper oil circulation, adequate refrigerant velocities must be maintained not only in the suction and discharge lines, but in the evaporator circuits as well.

Oil logging in the evaporator can be minimized, even at extremely low evaporating temperatures, with adequate refrigerant velocities and properly designed evaporators.

Normally, oil separators are necessary for operation at evaporating temperatures below –50°F in order to minimize the amount of oil in circulation.

Each valve, fitting, and bend in a refrigerant line contributes to the friction pressure drop because of its interruption or restriction of smooth flow.

For accurate calculations of pressure drop, the equivalent length for each fitting should be calculated.

Pressure drop in discharge lines is probably less critical than in any other part of the system.

As a general guide, for discharge line pressure drops up to 5 psi, the effect on the system performance should be so small that it would be difficult to measure.

Actually, a reasonable pressure drop in the discharge line is often desirable to dampen compressor discharge pulsations, and thereby reduce noise and vibration.

Oil circulation in discharge lines is normally a problem only on systems where large variations in system capacity are encountered.

The primary concern in liquid line sizing is to ensure a solid liquid column of refrigerant at the expansion valve.

For proper system performance, it is essential that liquid refrigerant reaching the flow control device be subcooled slightly below its saturation temperature.

Liquid line pressure drop causes no direct penalty in electrical power consumption, and the decrease in system capacity due to friction losses in the liquid line is negligible. Because of this, the only real restriction on the amount of liquid line pressure drop is the amount of subcooling available.

Suction line sizing is more important than that of the other lines from a design and system standpoint. Any pressure drop occurring due to frictional resistance to flow results in a decrease in the refrigerant pressure at the compressor suction valve, compared to the pressure at the evaporator outlet.

The maintenance of adequate velocities to return the lubricating oil to the compressor properly is also of great importance when sizing suction lines.

Nominal minimum velocities of 700 fpm in horizontal suction lines and 1500 fpm in vertical suction lines have been recommended and used for many years as suction line sizing design standards.

As a general approach, in suction line design, velocities should be kept as high as possible by sizing lines on the basis of the maximum pressure drop that can be tolerated. In no case, however, should the vapor velocity be allowed to fall below the minimum levels necessary to return oil to the compressor.

The two lines of a double riser should be sized so that the total cross-sectional area is equivalent to the cross-sectional area of a single riser that would have both satisfactory vapor velocity and an acceptable pressure drop at maximum load conditions.

Horizontal suction lines should be pitched downward in the direction of flow to aid in oil drainage, with a downward pitch of at least 1/2 in. in 10 ft.

Piping should be located so that access to system components is not hindered, and so that any components that could possibly require future maintenance are easily accessible.

Every vertical riser greater than 3 to 4 ft. in height should have a P trap at the base to facilitate oil return up the riser.

In general, trapped sections of the suction line should be avoided except where necessary for oil return.

Where multiple evaporators are connected to a common suction line, the connections to the common suction line must be made with inverted traps to prevent drainage from one evaporator from affecting the expansion valve bulb control of another evaporator.

Chapter 10

Duct Systems and Design

Air distribution—what does it mean to personal comfort? Does it mean a gale of air being blown into a space so strongly that the occupants feel like they are in a whirlwind? Does it mean that there is so little air motion in the space that the occupants feel sleepy and drowsy or stuffy and humid? Actually, it should be neither of these, but some happy medium between them.

INTRODUCTION

Local climatic conditions are a very important factor in the proper selection of an air distribution system. There are certain performance characteristics that are needed in certain areas of extreme climatic conditions which are required to maintain indoor comfort on a year-round basis. Areas in cold climates require warm floors in the winter. Hot summer areas require comfort cooling. In areas where both extremes occur, it may be required to have both warm floors and comfort cooling. An air distribution system that performs satisfactorily in one area may not perform properly in another area. A perimeter-type floor warming duct system might be recommended in the northern section of the country. At the same time an overhead distribution system without floor warming might be recommended for the same type of building in the South. While the winter and summer inside and outside design conditions determine the capacity requirements of the heating and cooling equipment, it is the overall season conditions that determine the type of air distribution system used.

To aid us in determining what type of air distribution system to select, climatic zones, defined in terms of degree days, have been established (Figure 10-1).

The U.S. continent has been divided into three areas:

Figure 10-1. Degree day zone map. (Courtesy of The Trane Company)

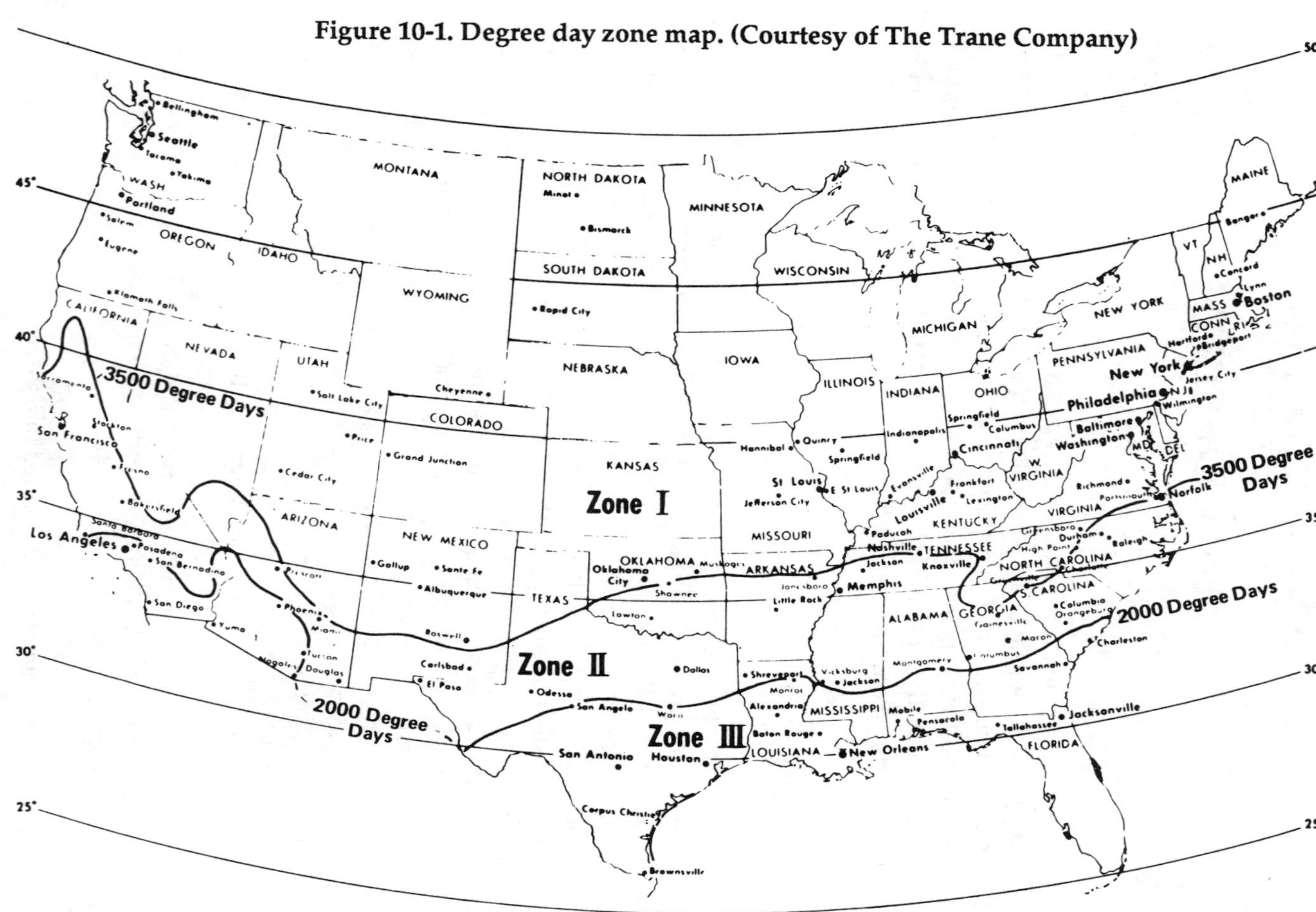

Zone I: More than 3500 degree days

Zone II: 2000 to 3500 degree days

Zone III: Fewer than 2000 degree days

Note: The degree day is the unit that represents one degree of difference between the inside temperature and the average outdoor temperature for one day. It is based on the temperature differential below 65°F. For example, on a day when the average outdoor temperature is 45°F, the degree days will be 65 – 45 = 20. Thus we should have an equivalent of 20 degree days for this one day.

SELECTING AIR DISTRIBUTION SYSTEMS

No single type of air distribution system is best suited for all applications. At times more than one type of system will be used in a single building. Air distribution systems are made up of ducts, supply air outlets, and return air intakes; therefore, they can be defined in terms of the location of these components.

The location of the supply ducts, supply air outlets, and return air intakes to maintain proper air circulation and uniform air temperatures should be based on certain factors, such as:

1. The type of residence
   - A. Slab floor on the ground
   - B. Crawl space beneath the floor
   - C. Split level
   - D. Multilevel
   - E. Apartment

2. Are warm floors necessary?

3. Will large heat storage in the floors be an advantage or disadvantage?

4. Will zoning be necessary to maintain comfort from top to bottom in split-level buildings?

5. Is some heat in the crawl space desirable?

6. Is it necessary to heat the full basement?

TYPES OF AIR DISTRIBUTION SYSTEMS

The perimeter system is the most popular and most widely used air distribution system for residential comfort conditioning systems. This system may be defined as a system is which the forced air is conveyed underneath the floor to warm the floor surface, during the heating cycle, and is introduced into the conditioned space near the floor. The air is delivered in an upward direction so as to blanket the outside walls and windows. Indoor comfort is provided with the system by counteracting heat losses or heat gains in the major heat transfer areas.

Perimeter Systems

There are several types of perimeter air distribution systems. The loop system is used to heat slab floors because it provides uniform floor temperatures and satisfactory floor-to-ceiling gradients and a satisfactory room-to-room temperature balance (Figure 10-2).

The radial system is popular in both slab floor and crawl space buildings (Figure 10-3).

This system also provides satisfactory temperature gradients, but with greater variation in floor temperature.

The trunk and branch system is used in buildings with basements (Figure 10-4). It also provides satisfactory temperature gradients.

Ceiling Diffuser Systems

In this method of conditioning a building, the forced air is directed both horizontal and parallel to the ceiling (Figure 10-5).

Figure 10-2. Perimeter loop system. (Courtesy of The Trane Company)

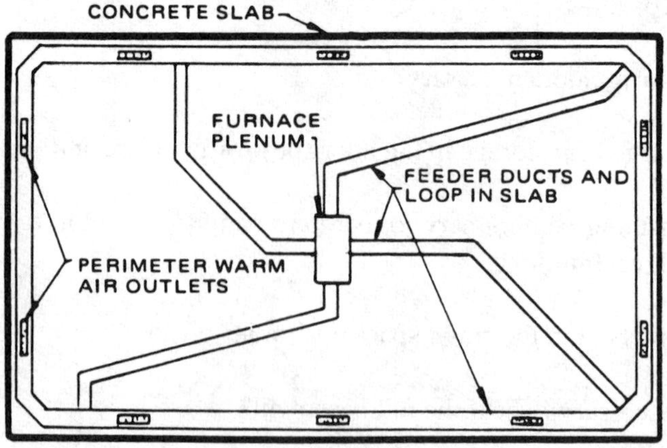

Figure 10-3. Perimeter radial system. (Courtesy of The Trane Company)

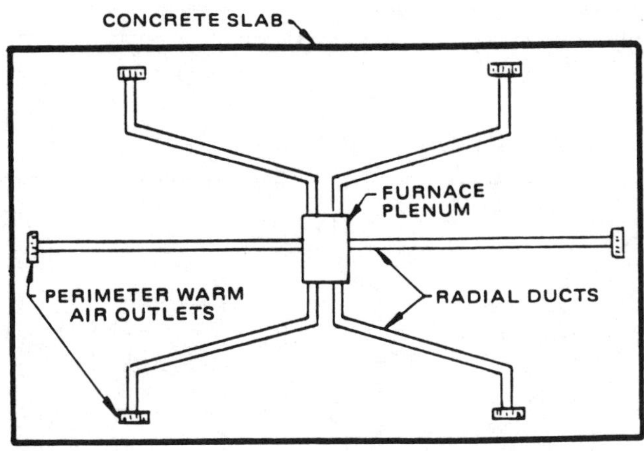

During the cooling cycle, air will drift down into the conditioned space. During the heating cycle the air has a tendency to stratify on the ceiling unless it is forced downward into the room. Occupant comfort is not satisfactory when the air does stratify at the ceiling. To aid in air circulation, the return air intakes should be located low on either the inside or outside wall.

High-Inside-Wall Furred Ceiling Systems

These types of systems are popular in mild climates (Figure 10-6).

Returns that are direct and close coupled at the floor level are preferred with these types of systems.

High-Inside-Wall Systems

In this method of comfort conditioning the forced air is delivered horizontally across the room toward the outside wall (Figure 10-7).

These systems work well for cooling but are less desirable for heating because of the drafts over the occupants. The return air intakes should be located low on the outside wall to aid in air circulation.

Overhead High-Inside-Wall Systems

These types of systems are popular in areas where the summers are hot (Figure 10-8).

Use close-coupled and direct floor-level returns with these systems, especially when the equipment is located within the building.

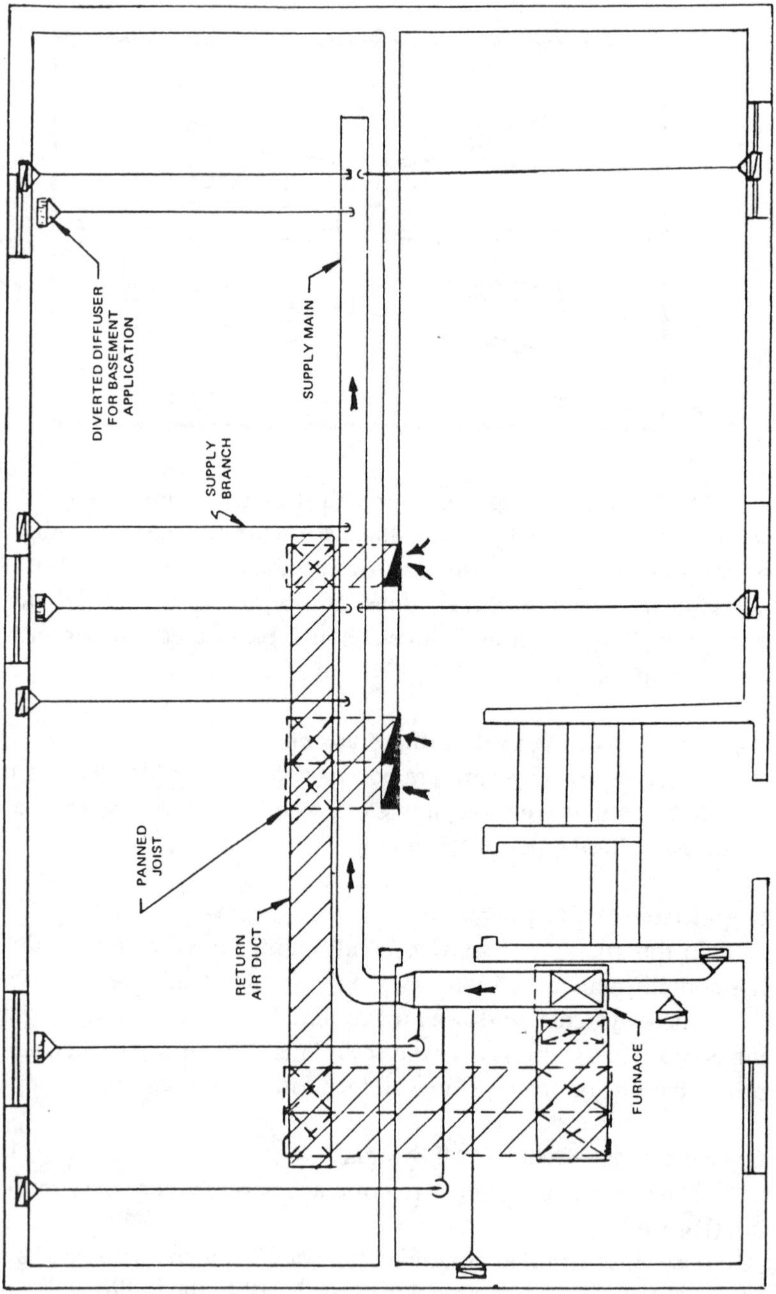

Figure 10-4. Perimeter trunk and branch system. (Courtesy of The Trane Company)

Figure 10-5. Ceiling diffuser system. (Courtesy of The Trane Company)

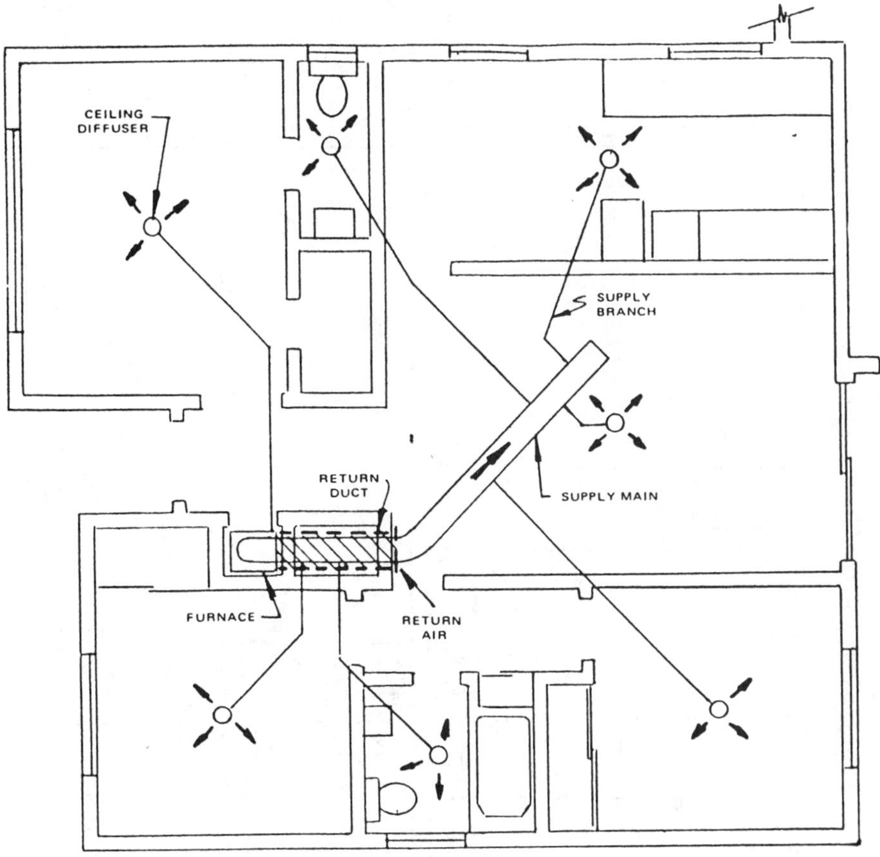

Parallel-Flow Systems

These systems may be defined as a method of comfort conditioning where the forced air is directed parallel to an outside wall. The supply air outlets are located high on the sidewall next to the outside wall. The air delivery is directed horizontally (Figure 10-9).

The return air intakes should be located on inside walls, but not directly opposite the supply outlet. The return air intakes preferably should be located in the baseboard but can be high in the sidewall. Low supply systems are generally not recommended because it is impossible to avoid drafts on the occupants.

Perimeter System Return Air Intakes

The return air intakes for perimeter residential air conditioning systems are usually located on the side walls. Even though the return air

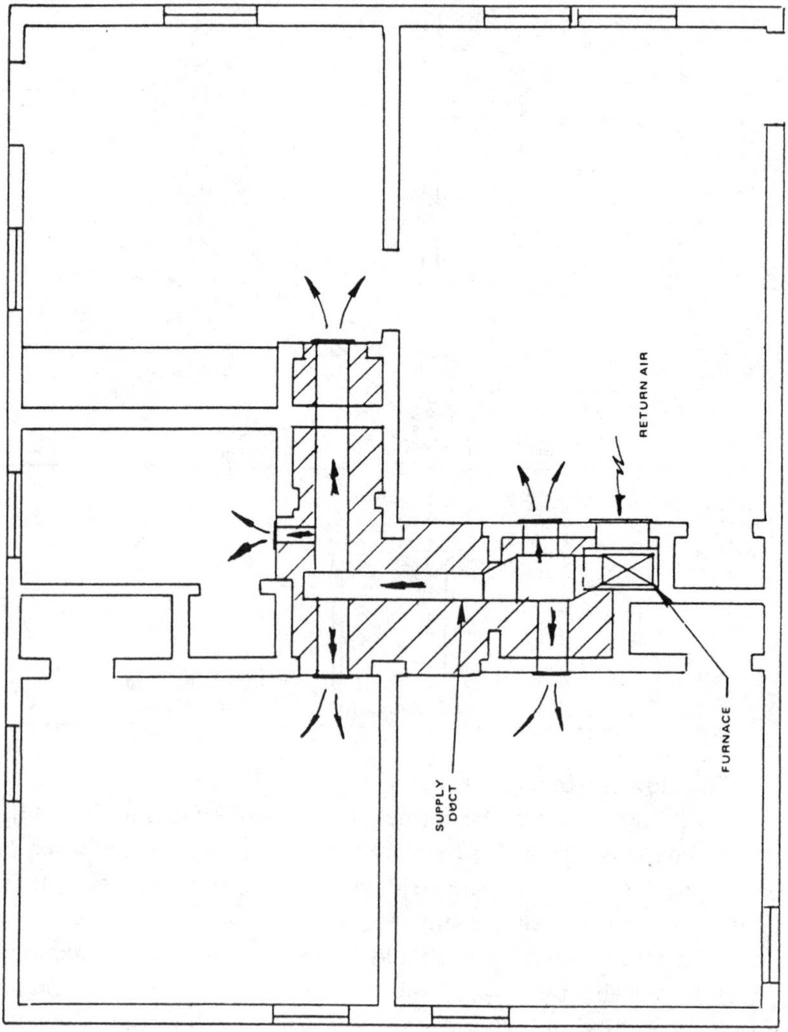

SUPPLY DUCT

RETURN AIR

FURNACE

Figure 10-6. High side wall furred ceiling system. (Courtesy of The Trane Company)

Figure 10-7. High inside wall system. (Courtesy of The Trane Company)

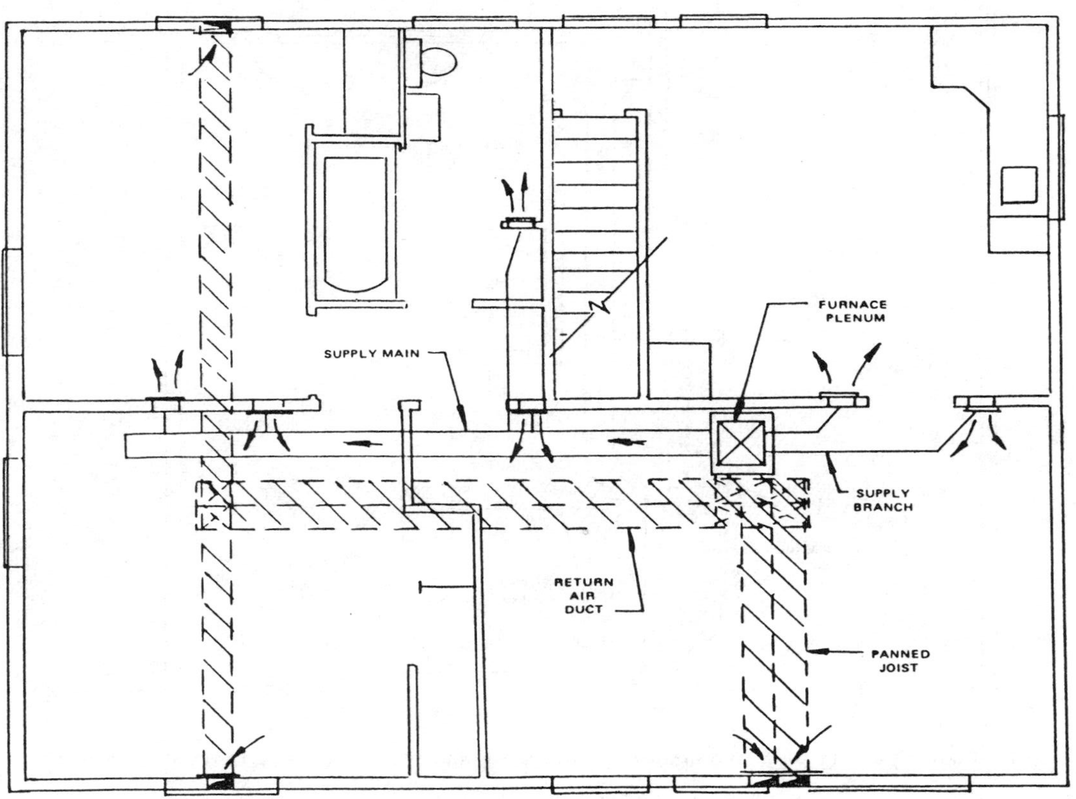

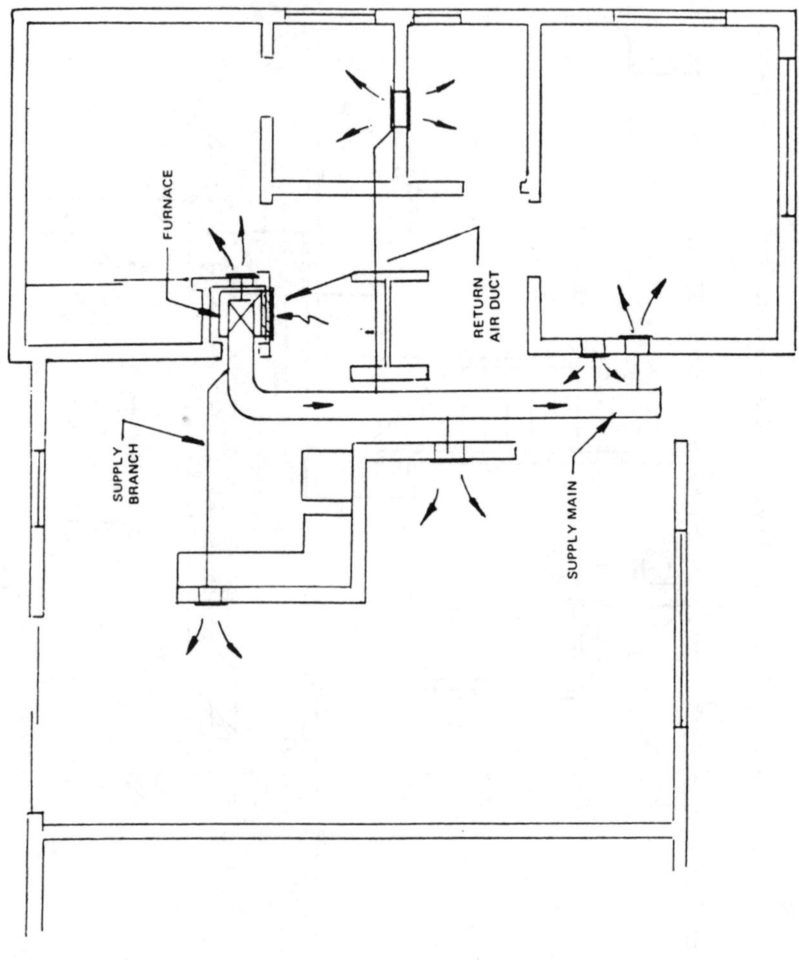

Figure 10-8. Overhead high side wall systems. (Courtesy of The Trane Company)

Figure 10-9. Parallel flow system. (Courtesy of The Trane Company)

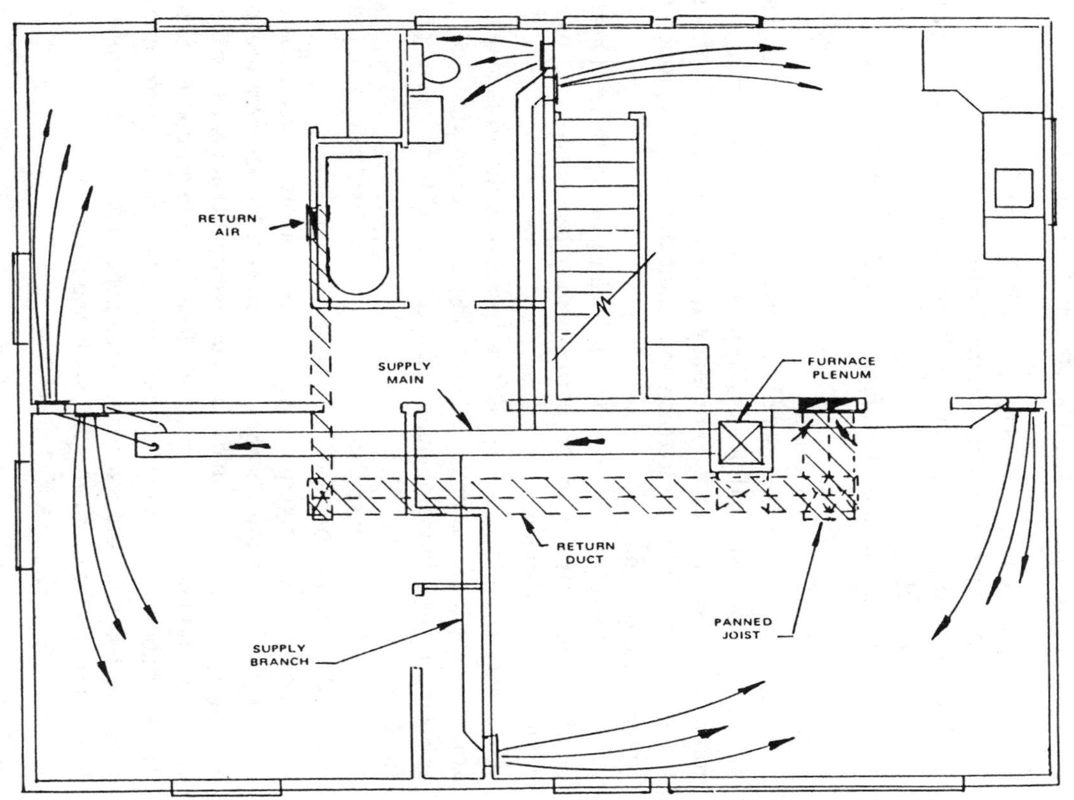

intakes are located at the baseboard level, the velocity of the air is so low that the air will not cause annoying drafts on the occupants. The return air intakes may be located in each room or provisions must be made to allow the air to pass from all rooms to the central return air intakes.

Listed below are some considerations for return air intakes:

1. Residential return air velocity should be around 500 fpm.
2. They should be located in a stratified zone.
3. They should be located on a wall. Avoid floor or ceiling locations when possible.
4. The grill location does not affect air movement in the room.

Special Zoning Considerations

Special considerations should be given to zone control to achieve satisfactory comfort conditions. Temperature variations from room to room should not exceed 3°F. In many types of construction this small variation may be difficult to reach and maintain without a properly controlled zone system.

In split level and tri-level homes, because of the different heat load characteristics in parts of the home, it may be necessary to use zone control for different levels.

Ranch homes that have large spacious rooms and areas may require zone control in some areas. Also, living areas over unheated spaces present problems in maintaining satisfactory conditions when zone control is not used.

Homes or buildings that are U-, L- or H-shaped cannot be properly balanced when only one thermostat is used.

Family activity rooms or living spaces that are located in basement areas are unique in their heating requirements. The temperature in such areas are not affected as rapidly by daily weather changes as are the areas above ground and they are greatly affected by the ground temperature. The areas above ground may require very little heat on a warm sunny day while the basement areas may require a good deal of heat because of the losses through the cold basement walls and floor areas.

TYPES OF STRUCTURES

As stated earlier, the air distribution system must be selected on the basis of its performance characteristics, the local climate conditions, and the type of structure. The type of structures and the recommended air distribution systems are listed on the following page.

SLAB FLOOR STRUCTURES

In Zone I (See Figure 10-1), practically all slab floor buildings must be provided with floor warming if the building is to be comfortable. A perimeter loop or radial system is recommended in this type of climate.

In Zone II, a perimeter system is recommended because of its floor warming characteristics. However, nonperimeter systems that do not provide floor warming are also used.

In Zone III, ceiling diffusers and high-side-wall supply air outlets are recommended.

CRAWL SPACE STRUCTURES

In Zone I, perimeter systems are recommended. Either radial or extended plenum systems can be used.

In Zone II, perimeter systems are also recommended. However, nonperimeter systems may be used when assisted with floor warming methods. Systems that use crawl space as a return air plenum are not acceptable.

In Zone III, all nonperimeter systems are recommended.

BASEMENT STRUCTURES

In Zone I, perimeter systems are recommended and the basement area must be heated to provide warm floors. Nonperimeter systems are acceptable when equipped with floor warming provisions.

In Zone II, perimeter systems are not recommended. Nonperimeter systems are acceptable without the use of floor warming provisions.

In Zone III, any of the air distribution systems described previously are recommended. The overhead ceiling or high-wall-supply openings are favored most.

SPLIT-LEVEL AND MULTILEVEL STRUCTURES

In Zone I, any slab floors at grade level should use a perimeter system in the floor. Crawl space areas must be equipped with floor warming provisions. Full perimeter systems are generally recommended in this zone.

In Zone II, perimeter systems are generally recommended. However, nonperimeter systems are acceptable.

In Zone III, nonperimeter systems are recommended.

APARTMENT STRUCTURES

In Zone I, perimeter systems are generally recommended. Overhead or high wall supply air outlets are acceptable for apartments on intermediate floors.

In Zone II, overhead or high-wall-supply outlets are acceptable, especially for intermediate- and top-floor apartments.

In Zone III, overhead or high-wall-supply outlets are recommended.

DUCT DESIGN

When a new air conditioning system is to be installed, it is customary for an engineer to make a layout of the job on paper. This layout, or arrangement, of the job is generally drawn on transparent tracing paper so that blueprints can be made from the drawing or tracing. Drawings may be made in detail where each fitting, duct, or other piece of the system is shown as it actually appears (Figure 10-10).

Notice that all of the fittings and ducts are drawn in considerable detail. The diagram illustrates a typical duct arrangement of an indirect heating and cooling system.

In order to save the time required in drawing and preparing tracings, a set of standard symbols has been adopted. By use of these symbols the drawings can be correctly interpreted (Table 10-1).

The plan of a heating installation in which the various pipes, fittings, and so on, are indicated by means of symbols instead of being drawn in full is shown in Figure 10-11.

To a person who is familiar with the meaning of the symbols, the drawing is entirely clear and a great deal of information is shown on it. A number of special items in the drawings are covered by special notes. For example, the figures around the outside edges of the drawing, followed by a square with the letter "S" drawn through it, indicate in each case the number of square feet of radiation heating surface of the radiator which the pipe serves. The letters "F" and "Q" indicate special fittings—a float vent and a quick vent, respectfully—which are not covered by standard symbols.

Figure 10-10. Supply air system.

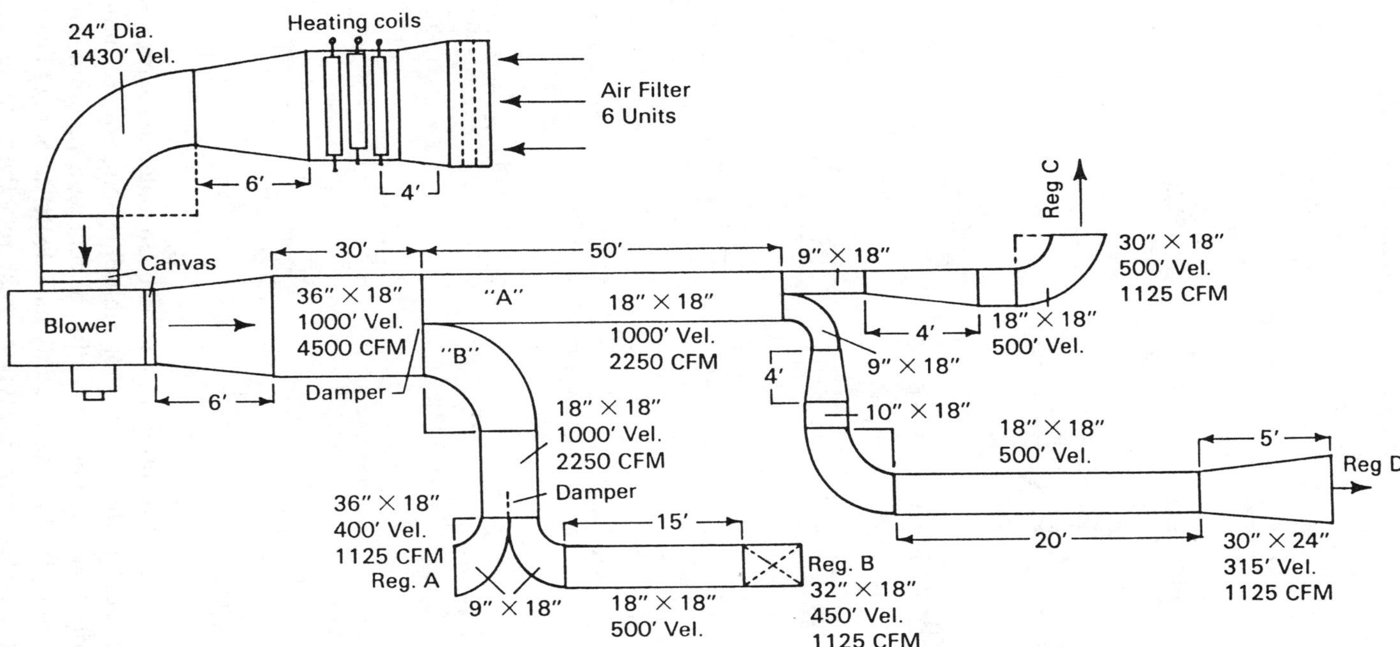

Table 10-1. Ductwork Symbols.

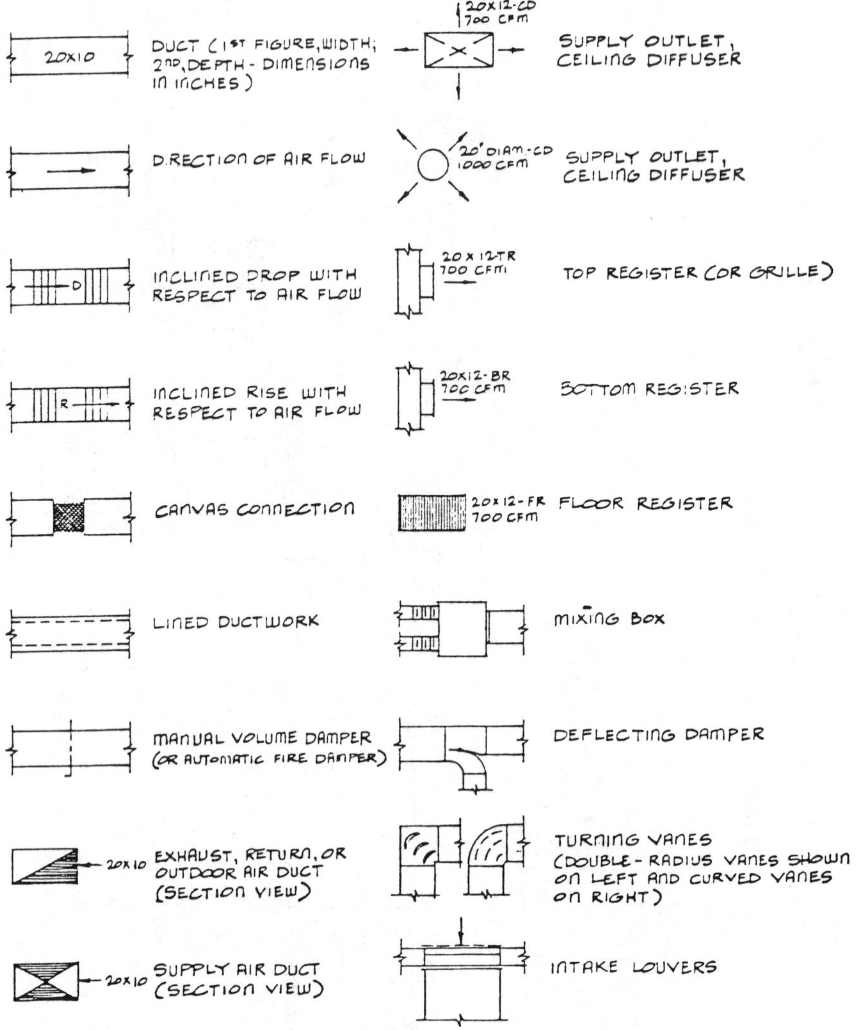

AIR SUPPLY

The total quantity of air to be circulated through any building is dependent on the necessity for controlling temperature, humidity, and air distribution when either heating or cooling is required. The factors that determine the total air quantity include the amount of heating or cooling to be done, as well as the type and nature of the building, locality, climate,

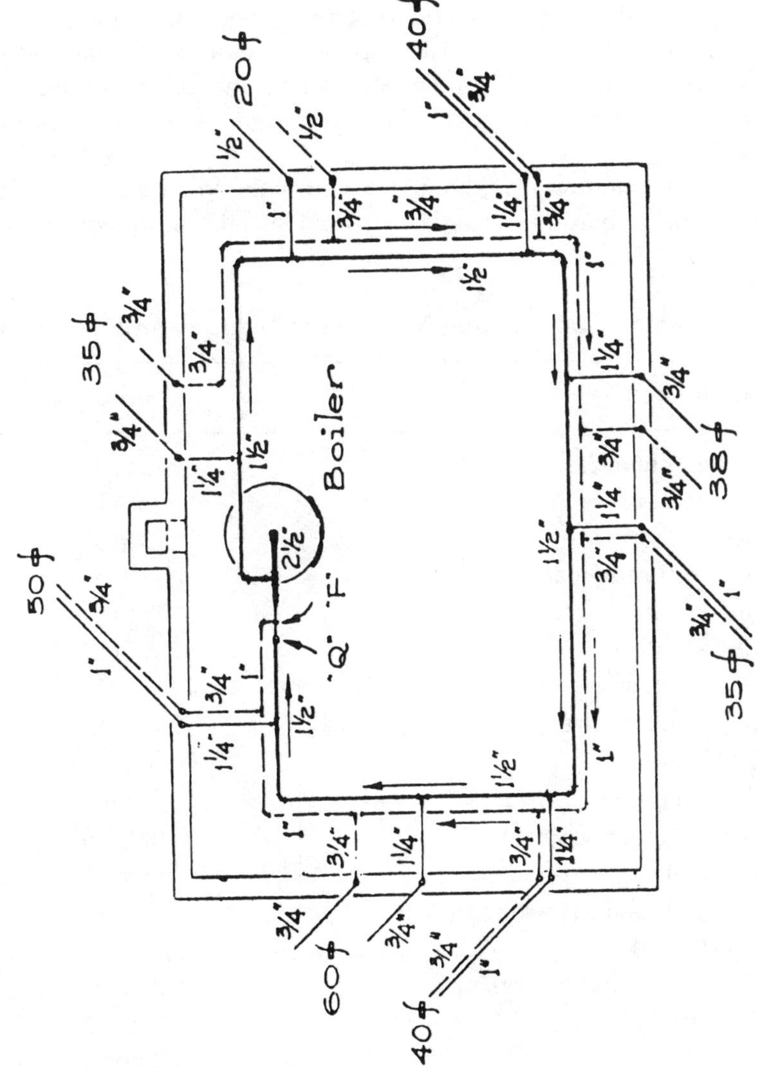

Figure 10-11. Piping diagram using symbols.

height of the room, floor area, window area, occupancy, and the method of air distribution.

The air supplied to the conditioned space must always be adequate to satisfy the ventilation requirements of the occupants. It must be without drafts and with a reasonable degree of uniformity. This total air quantity will ordinarily be composed of two parts: the outdoor fresh air supply and the recirculated air supply. The requirements as given by the American Society of Heating, Ventilating and Air-Conditioning Engineers call for 30 cfm to be circulated per person, at least two cfm of which is fresh air.

The minimum amount of fresh air required for ventilation for most common applications has been placed in table form for convenience (Table 10-2).

Table 10-2. Minimum Outdoor Air Required for Ventilation.

Application	CFM per person
Apartment or residence	10 to 15
Auditorium	5 to 7-1/2
Barber shop	10 to 15
Bank or beauty parlor	7-1/2 to 10
Broker's board room	25 to 40
Church	5 to 7-1/2
Cocktail lounge	20 to 30
Department store	5 to 7-1/2
Drugstore	7-1/2 to 10
Funeral parlor	7-1/2 to 10
General office space	10 to 15
Hospital rooms (private)	15 to 25
Hospital rooms (wards)	10 to 15
Hotel room	20 to 30
Night clubs and taverns	15 to 20
Private office	15 to 25
Restaurant	12 to 15
Retail shop	7-1/2 to 10
Theater (smoking permitted)	10 to 15
Theater (smoking not permitted)	5 to 7-1/2

As should be noted, the figures given in this table depend to a great extent on the degree of smoking to be expected within the conditioned space. The figures in this table should be used for estimating purposes where they do not conflict with existing local codes. Also, note that this table aims to provide minimum rather than adequate requirements.

DUCT CONSTRUCTION

Metal ducts are preferable because it is possible to obtain smooth surfaces, thereby avoiding excessive resistance to the airflow. Metal ducts can also be worked into compact sizes, shapes, and locations.

Metal ducts are usually made either of galvanized sheet metal or aluminum and the sheets are made in various weights or gauges. The diameter of the ducts or their width, if they are to be rectangular, determines the gauge of metal to be used (Table 10-3).

Table 10-3. Recommendations for Round Ducts.

Diameter (in.)	Gauge of Steel: U.S. Standard	Gauge of Aluminum: B&S
Up to 13	26	24
14 to 33-1/2	24	24
34 to 67-1/2	22	20

All metal duct work should be rigidly constructed and installed to eliminate possible vibration. All slip joints should be in the direction of airflow.

When designing a more extensive system, it is good practice to gradually lower the velocity both in the main duct and the remote branches. This scheme of design has the following advantages:

1. Enables the air to distribute in a uniform way.

2. Decreases the friction in the smaller ducts, where it otherwise would be the greatest.

3. When the velocity is lowered, some method should be provided for removing the dust and soot, particularly if the intake is close to the

street or ally. Air filters suitable for this purpose are used on the inlet side of the system. Where so used, additional resistance is added and must be taken into account.

Provisions might be made to heat or "temper" the cold incoming air. A blast heating coil may be used. This also adds to the system and must be considered in the calculations.

Dampers and deflectors should be placed at all points necessary to assure proper balance of the system. It is difficult to design a duct system so that the correct amount of air will be delivered at each outlet. Therefore, the use of dampers is essential so that the air supply may be directly proportioned after the system is placed in operation.

Changing Duct Sizes

In reducing the size or changing the shape of a duct, care must be taken that the angle of the slope is not too abrupt (Figure 10-12).

Figure 10-12. Changing duct sizes.

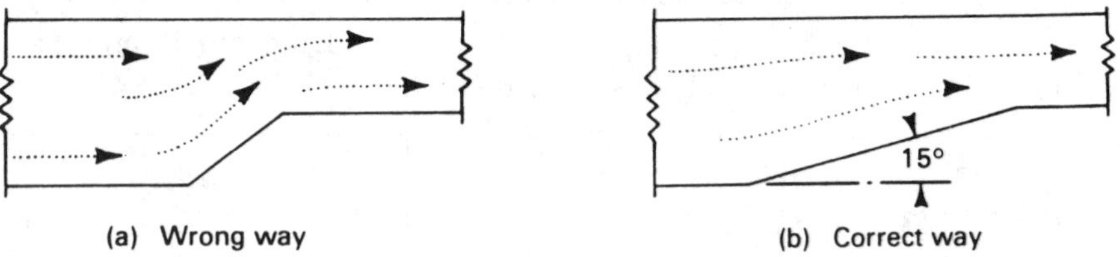

From this example it is seen that any sharp obtrusions in the path of the air through the ducts greatly increases the restriction and the static pressure. This should be kept in mind when designing an air distribution system.

Elbows

When designing a duct system, great care should be taken in shaping the elbows because sharp turns add greatly to the friction and lowers the efficiency of the entire system. Whenever possible, 90° bends should be made with a centerline radius equal to one and one-half times the diameter of the duct at the point of the bend, or one diameter at the very least in a tight corner. Figures 10-13 through 10-16 show four different 90° elbows. Figure 10-13 shows a centerline radius of one and one-half times the diameter; Figure 10-14, a centerline radius equal to the diameter; Figure 10-15, an inside throat square; and Figure 10-16, a square 90° turn.

Figure 10-13. Elbow, radius equal to 1-1/2 diameters.

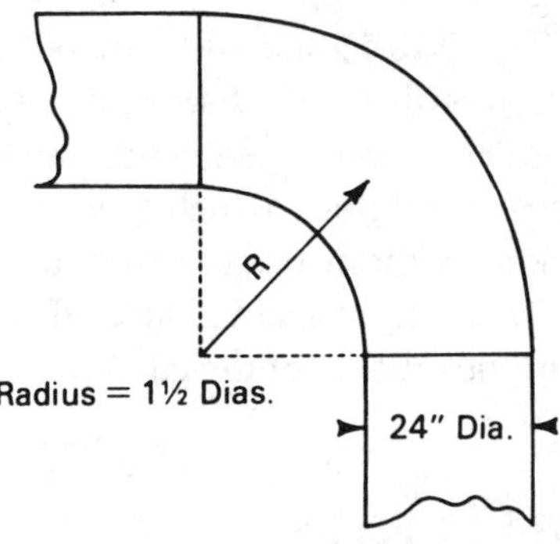

Figure 10-14. Elbow, radius equal to diameter.

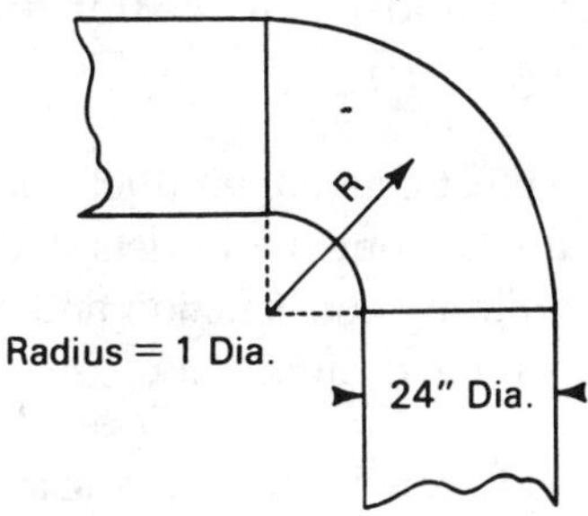

Figure 10-15. Elbow, inside throat square, with splitter vanes.

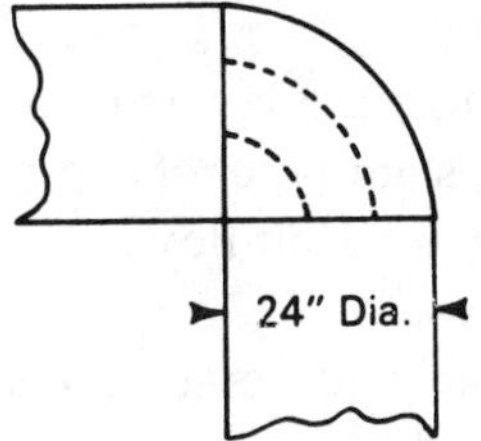

Figure 10-16. Elbow, square 90° turn, with turning vanes.

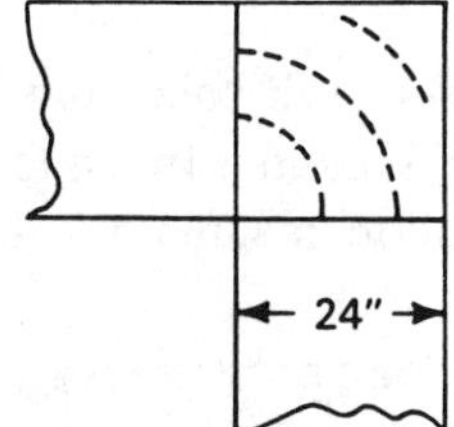

A bend similar to Figure 10-13 should be used whenever possible, although a bend similar to Figure 10-14 is permissible where available space will not allow a greater sweep.

Bends similar to Figure 10-15 and 10-16 are not desirable and should be avoided in connection with an air distribution system, as the power cost is greatly increased. However, these bends can be used when necessary with turning vanes or splitters, which tend to direct the air stream uniformly around bends. Without them, congestion and eddy currents would result and increase the resistance to airflow because the entire volume of air would hit the outside of the elbow.

PROCEDURE FOR DUCT DESIGN

The general procedure used in duct design is as follows:

1. Do a thorough study of the building plan and locate the supply air outlet positions to provide the proper air distribution within the conditioned space. Choose the outlet sizes from the grill manufacturer's catalog.

2. Make a sketch of the most convenient duct system. Include both the supply and return ducts from the outlets and intakes to the unit. Make a note of the building construction and avoid all obstructions while maintaining a simple design.

3. Calculate the main and branch duct sizes using one of the methods given in the section entitled "Design Methods."

4. Determine the total static pressure requirement of both the supply and the return duct systems. Ordinarily, only the pressure loss of the duct run having the greatest static resistance is considered as the pressure loss of the total system, even though the total loss in pressure of each duct run connecting the unit to each supply outlet, or return inlet, should be calculated and made the same for all runs.

5. The more self-balancing the duct system is, the less expensive the overall system is in the long run, from the standpoint of engineering, fabrication, installation, and the balancing of the air flow.

The static pressure regain should be given consideration regardless of the design method used.

Velocities

By velocity is meant the rate of speed of the air traveling through the ducts or openings. The following suggested velocities should be maintained because higher velocities will result in an increase in the noise level and electric power consumption. The velocities given in Tables 10-4, 10-5, and 10-6 are for general purposes.

Table 10-4. Perimeter trunk and branch system. (Courtesy of The Trane Company)

Location	*Residences*	*Schools and Public Buildings*	*Industrial Buildings*
Main ducts	700 900 (1000)	1000 1300 (1400)	1200-1800 (2000)
Branch ducts	600 (700)	600 900 (1000)	800 1000 (1200)
Branch risers	500 (650)	600 700 (900)	800 (1000)
Outside air intakes	700 (800)	800 (900)	1000 (1200)
Filters	250 (300)	300 (350)	350 (350)
Heating coils	450 (500)	500 (600)	600 (700)
Air washers	500	500	500
Suction connections	700 (900)	800 (1000)	1000 (1400)
Fan outlets	1000 1600 (1700)	1300 2000 (2200)	1600-2400 (2800)

Table 10-5. Ceiling diffuser system. (Courtesy of The Trane Company)

Application	*Normal Throw: 10-25 Ft/Min*	*Long Throw: 30-60 Ft*
Radio studios	300-400	Not recommended
Funeral homes	500-600	Not recommended
Residences	500-600	Not recommended
Private offices	650-750	Not recommended
General offices	750-850	1000
Theaters	800-900	1000
Small shops. etc.	800-900	1000
Cafes, bars, etc.	900-1000	1200
Department and grocery stores	900-1200	1500
Industrial plants	1000-1500	1800

Table 10-6. High side wall furred ceiling system. (Courtesy of The Trane Company)

Type	*Feet per Minute*
Baseboard registers	300-500
Wall registers	500-600
Ceiling registers	500-2500
Return grills	500-1000

Note that the maximum duct velocities listed in Table 10-4 are given in parentheses. Because the fan horsepower requirement increases as the square of the velocity, approximately, and the noise generated increases with an increase in static pressure, the velocities should be kept low for a quiet and economical operating system. However, it must be remembered that at a given airflow rate, the duct size increases as the velocity decreases. Sometimes it is possible to reduce the between-floor height by using very small ducts in multistory buildings, thus allowing a considerable reduction in the building costs.

DESIGN METHODS

There are three methods used in designing air conditioning duct systems: (1) equal friction method, (2) velocity reduction, and (3) static regain. These three methods represent different design levels of accuracy and complexity. Therefore, the method should be selected that will best suit the application. When simple duct systems are used, they may be designed as quickly and as easily as is possible. However, for large installations the system static-pressure requirement must be determined as accurately as possible.

Keep in mind, however, that none of the three design methods listed here will automatically produce the most economical air delivery system for all conditions. If maximum economy is to be reached, a careful evaluation and balancing of all the cost variables that enter into the design of a duct system should be considered with each design method. The main variables that affect the cost of a duct system are ductwork cost, duct insulation, fan horsepower, space requirements, and cost of sound attenuation.

Equal-Friction Method

This method uses the principle of making the pressure loss per foot of length the same throughout the entire system. Very little air balancing is necessary for duct systems in which all the runs have about the same amount of resistance.

This method can be modified, including the design of the longest duct run at two or more different friction values. When a relatively high friction rate must be used on the discharge side of the system, a lower rate can be used on the return duct system when conditions are less critical to maintain a total system static pressure within the required limits.

Generally, the velocity in the main duct is selected near the fan discharge velocity to provide a satisfactory noise level for a particular installation. Because the flow rate in cfm is generally known, this determines a friction loss per 100 ft. of duct (Figure 10-17 or 10-18).

This same amount of friction loss is maintained through the entire system.

The flow rate in the main duct after the first branch takeoff is reduced by the amount taken off by the branch. Thus, in Figures 10-17 and 10-18, proceed vertically downward to the new flow rate value, in cfm, and read the velocity and the duct diameter. Notice that the velocity has been reduced. A great advantage of this method is that it automatically reduces the velocities in the ducts in the direction of airflow, thus reducing air noise problems. The equivalent rectangular size for round pipes of any diameter can be found in Table 10-7.

By continuing to use this procedure, the design person can size all sections of an air duct system at the same friction loss per linear foot of ductwork.

When the system has been sized, the pressure loss in the duct run which apparently has the greatest resistance should be calculated. The pressure losses caused by all elbows and transitions are to be included and expressed in terms of equivalent length of straight pipe.

The equal-friction method has one main limitation: It does not differentiate between duct runs consisting of several elbows, transitions, and so on, and the runs which have very little resistance. The actual length of duct regulates the cfm flow and determines the duct size. Also, when the system resistance is being computed, the design person must calculate the pressure losses of the various fittings and add these calculations to the straight pipe losses.

When the available pressure for the ductwork is known, this available pressure can be divided by the total equivalent length of the duct run which apparently has the greatest resistance, to determine the friction loss

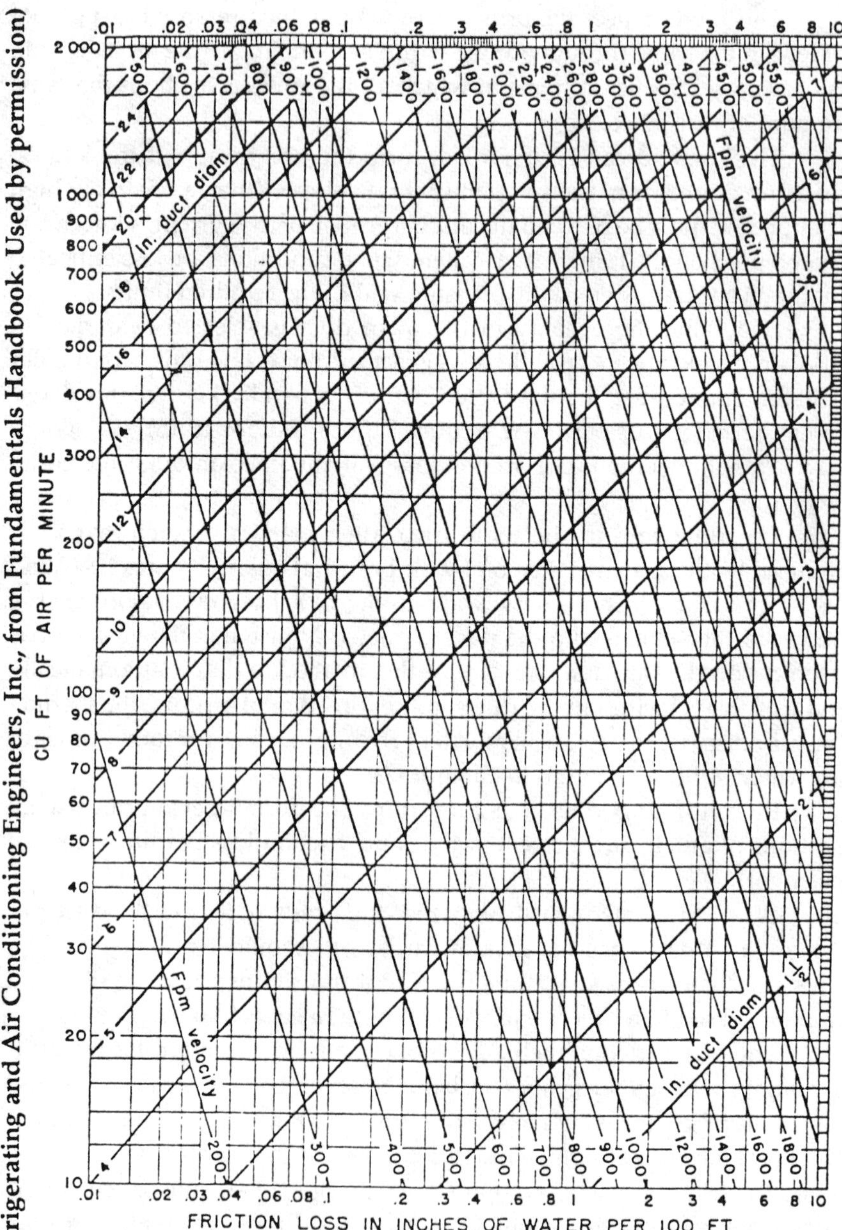

(Based on Standard Air of 0.075 lb per cu ft density flowing through average, clean, round, galvanized metal ducts having approximately 40 joints per 100 ft.) Caution: Do not extrapolate below chart.

Figure 10-17. Friction of air in straight ducts for volumes of 10 to 2000 cfm. (Copyright 1981 by the American Society of Heating, Refrigerating and Air Conditioning Engineers, Inc., from Fundamentals Handbook. Used by permission)

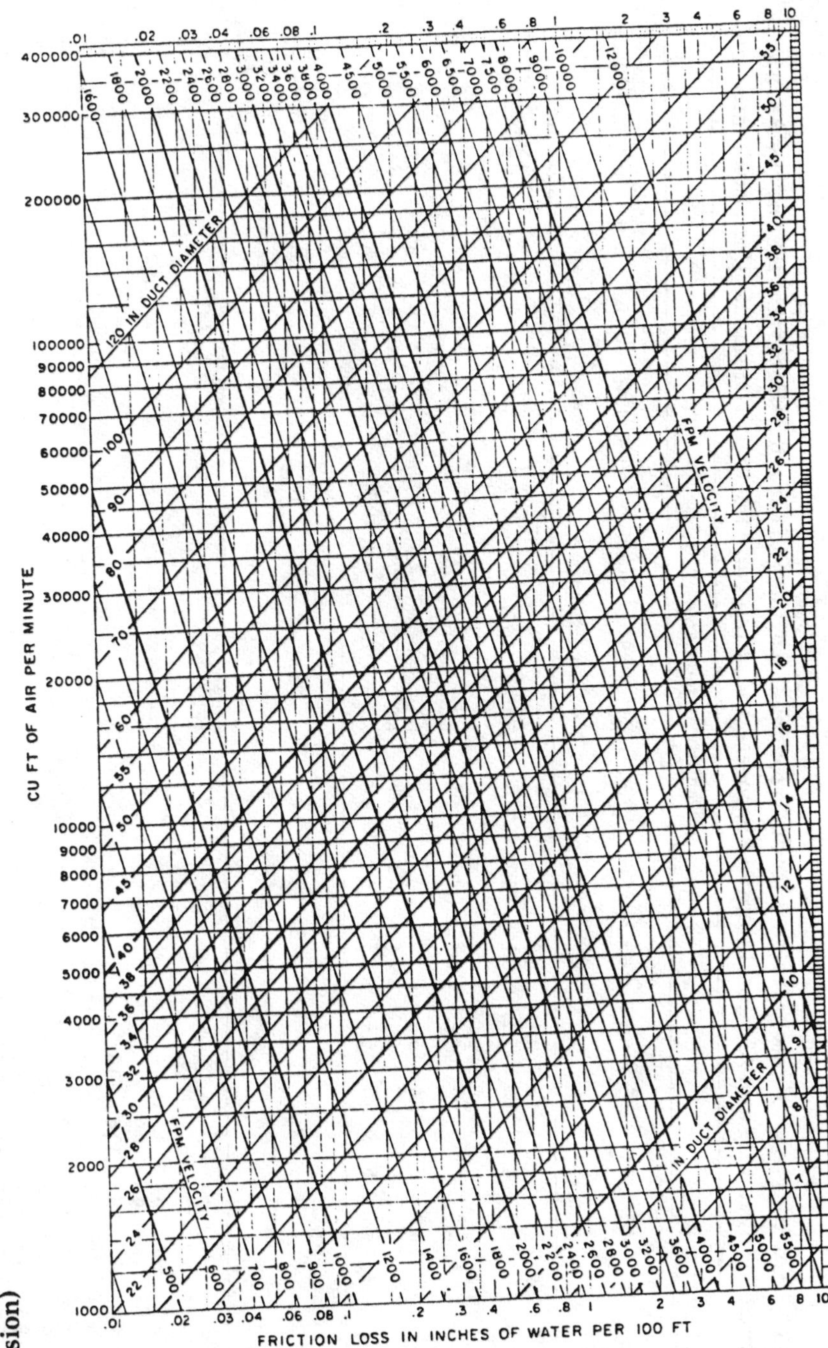

(Based on Standard Air of 0.075 lb per cu ft density flowing through average, clean, round, galvanized metal ducts having approximately 40 joints per 100 ft.)

Figure 10-18. Friction of air in straight ducts for volumes of 1000 to 400,000 cfm. (Copyright 1981 by the American Society of Heating, Refrigerating and Air Conditioning Engineers, Inc., from Fundamentals Handbook. Used by permission)

Table 10-7. Circular equivalents of rectangular ducts for equal friction and capacity. (Copyright 1981 by the American Society of Refrigerating and Air Conditioning Engineers, Inc., from Fundamentals Handbook. Used by Permission)

Dimensions in inches

Side Rectangular Duct	4.0	4.5	5.0	5.5	6.0	6.5	7.0	7.5	8.0	9.0	10.0	11.0	12.0	13.0	14.0	15.0	16.0
3.0	3.8	4.0	4.2	4.4	4.6	4.8	4.9	5.1	5.2	5.5	5.7	6.0	6.2	6.4	6.6	6.8	7.0
3.5	4.1	4.3	4.6	4.8	5.0	5.2	5.3	5.5	5.7	6.0	6.3	6.5	6.8		7.0	7.2	7.4
7.6																	
4.0	4.4	4.6	4.9	5.1	5.3	5.5	5.7	5.9	6.1	6.4	6.8	7.1	7.3	7.6	7.8	8.1	8.3
4.5	4.6	4.9	5.2	5.4	5.6	5.9	6.1	6.3	6.5	6.9	7.2	7.5	7.8	8.1	8.4	8.6	8.9
5.0	4.9	5.2	5.5	5.7	6.0	6.2	6.4	6.7	6.9	7.3	7.6	8.0	8.3	8.6	8.9	9.1	9.4
5.5	5.11	5.4	5.7	6.0	6.3	6.5	6.8	7.0	7.2	7.6	8.0	8.4	8.7	9.0	9.4	9.6	9.8

Side Rectangular Duct	6	7	8	9	10	11	12	13	14	15	16	17	18	19	20	22	24	26	28	30	Side Rectangular Duct
6	6.6																				6
7	7.1	7.7																			7
8	7.5	8.2	8.8																		8
9	8.0	8.6	9.3	9.9																	9
10	8.4	9.1	9.8	10.4	10.9																10
11	8.8	9.5	10.2	10.8	11.4	12.0															11
12	9.1	9.9	10.7	11.3	11.9	12.5	13.1														12
13	9.5	10.3	11.1	11.8	12.4	13.0	13.6	14.2													13
14	9.8	10.7	11.5	12.2	13.5	14.2	14.7	15.3													14
15	10.1	11.0	11.8	12.6	13.3	14.0	14.6	15.3	16.4												15

16	10.4	11.4	12.2	13.0	13.7	14.4	15.1	15.7	16.3	16.9	17.5										16
17	10.7	11.7	12.5	13.4	14.1	14.9	15.5	16.1	16.8	17.4	18.0	18.6									17
18	11.0	11.9	12.9	13.7	14.5	15.3	16.0	16.6	17.3	17.9	18.5	19.1	19.7								18
19	11.2	12.2	13.2	14.1	14.9	15.6	16.4	17.1	17.8	18.4	19.0	19.6	20.2	20.8							19
20	11.5	12.5	13.5	14.4	15.2	15.9	16.8	17.5	18.2	18.8	19.5	20.1	20.7	21.3	21.9						20
22	12.1	13.1	14.1	15.0	15.9	16.7	17.6	18.3	19.1	19.7	20.4	21.0	21.7	22.3	22.9	24.1					22
24	12.4	13.6	14.6	15.6	16.6	17.6	18.3	19.1	19.8	20.6	21.3	21.9	22.6	23.2	23.9	25.1	26.2				24
26	12.8	14.1	15.2	16.2	17.2	18.1	19.0	19.8	20.6	21.4	22.1	22.8	23.5	24.1	24.8	26.1	27.2	28.4			26
28	13.2	14.5	15.6	16.7	17.7	18.7	19.6	20.5	21.3	22.1	22.9	23.6	24.4	25.0	25.7	27.1	28.2	29.5	30.6		28
30	13.6	14.9	16.1	17.2	18.3	19.3	20.2	21.1	22.0	22.9	23.7	24.4	25.2	25.9	26.7	28.0	29.3	30.5	31.6	32.8	30
32	14.0	15.3	16.5	17.7	18.8	19.8	20.8	21.8	22.7	23.6	24.4	25.2	26.0	26.7	27.5	28.9	30.1	31.4	32.6	33.8	32
34	14.4	15.7	17.0	18.2	19.3	20.4	21.4	22.4	23.3	24.2	25.1	25.9	26.7	27.5	28.3	29.7	31.0	32.3	33.6	23.8	34
36	14.7	16.1	17.4	18.6	19.8	20.9	21.9	23.0	23.9	24.8	25.8	26.6	27.4	28.3	29.0	30.5	32.0	33.0	34.6	35.8	36
38	15.0	16.4	17.8	19.0	20.3	21.4	22.5	23.5	24.5	25.4	26.4	27.3	28.1	29.0	29.8	31.4	32.8	34.2	35.5	36.7	38
40	15.3	16.8	18.2	19.4	20.7	21.9	23.0	24.0	25.1	26.0	27.0	27.9	28.8	29.7	30.5	32.1	33.6	35.1	36.4	37.6	40
42	15.6	17.1	18.5	19.8	21.1	22.3	23.4	24.5	25.6	26.6	27.6	28.5	29.4	30.4	31.2	32.8	34.4	35.9	37.3	38.6	42
44	15.9	17.5	18.9	20.2	21.5	22.7	23.9	25.0	26.1	27.2	28.2	29.1	30.0	31.0	31.9	33.5	35.2	36.7	38.1	39.5	44
46	16.2	17.8	19.2	20.6	21.9	23.2	24.3	25.5	26.7	27.7	28.7	29.7	30.6	31.6	32.5	34.2	35.9	37.4	38.9	40.3	46
48	16.5	18.1	19.6	20.9	22.3	23.6	24.8	26.0	27.2	28.2	29.2	30.2	31.2	32.2	33.1	34.9	36.6	38.2	39.7	41.2	48
50	16.8	18.4	19.9	21.3	22.7	24.0	25.2	26.4	27.6	28.7	29.8	30.8	31.8	32.8	33.7	35.5	37.3	38.9	40.4	42.0	50
52	17.0	18.7	20.2	21.6	23.1	24.4	25.6	26.8	28.1	29.2	30.3	31.4	32.4	33.4	34.3	36.2	38.0	39.6	41.2	42.8	52
54	17.3	19.0	20.5	22.0	23.4	24.8	26.1	27.3	28.5	29.7	30.8	31.9	32.9	33.9	34.9	36.8	38.7	40.3	42.0	43.6	54
56	17.6	19.3	20.9	22.4	23.8	25.2	26.5	27.7	28.9	30.1	31.2	32.4	33.4	34.5	35.5	37.4	39.3	41.0	42.7	44.3	56
58	17.8	19.5	21.1	22.7	24.2	25.5	26.9	28.2	29.3	30.5	31.7	32.9	33.9	35.0	36.0	38.0	39.8	41.7	43.4	45.0	58
60	18.1	19.8	21.4	23.0	24.5	25.8	27.3	28.7	29.8	31.0	32.2	33.4	34.5	35.5	36.5	38.6	40.4	42.3	44.0	45.8	60
62	18.3	20.1	21.7	23.3	24.8	26.2	27.6	29.0	30.2	31.4	32.6	33.8	35.0	36.0	37.1	39.2	41.0	42.9	44.7	46.5	62
64	18.6	20.3	22.0	23.6	25.2	26.5	27.9	29.3	30.6	31.8	33.1	34.2	35.5	36.5	37.6	39.7	41.6	43.5	45.4	47.2	64
66	18.8	20.6	22.3	23.9	25.5	26.9	28.3	29.7	31.0	32.2	33.5	34.7	35.9	37.0	38.1	40.2	42.2	44.1	46.0	47.8	66
68	19.0	20.8	22.5	24.2	25.8	27.3	28.7	30.1	31.4	32.6	33.9	35.1	36.3	37.5	38.6	40.7	42.8	44.7	46.6	48.4	68
70	19.2	21.0	22.8	24.5	26.1	27.6	29.1	30.4	31.8	33.1	34.3	35.6	36.8	37.9	39.1	41.3	43.3	45.3	47.2	49.0	70

(Continued)

Table 10-7. (*Continued*)

Dimension in inches

Side Rectangular Duct	6	7	8	9	10	11	12	13	14	15	16	17	18	19	20	22	24	26	28	30	Side Rectangular Duct
															39.6	41.8	43.8	45.9	47.8	49.7	72
															40.0	42.3	44.4	46.4	48.4	50.3	74
															40.5	42.8	44.9	47.0	49.0	50.8	76
															40.9	43.3	45.5	47.5	49.5	51.5	78
															41.3	43.8	46.0	48.0	50.1	52.0	80
															41.8	44.2	46.4	48.6	50.6	52.6	82
															42.2	44.6	46.9	49.2	51.1	53.2	84
															42.6	45.0	47.4	49.6	51.6	53.7	86
															43.0	45.4	47.9	50.1	52.2	54.3	88
															43.4	45.9	48.3	50.6	52.8	54.8	90
															43.8	46.3	48.7	51.1	53.4	55.4	92
															44.6	47.2	49.5	52.0	54.4	56.3	96

Equation for Circular Equivalent of a Rectangular Duct.

$$d_c = 1.30 \frac{(ab)^{0.625}}{(a+b)^{0.250}} = 1.30 \sqrt[8]{\frac{(ab)^5}{(a+b)^2}}$$

where

a = length of one side of rectangular duct, inches.

b = length of adjacent side of rectangular duct, inches.

d_c = circular equivalent of a rectangular duct for equal friction and capacity, inches.

Side Rectangular Duct	32	34	36	38	40	42	44	46	48	50	52	56	60	64	68	72	76	80	84	88	Side Rectangular Duct
32	35.0																				32
34	36.0	37.2																			34
36	37.0	38.2	39.4																		36
38	38.0	39.2	40.4	41.6																	38
40	39.0	40.2	41.4	42.6	43.8																40

42	39.9	41.1	42.4	43.6	44.8	45.9															42
44	40.8	42.0	43.4	44.6	45.8	46.9	48.1														44
46	41.7	43.0	44.3	45.6	46.8	47.9	49.1	50.3													46
48	42.6	43.9	45.2	46.5	47.8	48.9	50.2	51.3	52.6												48
50	43.5	44.8	46.1	47.4	48.8	49.8	51.2	52.3	53.6	54.7											50
52	44.3	45.7	47.1	48.3	49.7	50.8	52.2	53.3	54.6	55.8	56.9										52
54	45.0	46.5	48.0	49.2	50.6	51.8	53.2	54.3	55.6	56.8	57.0										54
56	45.8	47.3	48.8	50.1	51.5	52.7	54.1	55.3	56.5	57.8	58.9	61.3									56
58	46.6	48.1	49.6	51.0	52.4	53.7	55.0	56.2	57.5	58.8	60.0	62.3									58
60	47.3	48.9	50.4	51.8	53.3	54.6	55.9	57.1	58.5	59.8	61.0	63.3	65.7								60
62	48.0	49.7	51.2	52.6	54.2	55.5	56.8	58.0	59.4	60.7	62.0	64.3	66.7								62
64	48.7	50.4	52.0	53.4	55.0	56.4	57.7	59.0	60.3	61.6	62.9	65.3	67.7	70.0							64
66	49.5	51.1	52.8	54.2	55.8	57.2	58.6	59.9	61.2	62.5	63.9	66.3	68.7	71.1							66
68	50.2	51.8	53.5	55.0	56.6	58.0	59.5	60.8	62.1	63.4	64.8	67.3	69.7	72.1	74.4						68
70	50.9	52.5	54.2	55.8	57.3	58.8	60.3	61.7	63.0	64.3	65.7	68.3	70.7	73.1	75.4						70
72	51.5	53.2	54.9	56.5	58.0	59.6	61.1	62.6	63.9	65.2	66.6	69.2	71.7	74.1	76.4	78.8					72
74	52.1	53.9	55.6	57.2	58.8	60.4	61.9	63.3	64.8	66.1	67.5	70.1	72.7	75.1	77.4	79.9					74
76	52.7	54.6	56.3	57.9	59.5	61.2	62.7	64.1	65.6	67.0	68.4	71.0	73.6	76.1	78.4	80.9	83.2				76
78	53.3	55.2	57.0	58.6	60.3	62.0	63.4	64.9	66.4	67.9	69.3	71.8	74.5	77.1	79.4	81.8	84.2				78
80	53.9	55.8	57.6	59.3	61.0	62.7	64.1	65.7	67.2	68.7	70.1	72.7	75.4	78.1	80.4	82.8	85.2	87.5			80
82	54.5	56.4	58.2	60.0	61.7	63.4	64.9	66.5	68.0	69.5	71.0	73.6	76.3	79.0	81.4	83.8	86.2	88.6			82
84	55.1	57.0	58.9	60.7	62.4	64.1	65.7	67.3	68.8	70.3	71.8	74.5	77.2	79.2	82.4	84.8	87.2	89.6	91.9		84
86	55.7	57.6	59.5	61.3	63.0	64.8	66.4	68.0	69.5	71.1	72.6	75.4	78.1	80.8	83.3	85.8	88.2	90.6	92.9		86
88	56.3	58.2	60.1	62.0	63.7	65.4	67.0	68.7	70.3	71.8	73.4	76.3	79.0	81.6	84.2	86.8	89.2	91.6	93.9	96.3	88
90	56.9	58.8	60.7	62.6	64.4	66.0	67.8	69.4	71.1	72.6	74.2	77.1	79.9	82.5	85.1	87.8	90.2	92.6	94.9	97.3	90
92	57.4	59.4	61.3	63.2	65.0	66.8	68.5	70.1	71.8	73.3	74.9	77.9	80.8	83.4	86.0	88.7	91.2	93.6	95.9	98.3	92
94	57.9	60.0	61.9	63.8	65.6	67.5	69.2	70.8	72.5	74.1	75.6	78.7	81.7	84.3	86.9	89.6	92.1	94.6	95.9	99.3	94
96	58.4	60.5	62.4	64.4	66.2	68.2	69.8	71.5	73.2	74.3	76.3	79.4	82.6	85.2	87.8	90.5	93.0	95.6	97.9	100.3	96

figure per foot. Thus it is not necessary to select an initial airflow velocity for these types of systems. There is a weakness of this method, however, in that the fitting resistances must be expressed as equivalent length. Transitions, elbows, and so on, have predominate dynamic losses; therefore, the equivalent length of a particular fitting varies considerably with its actual size. This method requires that the duct size be estimated in advance. If there is a considerable difference between the calculated duct size and the initial estimated size, the calculated size should be used.

Fewer volume dampers are required if this method is modified so that only the main duct is sized by the equal-friction method. The total duct resistance is used as described for the velocity-reduction method for fan selection. The available air pressure for each branch is divided by its equivalent length in hundreds of feet to calculate a design friction loss figure for use in Figures 10-17 and 10-18, together with the branch airflow in cfm. The branch ducts should be sized as close as possible to use all the available air pressure.

Care should be exercised when using the equal-friction method to prevent branch air velocities from getting too high, thus avoiding noise problems. This problem is easily avoided during the design phase, because the resulting velocity can be read directly on the air friction chart. When the velocity becomes too high, read horizontally to the left-hand column on the friction chart and choose a duct diameter that will provide the desired velocity. The air volume damper for this duct run will be used to reduce the excess pressure. Ductwork attenuates noise to some extent; thus the damper should be located as close to the main duct run as possible. An alternative solution to this type of problem may be to change the duct layout so the resistance on that duct run is increased. Perhaps this can be accomplished by changing the branch takeoff location to increase the total duct length.

Example

A duct layout has three outlets (Figure 10-19).

Outlets 1 and 2 have an air flow of 750 cfm each. Outlet 3 has an airflow of 1000 cfm. The air flows with a velocity of 1600 fpm in section A. Size the duct system and calculate the static-pressure requirement.

Solution

The total cfm to be delivered is 750 + 750 + 1000 = 2500 cfm. Using Figure 10-18, locate 2500 cfm at 1600-fpm velocity. Read a round pipe diameter of 17 inches with a friction loss of 0.2 in. of water column per 100 ft. of duct in section A. Now subtract the 750 cfm delivered to outlet 1, to

Figure 10-19. Duct layout for Example 1. (Copyright 1981 by the American Society of Heating, Refrigerating and Air Conditioning Engineers, Inc., from Fundamentals Handbook. Used by permission)

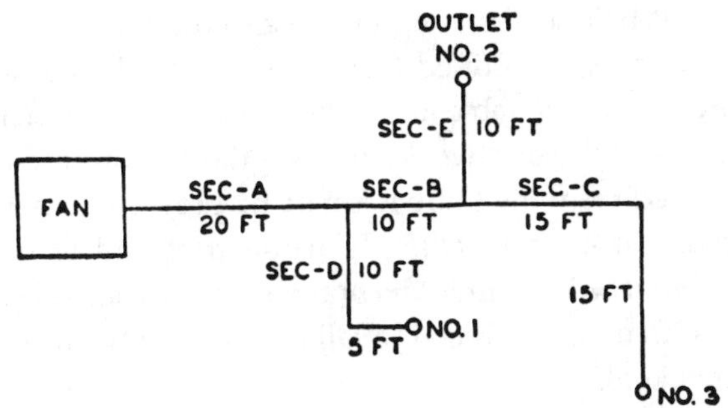

determine that there is 750 cfm of airflow in section B. Read along the 0.2 friction line. All the ducts can be sized now because all the flow rates are known (Table 10-8).

Table 10-8. Tabulation of Results. (Copyright 1981 by the American Society of Heating, Refrigerating and Air Conditioning Engineers, Inc., from Fundamentals Handbook. Used by permission)

Section	Flow Rate (Cfm)	Friction per 100 Ft (In H_2O)	Duct Diameter (In.)	Velocity (Fpm)	Rectangular Duct (In.)
A	2500	0.2	17.0	1600	20 × 12
B	1750	0.2	14.8	1480	15 × 12
C	1000	0.2	12.0	1290	15 × 8
D	750	0.2	10.7	1190	12 × 8
E	750	0.2	10.7	1190	12 × 8

Velocity-Reduction Method

The velocity-reduction method consists of choosing an air velocity at the fan discharge and designing the ducts for progressively lower air velocities in the main duct at each branch takeoff. For the chosen velocities from Figure 10-17 or 10-18, the equivalent rectangular duct sizes may be obtained from Table 10-7. The static pressure loss of the duct run which apparently has the greatest resistance is found by adding the straight pipe, elbow, and transition losses; this sum is representative of the static pressure required for the supply duct system. The return air duct system is then sized in a similar manner, starting with the lowest air velocities at the return air intakes and increasing them progressively toward the fan inlet. The system is balanced through the use of dampers.

This method may be refined to involve sizing the branch ducts to use the pressure that is available at the entrance to each branch duct. The static pressure loss of the ductwork between the fan and the first branch takeoff is subtracted from the known fan static pressure to determine the available pressure at each takeoff. By using trial and error, a branch velocity is found that will result in the branch pressure loss being equal to, or somewhat less than, that which is available. This procedure is repeated for each branch takeoff.

The fan is selected to provide a specific static pressure to the ductwork; the method then consists of finding, by trial and error, the main duct velocities that will result in a pressure loss that is equal to the available pressure. The branch ducts are then sized as discussed above.

The advantages of the velocity-reduction design method are: (1) the duct sizes are easily determined, and (2) the velocities can be limited to prevent noise problems. The disadvantages are: (1) the proper choice of air velocities requires experience, and (2) the design person cannot always determine, by inspection, which run will have the greatest amount of resistance.

Static-Regain Method

In the now popular methods of duct design—equal friction, velocity reduction, or static regain—for average conditions, air velocities in a duct run are progressively reduced, resulting in the conversion of velocity air pressure to static pressure. In these terms, these design methods are static-regain methods. In low-velocity air distribution systems, the conversion of velocity pressure is often disregarded, thus providing a narrow safety margin in the duct system design. However, when considering high-velocity systems, the failure to recognize the amount of static regain will often result in an overdesigned air distribution system with wasted fan motor horsepower and additional air noise problems.

In the ordinary sense, the term static-regain method refers to a duct design procedure in which the duct is sized to cause an increase in static pressure or regain at each branch takeoff which will offset any pressure loss because of the succeeding section of the run. When the system or part of the system is designed by the static-regain method only the initial velocity pressure to offset the friction and dynamic losses in the duct system will be required.

This method of designing air distribution systems is especially suited for large, high-velocity systems having several long runs of duct, with each run having branch takeoffs, terminal units, or supply outlets. When this method is used, essentially the same static pressure is available at the entrance to each branch takeoff, outlet, or terminal unit, thus simplifying the selection of outlets or terminals and system balancing.

AIR DISTRIBUTION

The possible comfort conditions in a room are, to a great extent, dependent on the type and location of the supply air grills and to some extent on the location of the return air intake grills.

In general, air supply grills fall into four classifications, which are governed by their discharge air pattern: (1) vertical spreading, (2) vertical nonspreading, (3) horizontal height, and (4) horizontal low. The various grill manufacturers have tables which include the performance of the different types of outlets for both cooling and heating. See Table 10-9.

One advantage of the forced air system is that the same grill can be used for both heating and cooling. It should be kept in mind that no single outlet grill is best for both applications.

The best types of outlets for heating are those providing a vertical spreading air jet located in, or near, the floor next to an outside wall at the point of greatest heat loss, such as under a window. Air delivered at these points will blanket the cold areas and counteract any cold drafts that might occur. This method of air distribution is called perimeter heating.

The best types of outlets for cooling are located in the ceiling and provide a horizontal discharge air pattern. The air being discharged across the ceiling blankets this hot area. Also, the air bathes the outside walls to pick up any heat that has entered the building through these areas.

When year-round operation of the system is desired, the type of system chosen depends on the principal application. If heating is to be the major use of the unit, perimeter-type diffusers should be used with the duct system designed to supply maximum supply air velocity for the cooling season. When cooling is to be the major use, ceiling diffusers should be chosen.

The location of return air grills is much more flexible than the supply air grill. They may be located in hallways, under windows, in exposed corners, near entrance doors, or on an inside wall, depending on the supply air grill location. Baseboard returns are preferred to floor

Table 10-9. Modulaire Series Performance Data—Example of Grill Performance Table. (Courtesy of Standard Perforating & Mfg., Inc.)

C.F.M.	SIZES	8 × 4			10 × 4			10 × 5 12 × 4			10 × 6 12 × 5 14 × 4 16 × 4			12 × 6 14 × 5 18 × 4			14 × 6 16 × 5 18 × 5 20 × 4 22 × 4		
		0	22	45	0	22	45	0	22	45	0	22	45	0	22	45	0	22	45
50	Throw	8	6	5															
	Drop	4.5	2.5	1.0															
	Velocity	335	385	420															
	Static Pressure	009	009	009															
100	Throw	16	13	11	14	12	10	12	11	9	11	10	8	10	9	7			
	Drop	6.0	4.0	3.5	6.0	4.5	4.0	7.0	5.0	3.5	7.0	4.5	3.5	6.0	.50	3.5			
	Velocity	670	770	840	530	610	660	440	505	550	375	430	470	290	335	365			
	Static Pressure	.028	.037	.044	.017	.023	.027	.012	.016	.019	.009	.011	.013	.009	.009	.009			
150	Throw	23	20	17	20	18	15	19	16	13	17	14	12	15	13	11	14	12	10
	Drop	7.0	5.5	4.5	7.0	5.5	4.5	7.5	5.5	4.5	7.5	5.5	4.5	7.5	5.5	4.5	8.0	5.5	4.5
	Velocity	1000	1150	1260	800	915	990	660	760	820	560	645	705	435	500	545	390	450	485
	Static Pressure	.062	.081	.098	.040	.052	.061	.027	.036	.042	.019	.026	.031	.012	.016	.018	.009	. 013	.015
200	Throw	31	27		27	24	20	25	22	!8	22	19	16	20	18	15	18	16	14
	Drop	8.5	7.0		8.5	7.0	5.5	8.5	6.5	5.5	8.0	6.5	5.0	8.5	7.0	5.5	8.5	65	5.5
	Velocity	1340	1540		1060	1220	1320	880	1010	1090	750	860	940	580	670	730	520	600	650
	Static Pressure	.111	.147		.070	.092	.107	.048	.063	.074	.035	.046	.055	.021	.028	.033	.017	.022	.026

250	Throw	35	30		32	38	23	28	24	20	26	22	18	24	20	17
	Drop	10.0	7.5		10.0	7.5	6.0	9.5	7.5	6.0	10.0	7.5	6.0	10.0	7.5	6.0
	Velocity	1330	1525		100	1265	1360	940	1080	1170	725	833	910	650	750	810
	Static Pressure	.109	.146		.075	.099	.115	.055	.072	.084	.032	.043	.051	.026	.035	.041
300	Throw				38	33		34	29	25	30	26	22	20	24	21
	Drop				11.0	8.5		10.5	8.0	6.5	10.5	8.0	6.5	10.5	8.5	6.5
	Velocity				1320	1420		1120	1290	1405	870	1000	1090	780	900	972
	Static Pressure				.107	.144		.077	.104	.121	.047	.062	.074	.038	.050	.058
350	Throw							40	35		36	30	25	32	28	25
	Drop							11.5	9.0		11.5	9.0	7.0	11.0	8.5	7.5
	Velocity							1310	1510		1020	1165	1275	910	1050	1135
	Static Pressure							.106	.142		.065	.083	.101	.051	.068	.078
400	Throw							45			41	36	29	38	33	31
	Drop							12.0			12.5	10.0	7.5	12.5	9.5	8.5
	Velocity							1500			1160	1330	1455	1040	1200	1295
	Static Pressure							.140			.083	.109	.131	.066	.088	.104
450	Throw										46	40		42	37	35
	Drop										13.0	10.5		13.0	10.0	9.0
	Velocity										1310	1500		1170	1350	1460
											.106	.140		.084	.113	.132
500	Throw										50			47	40	
	Drop										14.0			14.0	10.5	
	Velocity										1450			1300	1500	
	Static Pressure										.130			.150	.140	
550	Throw													51		
	Drop													14.5		
	Velocity													1365		
	Static Pressure													.116		

(Continued)

Table 10-9 (*Continued*)

C.F.M.	Sizes	12 × 8 16 × 6 20 × 5 24 × 5			14 × 8 18 × 6 20 × 6 22 × 5 24 × 5 28 × 4 30 × 4			16 × 8 18 × 8 22 × 6 24 × 6 28 × 5 30 × 5 36 × 4			18 × 10 20 × 8 22 × 8 28 × 6 30 × 6 36 × 5 48 × 4			18 × 12 10 × 10 ×22 × 10 24 × 8 38 × 6 48 × 5			20 × 12 22 × 12 24 × 10 28 × 8 30 × 8 48 × 6		
		0	22	45	0	22	45	0	22	45	0	22	45	0	22	45	0	22	45
	Throw	13	11		12														
	Drop	8.0	6.0		8.5														
150	Velocity	325	365		270														
	Static Pressure	.009	.009		.009														
	Throw	17	15	12	15	14	11	14											
200	Drop	9.0	7.0	5.5	9.0	7.5	5.5	9.5											
	Velocity	430	490	535	360	415	450	280											
	Static Pressure	.011	.015	.018	.009	.010	.012	.009											
	Throw	21	19	15	19	17	14	18	15	12	16								
250	Drop	9.5	8.0	6.0	10.0	8.0	5.5	11.0	8.5	6.0	10.5								
	Velocity	540	610	670	450	520	560	350	400	440	300								
	Static Pressure	.018	.023	.028	.012	.017	.019	.009	.010	.012	.009								
	Throw	26	22	18	23	20	16	21	18	15	19	16	14	18					
300	Drop	11.0	8.0	6.0	11.0	8.5	6.0	11.5	9.0	7.0	11.5	9.0	7.0	12.0					
	Velocity	645	735	805	540	625	675	420	480	530	360	415	450	305					
	Static Pressure	.026	.033	.040	.018	.024	.028	.011	.014	.017	.009	.011	.012	.009					

350	Throw	30	26	21	27	24	19	25	22	18	22	19	16	20	18	15			
	Drop	11.5	9.0	6.5	11.5	9.0	7.0	12.5	9.5	7.5	12.5	9.5	7.5	12.5	10.0	7.5			
	Velocity	755	855	940	630	725	785	490	560	615	420	485	525	355	410	450			
	Static Pressure	.035	.045	.055	.025	.033	.038	.015	.019	.023	.011	.014	.017	.009	.010	.012			
400	Throw	35	30	25	31	27	22	28	24	20	26	23	18	24	21	18	22		
	Drop	12.5	9.5	7.5	12.5	9.5	7.0	13.5	10.0	7.5	13.5	10.5	7.5	13.5	11.0	8.0	15.0		
	Velocity	865	975	1075	720	830	895	560	640	705	480	550	600	405	465	525	315		
	Static Pressure	.046	.059	.071	.032	.043	.049	.019	.026	.030	.014	.018	.022	.010	.013	.017	.009		
450	Throw	39	34	28	36	30	25	32	27	22	28	26	21	27	24	20	25	22	17
	Drop	13.5	10.5	8.0	13.5	10.5	8.0	14.5	11.0	8.0	14.0	11.0	8.0	15.0	11.5	8.5	16.0	12.5	9.0
	Velocity	970	1100	1205	810	935	1010	630	720	790	540	620	675	455	525	595	355	405	445
	Static Pressure	.058	.075	.088	.041	.054	.063	.025	.032	.039	.018	.024	.028	.0113	.017	.022	.009	.010	.012
500	Throw	42	38	30	39	34	27	35	30	25	32	29	23	30	26	22	28	24	19
	Drop	13.5	11.5	8.0	14.0	10.5	8.0	14.5	11.0	8.5	14.5	11.5	8.5	15.5	12.0	8.5	17.0	12.5	9.5
	Velocity	1080	1220	1340	900	1040	1120	700	800	880	600	690	750	505	585	670	395	455	495
	Static Pressure	.072	..092	.1 11	.050	.066	.077	.030	..040	..048	.022	.029	.035	.016	.021	..028	.010	.012	.015
550	Throw	46	40	32	41	36	29	37	32	27	34	30	25	32	27	24	30	25	21
	Drop	14.0	11. 5	8.5	14.0	11.0	8.0	14.5	11.5	8.5	14.5	11.5	8.5	15.5	12.0	9.0	17.5	12.5	9.5
	Velocity	1190	1345	1475	990	1145	1235	770	880	970	660	760	830	560	640	740	435	500	545
	Static Pressure	.087	.111	.135	.061	.080	.093	.037	.048	.058	.027	.036	.043	.019	.026	.034	.012	.016	.018

(Conclusion)

grills. It is good practice to have them located on an outside wall. When a perimeter supply air system is used, the return air location is not extremely important because with these systems the return has little effect on the air distribution. Returns that are located centrally on all types of perimeter systems provide satisfactory operation. In large or multilevel buildings it may be advisable to locate a return air grill in each level or group of rooms. Also, some provision should be made for the return air to move from all rooms on one level, or group, to the desired return air grill. It is usually desirable to have individual room return air grills to avoid potential problems with airflow under doors or through adjacent rooms and hallways.

When basement-less buildings are to be air conditioned, it is good practice to locate the return air grills either in the ceiling or in the wall. For heating, however, this is simply a matter of preference because returns that are located at the baseboard perform as well as those located in the ceiling.

SIZING A DUCT SYSTEM

When designing a duct system it is often convenient to use a trunk duct, or extended plenum, to supply the feeders (branches) to different rooms.

When sizing a duct system, start at the grill located the farthest from the equipment and work back to the unit. When there are two or more feeders that are the same distance from the unit, it makes no difference which one is considered farther from the unit.

Example

We have determined from the heat load calculation that we need a duct system made up of two 6-in. feeder ducts, one 7-in. duct, and two 8-in. feeders. See Figure 10-20.

Solution

Either of the two 6-in. feeders may be considered the trunk. When two ducts join, a different size duct results. In this example, when the two 6-in. ducts join, the resulting duct size is 8-in. See Table 10-10.

To read the table, find the trunk size in the left-hand column. We are connecting one 6-in. feeder duct to the 6-in. trunk. Find "1-6" in the center column under "size of feeder joining trunk." In the right-hand column across from the 1-6 entry find 8 in. trunk duct picking up a 7-in. feeder

Figure 10-20. Round trunk duct system.

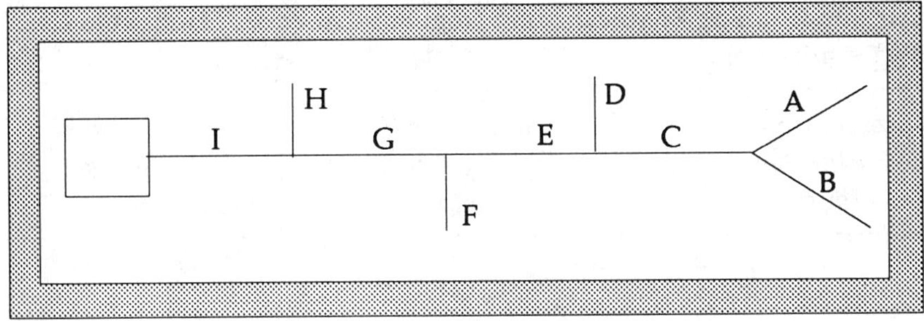

(section D). The resulting trunk size is 10-in. (section E). Next, we have a 10-in. trunk duct picking up an 8-in. feeder (section F). The resulting trunk size is 12-in. (section G). The 12-in. trunk picks up an 8-in. feeder (section H), resulting in a 14-in. trunk duct (section I) back to the unit plenum.

DESIGNING A DUCT SYSTEM

We will now design a complete air distribution system for a residence on which we made the heat load calculations in Chapter 3. Since this home is located in Fort Worth, Texas, the major use of the unit will be for cooling. Therefore, we will need to design the duct system to handle the correct volume of cooling air. We will use rectangular ceiling diffusers in each room. See Figure 10-21.

First, we must determine the supply air grill locations. We want the air to be blown toward the point of greatest heat gain. We now need to determine the size of each supply air grill. To do this we must determine the required amount of air for each room. We will use the percentage method for calculating the volume of air in cfm for each room. We will do this by determining what percent of the total heat gain each room represents. The air requirement for the living room will be: 9672 ÷ 41,046 × 100 = 23% of the total heat load. The unit we selected has a rated cfm of 1200. The cfm will be 1200 × 23% = 276 cfm.

SELECTING THE PROPER SUPPLY AIR GRILLS

The proper selection of the supply air outlet is very important for operation of the air conditioning system. As a part of the total air condi-

Table 10-10. Diameters of Round Trunks for Perimeter Loop System. (Copyright 1981 by the American Society of Heating, Refrigerating and Air Conditioning Engineers, Inc., from Fundamentals Handbook. Used by permission)

Size of Trunk before Junction (in.)	*Size of Feeder Joining Trunk (in.)*	*Size of Trunk after Junction (in.)*
6	1-6 1-7 or 1-8	8 9
	2-6 2-7 2-8	9 10 12
	1-6& 1-7 1-6 & 1-8 1-7& 1-8	10 10 10
7	1-6 or 1-7 1-8	9 10
	2 6 2-7 or 2-8	10 12
	1-6& 1-7 1-6& 1-8 1-7& 1-8	10 12 12
8	1-6 1-7 or 1-8	9 10
	2-6 2-7 or 2-8	10 12
	1-6& 1-7 6 & 1-8 1-7 & 1-8	12 12 12
9	1-6 or 1-7 1-8	10 12
	2-6 or 2-7 2-8	12 14
	1-6 & 1-7 1-6& 1-8 1-7& 1-8	12 12 12

Table 10-10 (*Continued*)

Size of Trunk before Junction (in.)	*Size of Feeder Joining Trunk (in.)*	*Size of Trunk after Junction (in.)*
10	1-6, 1-7, or 1-8	12
	2 6 or 2 7	12
	2-8	14
	1-6 & 1-7	12
	1-6 & 1-8	12
	1-7 & 1-8	14
12	1-6, 1-7, or 1-8	14
	2-6 or 2-7	14
	2-8	16
	1-6 & 1-7	14
	1-6 & 1-8	14
	1-7 & 1-8	14
14	1-6, 1-7, or 1-8	16
	2-6, 2-7, or 2-8	16
	1-6& 1-7	16
	1-6 & 1-8	16
	1-7 & 1-8	16

tioning system, the air outlet maintains air motion without objectionable drafts in the occupied space.

The air velocities in the conditioned space are a major factor in the comfort of the occupants. Generally, these air velocities should not exceed 50 fpm in the occupied portion of the conditioned space.

Because of the difference in temperature between the supply air and the air in the room, the supply air stream will drop below the level it is being discharged into the room. Caution and good judgement are necessary in the selection and location of the conditioned air supply grills to prevent the conditioned air stream from entering the occupied zone at a critical velocity which would cause discomfort due to excessive drafts. If the supply air is overthrown and strikes the wall opposite the grill location, objectionable down drafts will occur.

Figure 10-21. Model house with duct layout.

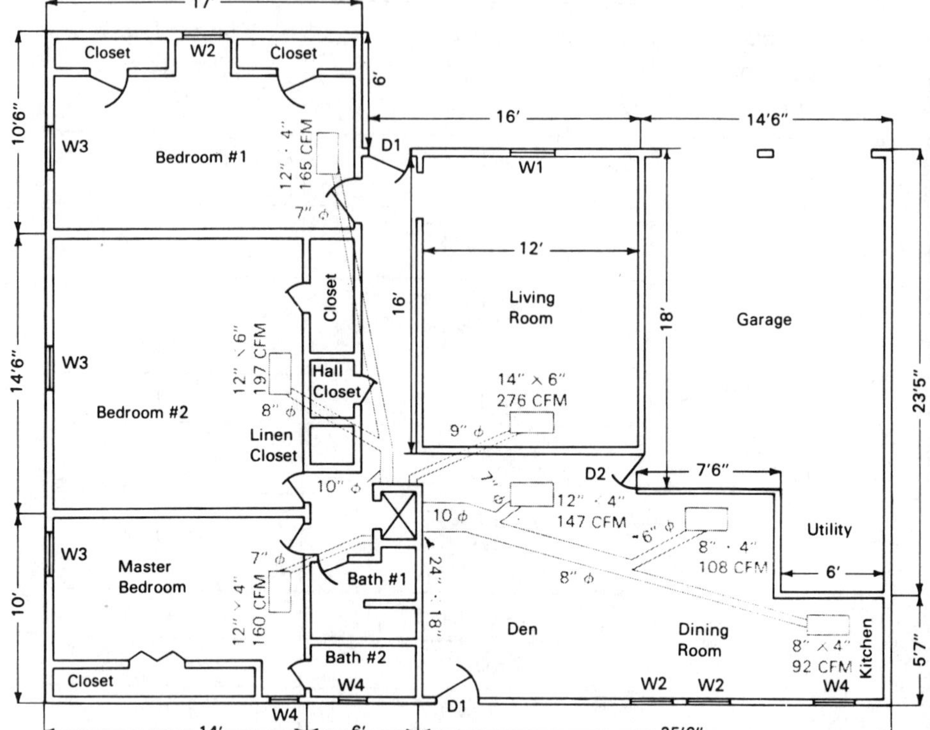

The prescribed general rule is to select a grill with a throw equal to three-fourths the distance to the opposite wall. The terminal velocity at this point should be approximately 6 ft. above the floor. See Figure 10-22.

In practice the physical room size and the heating and cooling loads have such a wide variation that grills with adjustable deflecting blades are used to help in compensating for these variables.

When applications require spot heating or cooling, grills with a single row of adjustable blades are generally selected. This type of grill will direct the air in one place, either vertical or horizontal, as desired.

Grills that are manufactured with two rows of adjustable blades direct the air in two planes. This type of grill allows complete flexibility in the adjustment for meeting the throw and drop requirements of the space. These blades are used as fine adjustment after the best possible selection has been made. If the air stream should prove to be short, the vertical blades may be adjusted to increase the distance of the throw. If the supply air stream enters the occupied zone at too high a velocity, the rear horizontal blades may be adjusted to arc the air above the occupied zone, thus

Figure 10-22. Recommended length of throw. (Courtesy of Standard Perforating & Mfg., Inc.)

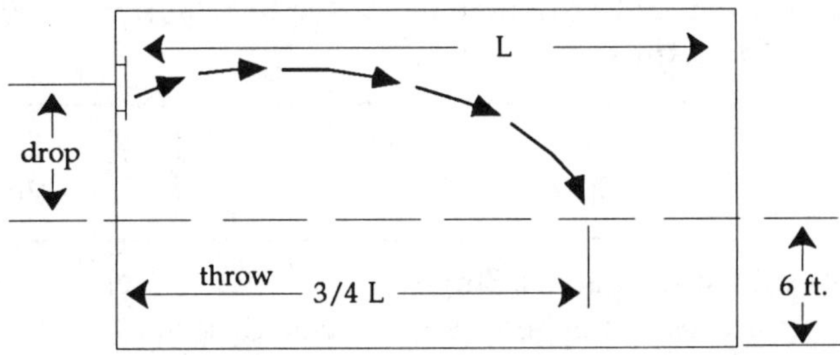

reducing the terminal velocity at this point to help in reducing objectionable drafts.

There are three different settings on the adjustable blades of the supply air grill (0°, 22°, and 45°), which provide different air patterns. See Figure 10-23.

The amount of noise caused by the supply air stream entering the room through the supply air grill is so extremely complicated that manufacturers seldom include decibel ratings in their performance tables. The noise created by the supply air entering the room is directly proportional to its velocity when leaving the supply air grill. The sound level may be

Figure 10-23. Different air patterns. (Courtesy of Standard Perforating & Mfg., Inc.)

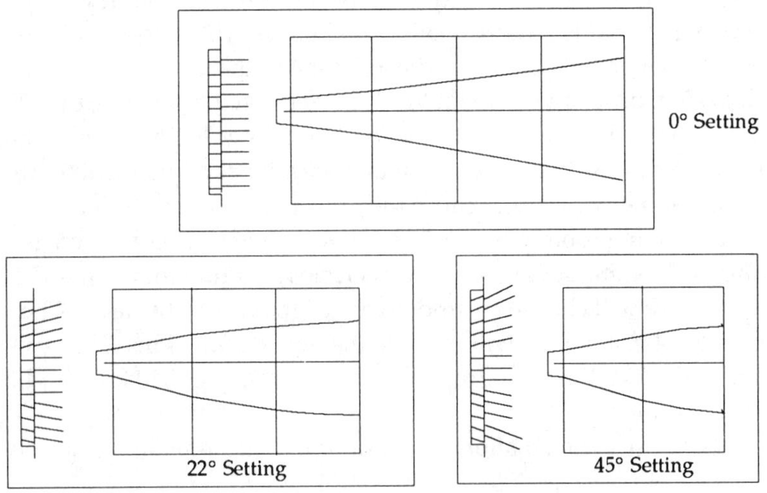

kept within satisfactory limits by selecting supply air grills with an outlet velocity in recognized approximate ranges. See Table 10-11.

Table 10-11. Maximum Recommended Velocity. (Courtesy of Standard Perforating & Mfg., Inc.)

Space	Maximum Velocity
Libraries, Broadcasting Studios, Surgery	500
Residences, Churches, Hotel Bedrooms, Private Offices	750
Banks, Theaters, Cafeterias, School Classrooms, General Office, Public Buildings	1000
Factories, Gymnasiums, Warehouses, Arenas, Department Stores	1500

We have determined previously the percentage of the total heat load that each room represents and have determined the volume of air in cfm required to handle this load. These figures should have been entered on the load estimate worksheet.

We can now choose the grills that best fill our needs for the model home. The living room needs 276 cfm to provide the required conditions. We can see on Table 10-9 that no grill is listed that exactly meets our needs. We will choose a cfm of 300. When we follow the numbers indicated for the distance of throw, horizontal line, the size of grill that will deliver the throw nearest our needs is found in the right-hand column. The grill sizes are located at the top of the column. The next step is to choose a grill that is readily available. Thus we will choose a 14 × 6 grill. When the deflector blades are adjusted for 45°, the air throw will be 21 ft. with a 6.5 ft. drop, a velocity of 972 fpm, and a static pressure of 0.058 in. water column. The throw is a little long and the velocity is a little higher than desired. However, we can make additional adjustments with the volume damper which will cause the grills to perform satisfactorily.

The dining room requires 108 cfm to satisfy its needs. There is not a listing for this amount of cfm. It is much closer to 100 cfm than to 150 cfm; therefore, we will choose an air delivery of 100 cfm on the table. The table lists an 8 × 4 grill which will provide the desired functions. This grill will provide an 11-ft. throw, a drop of 3.5 ft., a velocity of 840 fpm, and 0.044 static pressure.

Using this information and examples, the remaining supply air grills can be selected. Write the grill size on the house plans beside the grill. As a general rule when the desired air volume is greater than halfway between two listed volumes, choose the next higher listing. If it is less than halfway, choose the lower listing. As can be seen, selecting grills is at best an educated compromise.

SELECTING THE RETURN AIR GRILL

The most important points in selecting a return air grill are (1) it must handle the required cfm, and (2) it must handle the required cfm at the desired velocity. There are tables provided by grill manufacturers to aid in their selection (Table 10-12).

In our model home we will use a central return under the furnace. The furnace platform is 12 in. above the floor. Thus, the maximum height of our return air grill will be 12 in. The unit will move 1200 cfm of air. Checking Table 10-12, the horizontal line along the top of the table represents the air velocity. Since the required air velocity in a residence is 500 fpm, we can follow the 500 fpm column down until we find an allowed volume that is equal to or greater than the 1200 cfm our unit will move. There is no single 12" grill listed in the table that will allow the required cfm to flow through it. Thus, we must choose two grills. We can use a 24 × 12 grill that will allow 825 cfm at a velocity of 500 fpm and a 12 × 12 grill that will allow 400 cfm to flow through. These grills may be located on different walls that lead into the return air plenum.

Now that the grills have been selected and their location determined, we can make a duct layout for the building. This is accomplished by connecting all the supply air grills to the plenum by ductwork (Figure 10-22). Use the shortest route possible in connecting the ducts. It must be remembered that structural components will dictate the duct route taken. It is usually cheaper and easier to use a trunk duct system where possible, such as feeding the den, dining room, and kitchen with one connection to the discharge air plenum, as shown in Figure 10-21.

The next step is to size the ducts to convey the proper amount of air to each room. Referring to Figures 10-17 and 10-21, we size the ducts leading to each grill. We will use a 0.1 static pressure on the supply air system. We begin by sizing at the grill the farthest from the unit in each duct run. Starting with the kitchen, we need 92 cfm. We follow the 0.1 static pressure line, located on the bottom of the table, upward until it intersects with the approximate 92-cfm line, extending from the left side of the table. We find that this is closer to the 6-in. duct than the 5-in. duct; therefore, we will use the 6-in. duct. The duct size for the dining room is

Table 10-12. Fixed Air Return Grills—HFD, Air Capacities CFM. (Courtesy of Standard Perforating & Mfg., Inc.)

Size	Free Area Sq. Ft.	Free Area Velocity FPM							
		300	400	500	600	700	800	900	1000
10×6	.31	93	124	155	186	217	248	279	310
12×6	.37	111	148	185	222	259	296	333	370
10×8	.41	123	164	205	246	287	328	369	410
12×8	.52	156	208	260	312	364	416	468	520
18×6	.57	171	228	285	342	399	456	513	570
12×12	.80	240	320	400	480	560	640	720	800
18×12	1.23	369	492	615	738	861	984	1107	1230
24×12	1.65	495	660	825	990	1155	1320	1485	1650
18×18	1.87	561	748	935	1122	1309	1496	1683	1870
30×12	2.07	621	828	1035	1242	1449	1656	1863	2070
24×18	2.53	759	1012	1265	1618	1771	2024	2277	2530
30×18	3.19	957	1276	1595	1914	2233	2552	2871	3190
24×24	3.40	1020	1360	1700	2040	2380	2720	3060	3400
36×18	3.84	1152	1536	1920	2304	2688	3072	3456	3840
30×24	4.28	1284	1712	2140	2568	2996	3424	3852	4280
36×24	5.17	1551	2068	2585	3102	3619	4136	4653	5170
36×30	6.56	1968	2624	3280	3936	4592	5248	5904	6560
48×24	6.91	2073	2764	3455	4146	4837	5528	6219	6910
48×30	8.68	2604	3472	4340	5208	6076	6944	7812	8680
48×36	10.50	3150	4200	5250	6300	7350	8400	9450	10500
Press. Drop		.006	.010	.016	.023	.031	.040	.051	.063

found to be almost exactly 6-in. The den requires a 7-in. duct. These should be indicated on the duct layout (Figure 10-21). We can now determine the trunk duct size by using Table 10-12. Again, starting with the kitchen, this can be considered the trunk duct. Here the kitchen duct and the dining room duct intersect, we have two 6-in. ducts joining. From the table the resulting trunk duct would be 8-in. in diameter. The result of the 8-in. trunk duct connecting with the 7-in. feeder duct for the den is a 10-in. trunk duct connected to the discharge air plenum.

The duct to the master bedroom is a 7-in. pipe from the plenum to the supply air grill. There are no feeder ducts in this run.

The remainder of the duct runs can be sized in the same manner. The duct sizes should be indicated on the layout for future reference. Some designers include the cfm of each outlet on the plan. This is for future reference and convenience.

SUMMARY

The local climate conditions are very important in making the proper selection of an air distribution system.

Areas in cold climates require warm floors in the winter. Hot summer areas require comfort cooling.

An air distribution system that performs satisfactorily in one area may not perform properly in another area.

While the winter and summer inside and outside design conditions determine the capacity requirements of the heating and cooling equipment, it is the overall season conditions that determine the type of air distribution system used.

To aid us in finding the type of air distribution system to select, climatic zones, defined in terms of degree days, have been established.

No single type of air distribution system is best for all applications.

The location of the supply air outlets, and return air intakes, to maintain proper circulation and proper air temperatures should be based on given criteria.

The perimeter system is the most popular and most widely used air distribution system for residential comfort conditioning systems.

The radial system is popular in both slab floor and crawl space buildings.

The trunk and branch system is used in buildings with basements. It also provides satisfactory temperature gradients.

The ceiling diffuser system is popular in buildings with slab floors and where comfort cooling is the major equipment load.

The high-inside-wall furred ceiling system is popular in mild climates.

The overhead high-inside-wall system is popular in areas where the summers are hot.

The return air velocity in a residence should be around 500 fpm.

The return air intake should be located in a stratified zone and on a wall. Avoid floor or ceiling locations when possible.

Temperature variations from room to room should not exceed 3°F.

In split-level and tri-level homes, because of the different heat load characteristics in parts of the home, it may be necessary to use zone control for different levels.

Homes or buildings that are built in a U, L, or H shape cannot be properly balanced when only one thermostat is used.

The total quantity of air to be circulated through any building is dependent on the necessity for controlling temperature, humidity, and air distribution when either heating or cooling is required.

The total air quantity includes the amount of heating or cooling to be done as well as the type and nature of the building, locality, climate, height of the room, floor area, window area, occupancy, and method of distribution.

The air supplied to the conditioned space must always be adequate to satisfy the ventilation requirements of the occupants.

Metal ducts are preferable because it is possible to obtain smooth surfaces, thereby avoiding excessive resistance to the airflow.

When designing a more extensive system, it is good practice to gradually lower the velocity both in the main duct and in the remote branches.

Dampers and deflectors should be placed at all points necessary to assure a proper balance of the system.

In reducing the size or changing the shape of a duct, care must be taken that the angle of the slope is not too abrupt.

When designing a duct system, great care should be taken in shaping the elbows because sharp turns add greatly to the friction and lower the efficiency of the entire system.

By velocity is meant the rate of speed of the air traveling through the ducts or openings.

There are three methods used in designing air conditioning duct systems: (1) equal friction, (2) velocity reduction, and (3) static regain.

If maximum economy is to be reached, a careful evaluation and balancing of all the cost variables that enter into the design of a duct system should be considered with each design method.

Care should be exercised when using the equal-friction method to prevent branch air velocities from getting too high, thus avoiding noise problems.

The possible comfort conditions in a room are, to a great extent, dependent on the type and location of the supply air outlet grills and to some extent on the location of the return air intake grills.

The best types of outlets for heating use are those providing a vertical spreading air jet located in, or near, the floor next to an outside wall at the point of greatest heat loss, such as under a window.

The best types of outlets for cooling use are in the ceiling and provide a horizontal discharge air pattern.

The location of the return air grill is much more flexible than the supply air grill.

When designing a duct system it is often convenient to use a trunk duct, or extended plenum, to supply the feeders (branches) to different rooms.

When applications require spot heating or cooling, grills with a single row of adjustable blades are generally selected.

Grills that are manufactured with two rows of adjustable blades will direct the air in two planes.

The noise caused by the supply air entering the room is directly proportional to its velocity when leaving the supply air grill.

The most important points in selecting a return air grill are: (1) it must handle the required cfm, and (2) it must handle the desired velocity.

Chapter 11

Duct Pipe, Fittings, and Insulation

To direct the conditioned air to the desired space, some type of carrier is needed. These air carriers are referred to as ducts. They are made of many different types of materials which are fire resistant.

Air ducts work on the principle of air pressure differential. When a difference in pressure is present, air will move from the place of high pressure to the low-pressure area. When this pressure difference is great, the airflow will be faster than when the pressure areas are nearer the same.

The types of duct systems that we are concerned with in most air conditioning installations are low pressure. A low pressure duct system is one that conveys air at velocities less than 2000 fpm and a static pressure in the duct of 2 in. water column or less.

INTRODUCTION

Duct systems can be put into two general classifications when used for ventilation: (1) systems where the movement of air is of prime importance, without regard to quiet operation or power economy. In these systems the air velocity will be high. (2) systems where air must be moved quietly and with power economy. In these systems the air velocity will be low.

Pressure losses in duct systems are caused by the velocity of the airflow, number of fittings, and the friction of the air against the sides of the duct. The use of heating and cooling coils, air filters, air washers, dampers, and deflectors all increase the resistance to air flow.

DUCT PIPE SHAPES

Duct pipe is available in round, rectangular, or square shapes. Round duct is generally preferred because it can carry more air in less space than the other shapes. Also, less material is used, and therefore there is less surface, reducing the amount of friction and heat transfer through the duct, and less insulation is required when compared to the other popular shapes. The rectangular shape is generally used where appearance is a factor because a flat surface is easier to work with in regard to the surface of the room or space. Rectangular duct is also sometimes preferred when space is limited or of unusual shape, such as between floor joists or wall studs (Figure 11-1).

Figure 11-1. Square and rectangular duct.

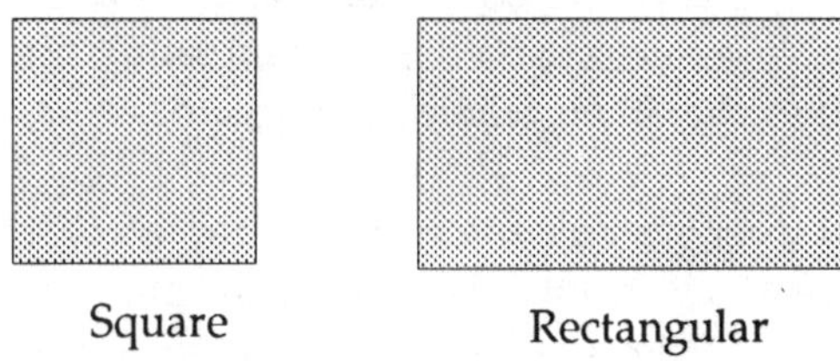

In practice, rectangular duct is sometimes used for plenums and round duct is used for the branch runs and usually the trunk line. Usually, a combination of rectangular duct and round duct is used to make take-offs from the rectangular plenum or trunk duct to the round duct branch runs (Figure 11-2).

Also, rectangular plenum or trunk ducts or rectangular branch take-offs and runs are used (Figure 11-3). When used in this manner the takeoff fittings are also completely rectangular.

The space between floor joists is sometimes used as an integral part of the air distribution system. In these cases, the joist space must be made airtight. In some installations, a space above the ceiling or the crawl space beneath the floor may be used as discharge or return air plenums. When these spaces are used for this purpose, they must be sealed airtight, made vapor tight, and insulated. Be sure to check the local codes and ordinances when using this method. The space between the floor joists may be used as part of the return air system when properly sealed. This space is seldom used for supply air because of the heat transfer through the wall and the possibility of condensation during the cooling cycle.

Figure 11-2. Extended plenum round and rectangular duct takeoff connections.

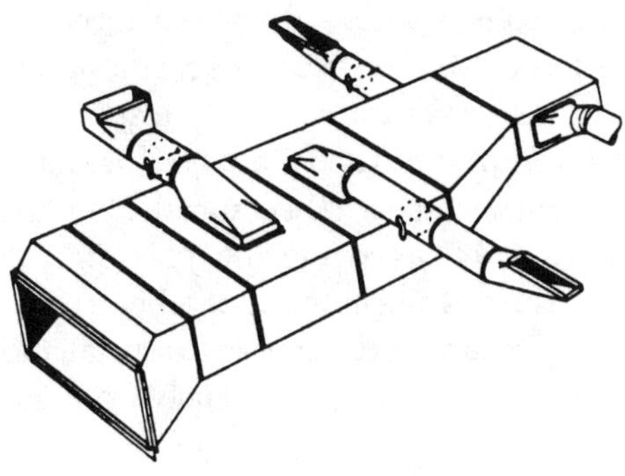

(a)Round pipe takeoff

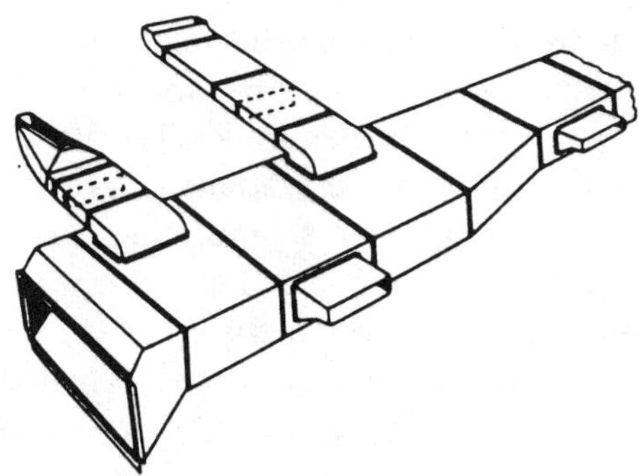

(b) Rectangular pipe takeoff

Figure 11 3. Rectangular duct and branch ducts.

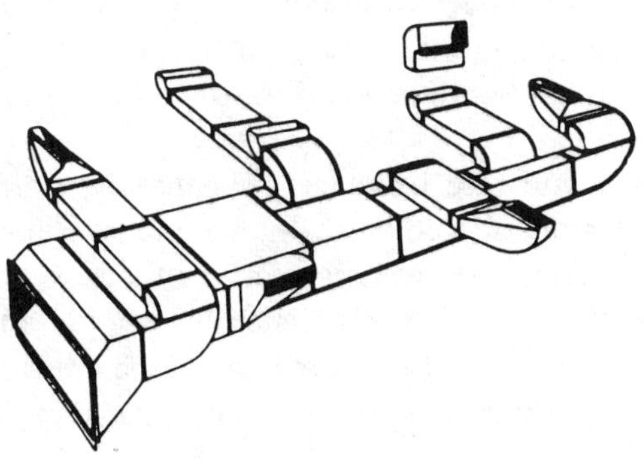

DUCT MATERIALS

Air ducts are made from a wide variety of materials, such as sheet metal, aluminum, flexible glass fiber, duct board, cement, and tile. Each type of material has certain advantages in a specific application; however, some have equal advantages for the same application and the one used is a matter of preference. Aluminum and glass fiber are fairly light in weight and are less subject to corrosion than sheet metal. However, they both cost more and are more easily damaged. Flexible glass fiber ducts are becoming more common because of the amount of labor saved during installation. Tile and cement pipe are more suitable for installation in a slab floor than other types because they will not crush from the weight of the concrete.

DUCTWORK FITTINGS

Fittings for ductwork are available in almost any shape and any size; adjustable and nonadjustable, round or rectangular in cross section, factory made and shop made are some examples. The fittings illustrated here include elbows, bends, turning vanes, reducers, takeoffs, collars, flexible connections, dampers, end caps, boots, register heads, offsets, floor pans, transitions, and combinations. When a duct system is being designed, careful consideration should be given to the use of fittings that change the direction of air flow or change the duct size. These changes should be kept to a minimum because each change of direction or reduction increases the resistance to the airflow. It is not meant that these fittings should not be used but it does mean that their use should be required for the desired system. Thus, the duct system should be kept as simple as possible to avoid increasing the resistance unnecessarily. Each one of these fittings add a resistance equal to a given number of straight pipe. Complete tables that give the resistance can be found in duct manufacturing manuals.

The centerline radii of elbows should be a minimum of one and one-half times the pipe diameter for round ducts. For rectangular ducts the radii should be a minimum of one-half the duct dimension in the turning plane.

All ductwork must be installed permanently, rigid, nonbuckling and rattle free. All joints should be airtight. Standards 90A and 90B of the National Board of Fire Underwriters specifies the type of materials that may be used in duct manufacture. Generally, supply air plenums and ducts should be constructed of noncombustible materials that are equivalent in strength and durability to those recommended in Table 11-1.

Table 11-1. Recommended Thickness for Duct Materials. (Copyright 1981 by the American Society of Heating, Refrigerating and Air Conditioning Engineers, Inc., from Fundamentals Handbook. Used by permission)

Round Ducts Diameter, In.	Minimum Thickness Galv. Iron, U.S. Gage	Minimum Thickness Aluminum B&S Gage	Minimum Weight of Tin-Plate
Less than 14	30	26	
14 or more	28	24	IX (135 lb)

Rectangular Ducts Diameter, In.	Minimum Thickness Galv. Iron, U.S. Gage	Minimum Thickness Aluminum B&S Gage	Minimum Weight of Tin-Plate
Ducts Enclosed in Partitions			
14 or less	30	26	
Over 14	28	24	IX (135 lb)
Ducts Not Enclosed in Partitions			
Less than 14	28	24	—
14 or more	26	23	—

Note: The table is in accordance with Standard 90B of the National Board of Fire Underwriters. Industry practice is to use heavier gage metals where maximum duct widths exceed 24 in.

The supply air ducts that serve single-family residences are not required to meet these requirements, except for the first 3 ft. from the unit. The unit must be a listed unit. They must be constructed from a base material or mineral and be properly applied. Also, combustible material that is otherwise suitable for the given application may be used as supply air ducts which are completely encased in concrete.

All supply air ducts should be supported and secured by metal hangers, metal straps, lugs, or brackets. Nails should not be driven through the duct walls and no unnecessary holes should be made in them.

Supply air stacks should not be installed in the outside walls, unless it would be impractical to place them elsewhere. When it is necessary to place them in the outside wall, the stack must be completely insulated against outside temperatures. Omission of the insulation will greatly reduce the heating or cooling capacity of that branch duct, resulting in unsatisfactory conditions in the area supplied by that duct.

All supply air ducts should be equipped with an adjustable locking volume damper for air control. This damper should be installed in the branch duct as far from the outlet grill as possible and be completely accessible for adjustment. Return air systems that have more than one return air intake may be equipped with volume dampers.

Noise should be eliminated when possible. When metal ducts are used, they should be connected to the unit by pieces of flexible, fire resistant fabric (flexible connectors). The electrical conduit and piping may increase noise transmission when connected directly to the unit. Return air intakes located next to the unit may also increase noise transmission. Fans located directly beneath a return air grill should be avoided.

INSULATION

When choosing and estimating the insulation to be included in the air distribution system, certain considerations must be given to the code requirements for insulations. ASHRAE Standard 90-75, "Energy Conservation in New Building Design," was developed by the American Society of Heating, Refrigerating and Air Conditioning Engineers in 1975 "to address new building design for effective utilization of energy." The standard establishes energy-efficient design requirements for:

1. Heating, ventilating, and air conditioning systems and equipment
2. Building exterior envelopes
3. Electrical distribution systems

These design standards have been adopted by most states as mandatory building codes and standards, either by legislative requirement or executive order. Those states that have not already made this adoption are moving in that direction. A strong push for this acceptance has been given by appropriate federal agencies, which could prevent funding assistance for any new building construction for states that have not established energy codes and regulations.

Listed below are the particular sections of ASHRAE 90-75 that deal with air distribution system insulation:

5-11. Air Handling Duct System Insulation: All ducts, plenums, and enclosures installed in or on buildings shall be thermally insulated as follows:

5-11.1. All duct systems, or portions thereof, shall be insulated to provide a thermal resistance, excluding film resistances, of

$$R = \Delta T \div 15 \text{ ft} - {}^\circ\text{F/Btu or } R = \Delta T \div 47.3 \text{ m} - \text{K/W}$$

where ΔT is the design temperature differential between the air in the duct and the surrounding air in °F.

Exceptions. Duct insulation is not required in any of the following cases:

a. Where the ΔT is 25°F or less.

b. For supply or return air ducts installed in basements, cellars, or unvented crawl spaces with insulated walls in one- and two-family dwellings.

c. When the heat gain or loss of the ducts, without insulation, will not increase the energy requirements of the building.

d. Within the HVAC equipment.

e. For exhaust air ducts.

5.11.2. Uninsulated ducts in uninsulated sections of exterior walls and in attics above the insulation might not meet the requirements of this standard.

5.11.3. The required thermal resistances do not consider condensation. Additional insulation with vapor barriers may be required to prevent condensation under some conditions.

5.12. Duct Construction: All ductwork must be constructed and erected in accordance with Chapter 1 of the 1975 ASHRAE Handbook and Product Directory, Equipment Volume, or the following NESCA/SMACNA or SMACNA standards.

a. Residential Heating and Air Conditioning Systems—Minimum Installation Standards, August 1973, NESCA/SMACNA.

b. Low Velocity Duct Construction Standards, 4th edition, 1969.

c. High Velocity Duct Construction Standards, 2nd edition, 1969.

d. Fibrous Glass Duct Construction Standards, 3rd edition, 1972.

e. Pressure Sensitive Tape Standards, 1973 (for fibrous glass ducts only).

5.12.1. High-pressure and medium-pressure ducts must be leak-tested in accordance with the applicable SMACNA standard, with the rate of leakage not to exceed the maximum rate specified in that standard.

5.12.2. There are no standards at this time for leak testing of low-pressure ducts. When low-pressure supply air ducts are located outside the conditioned space (except return air plenums), all traverse joints must be sealed using mastic or mastic plus tape. For fibrous glass ductwork, pressure sensitive tape is acceptable.

5.12.3. There is no standard at this time for damper leakage. Automatic or manual dampers installed for the purpose of shutting off outside air intakes for ventilation air must be designed with tight shut-off characteristics to minimize air leakage.

INSULATION TYPES AND USES

There are many types of insulation available for increasing the performance of air distribution systems. The type and thickness required for the particular installation depends on the local and national codes as well as system performance.

The following items explain the various types of insulation available and their uses.

Glass Fiber Duct Wrap

This is a resilient blanket of glass fiber insulation, available unfaced or faced with a reinforced foil kraft paper barrier facing (FRK) (Figure 11-4).

It is used to insulate residential and commercial air conditioning or dual-temperature sheet metal ducts operating at temperatures from 40 to 250°F.

Glass Fiber Duct Board

A complete duct product with glass fiber thermal and acoustical insulation bonded to a tough, flame-resistant aluminum foil barrier facing, for fabrication of rectangular ductwork and fittings (Figure 11-5).

Figure 11-4. Fiberglas duct wrap. (Courtesy of Owens-Corning Fiberglas Corp.)

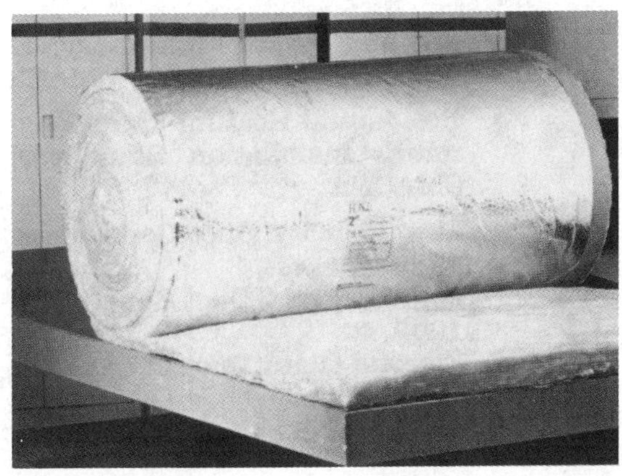

Figure 11-5. Fiberglas duct board. (Courtesy of Owens-Corning Fiberglas Corp.)

It is used for low-velocity heating, ventilating, and air conditioning duct systems in residential and commercial construction operating at temperatures up to 250°F, velocities to 2400 fpm, and 2-in. static pressure.

Flexible Duct Liner

Duct liner is a bonded mat of glass fiber coated with a black pigmented coating on the side toward the airstream. This coating tightly bonds the surface fibers to resist damage during installation and in service, and provides a uniquely tough airstream surface. It is available in three types, and in thicknesses of 1/2, 1, 1-1/2, and 2 in. (Figure 11-6).

Figure 11-6. Duct liner. (Courtesy of Owens-Corning Fiberglas Corp.)

Duct liner is designed for use as an acoustical and thermal insulation for sheet metal heating, cooling, and dual-temperature ducts and plenums operating at velocities up to 6000 fpm and temperatures to 250°F. The product is applied to the interior of the ductwork or plenum.

Duct Liner Board

This is a semirigid bonded board of glass fiber, coated with a flame-resistant coating to resist damage during installation and in service (Figure 11-7).

Figure 11-7. Duct liner board. (Courtesy of Owens Corning Fiberglas Corp.)

It is an acoustical and thermal insulation for sheet metal heating, cooling, and dual-temperature ducts and plenums operating at temperatures to 250°F, and velocities to 6000 fpm. It is applied to the interior of the duct.

Flexible Duct

Flexible duct is a lightweight flexible duct formed with a resilient inner air barrier, glass fiber insulation, and a reinforced vapor barrier jacket (Figure 11-8).

It is used as an air duct or connector on supply and return air ducts in residential, industrial, and commercial heating, ventilating, and air conditioning systems operating at temperatures to 250°F. In addition, it can be used on runouts to diffusers, registers, and mixing boxes. The flexible feature allows it to conform to gradual bends necessary when connecting air ducts to diffusers, or when routing air ducts through spaces with many obstructions.

INSTALLING DUCT WRAP INSULATION

Duct wrap insulation is used to reduce heat loss or gain through sheet metal heating and air conditioning ductwork. Properly insulated sheet metal ducts can mean that warmer air in the winter and cooler air in the summer will reach the rooms served by the ductwork, rather than losing or gaining heat energy through the duct walls.

Figure 11-8. Flexible duct. (Courtesy of Owens-Corning Fiberglas Corp.)

It is important to know that the higher the R value of the insulation, the greater the insulating value. It is equally important to realize that the R value of duct wrap insulation, once it is installed, depends on how much it is compressed during installation. If it is compressed too much, it will lose some of its insulating value.

Full Installed R Value

The instructions outlined here, if followed exactly, will assure you of getting the full R value of the duct wrap insulation. The installed R value is printed on the facing of the insulation by most manufacturers. It also appears on the product labels in each package of both faced and unfaced duct wrap insulations.

The installed R value may be obtained when the duct wrap is compressed to no less than 75% of its "as manufactured" thickness after being installed.

The key to correct installation is to cut each piece of duct wrap insulation to the stretch-out length in Table 11-2. This will prevent the insulation from being overcompressed (Figure 11-9).

Table 11-2. Length of Duct Wrap. (Courtesy of Owens-Corning Fiberglas Corp.)

If you are using duct wrap insulation which has these thicknesses printed on label:		Cut the duct wrap insulation to the lengths shown below, depending on shape of duct:		
Nominal (As-manufactured) Thickness (in.)	Average Installed Thickness (in.)	Round and Oval Ducts	Square Ducts	Rectangular Ducts
1.0*	3/4	P + 7.0	P + 6.0	P + 5.0
1.5	1-1/8	P + 9.5	P + 8.0	P + 7.0
2.0	1-1/2	P + 12.0	P + 10.0	P + 8.0
3.0	2-1/4	P + 17.0	P + 14.5	P + 11.5
4.0	3	P + 22.0	P + 18.5	P + 14.5

*Available in California only.

To obtain full installed R value of insulation, use the following steps:

1. Check the product label to determine the type and thickness of the duct wrap insulation you have bought (Figure 11-10).

Figure 11-9. Proper insulation tightness. (Courtesy of Owens-Corning Fiberglas Corp.)

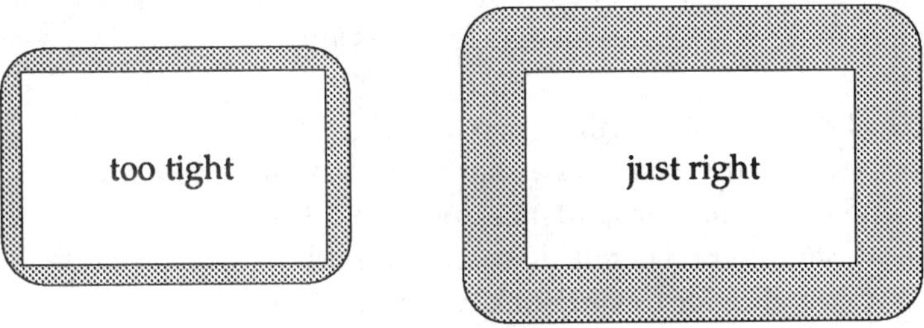

Figure 11-10. Insulation label. (Courtesy of Owens-Corning Fiberglas Corp.)

Ⓐ Wrap type
Ⓑ Installed R-value
Ⓒ Out-of-package R-value, thickness, width, and amount

2. Determine the shape and perimeter (P) dimension of the bare sheet metal duct. This dimension will be the circumference of round and oval ducts, or twice the height plus twice the width of square and rectangular ducts. Use Table 11-2 to determine the amount of duct wrap you must cut from the roll. (The stretch-out length).

Example

Your sheet metal ducts measures 8 in. high and 12 in. wide. The perimeter (P) dimension is twice the 8-in. height plus twice the 12-in. width: 2(8) + 2(12) = 16 + 24 = 40 in. total. If you are using 2-in.-thick duct wrap insulation, you should cut it to P = 8-in., or 48 in., to obtain the 1.5-in. average installed thickness and the full installed R value for the type of insulation that you bought.

Make the following checks before wrapping the ducts:

1. Check to see that the sheet metal ducts are tightly sealed. Air leakage at sheet metal joints wastes energy and will reduce the effectiveness of any insulation installed on the ducts. If the duct joints are sealed, it will help the duct wrap insulation do its job.

2. Collect the tools that will be needed. These are:
   a. Large utility shears, or a sharp serrated kitchen knife.
   b. Tape measure, or yardstick and some string.
   c. A marking pen (a felt-tip marker is best).
   d. A straightedge (a yardstick will serve here).
   e. 2-1/2- or 3-in. pressure-sensitive tape (foil or vinyl, to match the facing on the duct wrap insulation).

Now you are ready to cut, wrap, and tape the insulation. There are three simple steps to install the duct wrap insulation:

1. Cut:

- For faced duct wrap insulation:
   a. Use the felt-tipped marker and the straightedge to mark a fine line across the facing of the duct wrap for cutting a piece to the correct stretch-out dimension (Figure 11-11).
   Cut along this line with the facing up, using the utility shears or the serrated kitchen knife.
   b. Cut away a 2-in.-wide piece of the fiberglass insulation, being careful not to cut through the facing. This will provide a 2-in. overlapping tape flare (Figure 11-12).

Figure 11-11. Measuring insulation. (Courtesy of Owens-Corning Fiberglas Corp.)

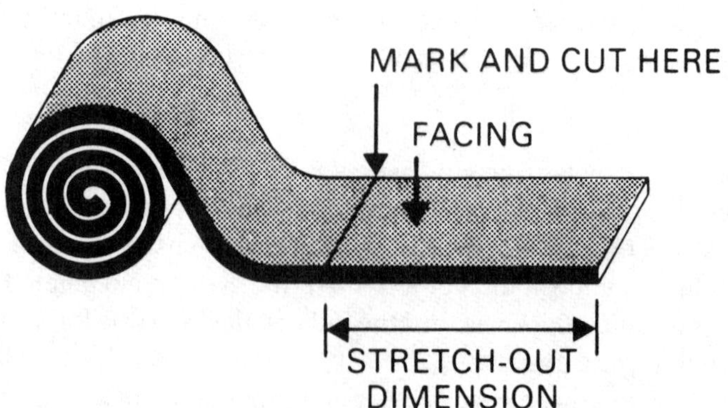

- For unfaced duct wrap insulation:
    a. Use the felt-tip marker and the straightedge to mark a line across the duct wrap for cutting a piece to the correct stretch-out dimension (Figure 11-13).
    b. Cut along this line using the utility shears or the serrated kitchen knife.

2. Wrap:

- For faced duct wrap insulation:
    a. Wrap the insulation around the duct, with the facing outside and the glass fiber insulation against the duct. The tape flap should overlap the insulation and the facing at the other end of the piece of duct wrap. The insulation should be tightly buttoned (Figure 11-14).

Figure 11-12. Cutting insulation. (Courtesy of Owens-Corning Fiberglas Corp.)

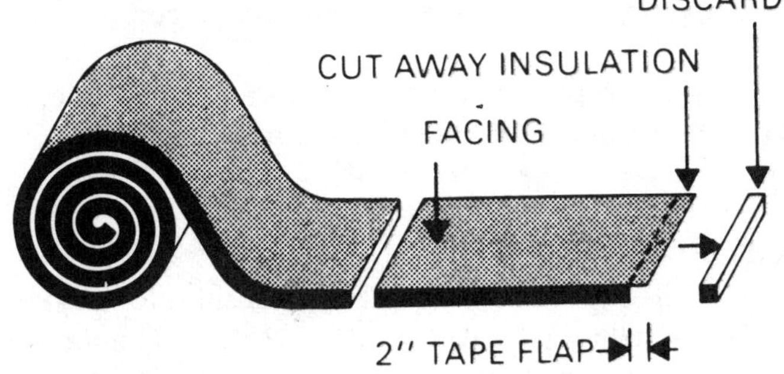

Figure 11-13. Cutting for unfaced duct wrap. (Courtesy of Owens-Corning Fiberglas Corp.)

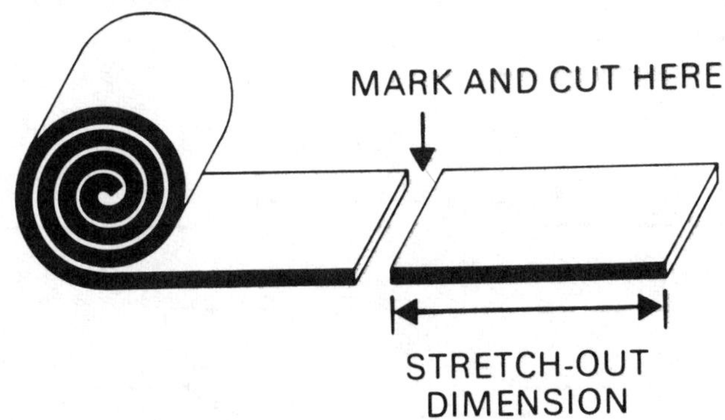

Figure 11-14. Wrapping faced duct wrap insulation. (Courtesy of Owens-Corning Fiberglas Corp.)

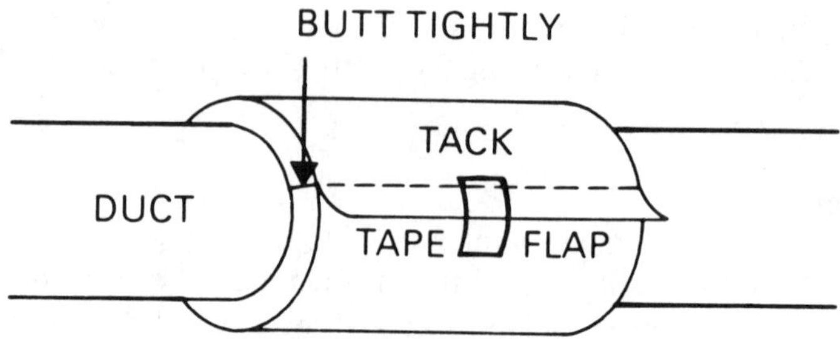

b. Next tape the flap with a short piece of tape in the center of the flap. Be careful not to pull too hard on the tape flap.

- For unfaced duct wrap insulation:
  a. Wrap the insulation around the duct. One end of the duct wrap should overlap the other end by about 2 in. (Figure 11-15).

  Next, tuck the overlap by stitching the insulation with finishing nails. Do not pull the insulation too tightly around the duct.

3. Tape:

- For faced duct wrap insulation:
  a. First tape the flap seam, using pressure-sensitive foil or vinyl tape compatible with the insulation facing. Rub the tape firmly with your hand. Be careful not to puncture the facing. Patch any holes or tears with tape (Figure 11-16).

Figure 11-15. Wrapping unfaced duct wrap. (Courtesy of Owens-Corning Fiberglas Corp.)

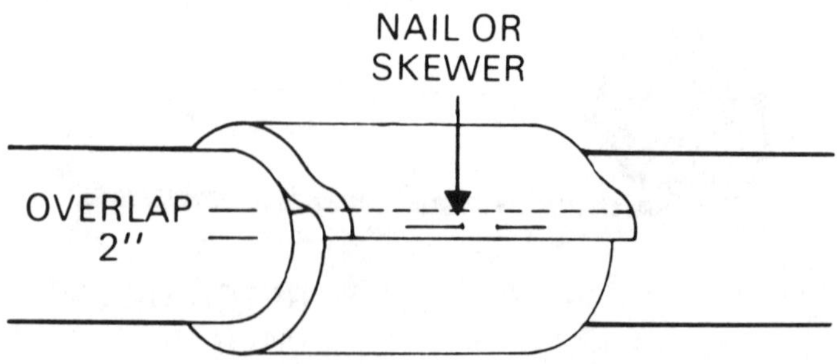

Figure 11-16. Taping faced duct wrap. (Courtesy of Owens-Corning Fiberglas Corp.)

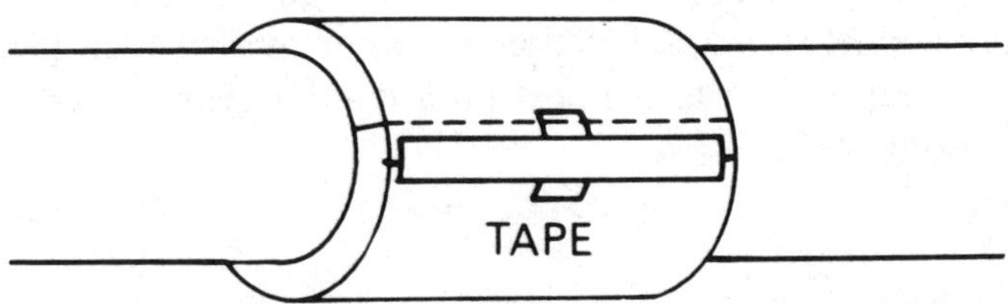

b. As you move along the duct, repeat the foregoing steps, butting each piece of duct wrap tightly against the previously installed piece so that the facing flap that runs the length of the roll the wrap is overlapping (Figure 11-17).
c. Next, tape all the way around each butt joint, continuing the steps until the duct is completely wrapped and taped (Figure 11-18).

Figure 11-17. Taping faced duct wrap (Continued). (Courtesy of Owens-Corning Fiberglas Corp.)

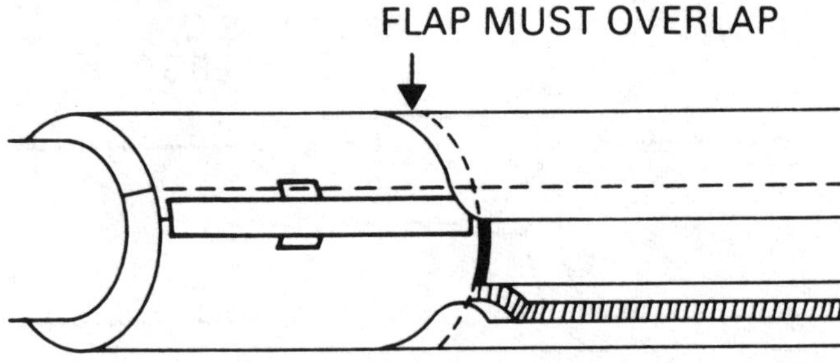

Figure 11-18. Taping all joints of faced duct wrap. (Courtesy of Owens-Corning Fiberglas Corp.)

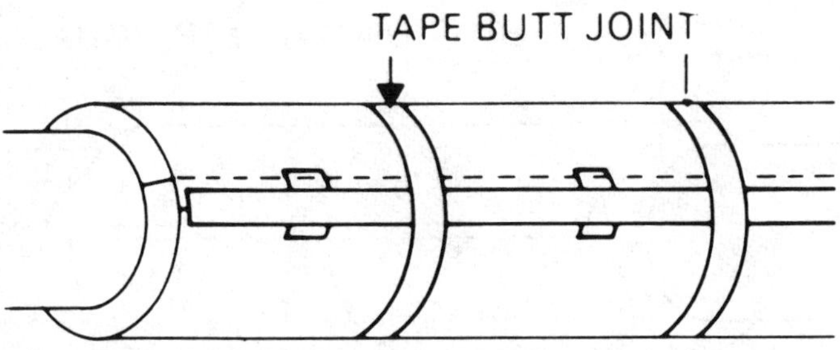

- For unfaced duct wrap insulation:
  a. Secure the duct wrap insulation overlap by stitching it together using nails or skewers 4-in. apart (Figure 11-19).
  b. Repeat step a, overlapping 2-in. or butting each piece of duct wrap tightly to the next until the duct is completely wrapped (Figure 11-20).

SUMMARY

Duct systems can be put into two general classifications when used for ventilation: (1) systems where the movement of air is of prime importance without regard to quiet operation or power economy, and (2) systems where air must be moved quietly and with power economy.

Pressure losses in duct systems are caused by the velocity of the airflow, number of fittings, and friction of the air against the sides of the duct.

Figure 11-19. Securing unfaced duct wrap. (Courtesy of Owens-Corning Fiberglas Corp.)

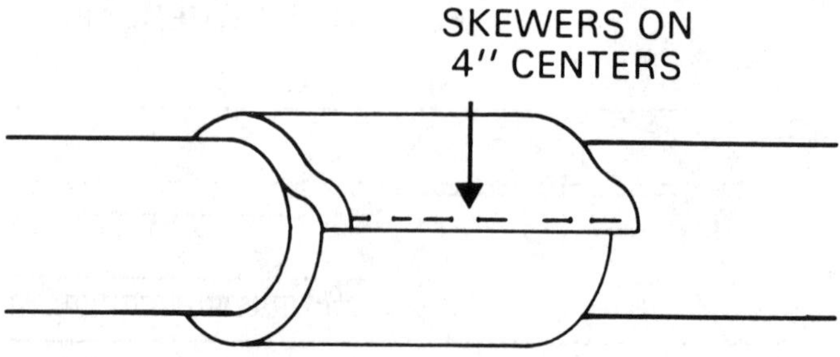

Figure 11-20. Joining unwrapped duct wrap. (Courtesy of Owens-Corning Fiberglas Corp.)

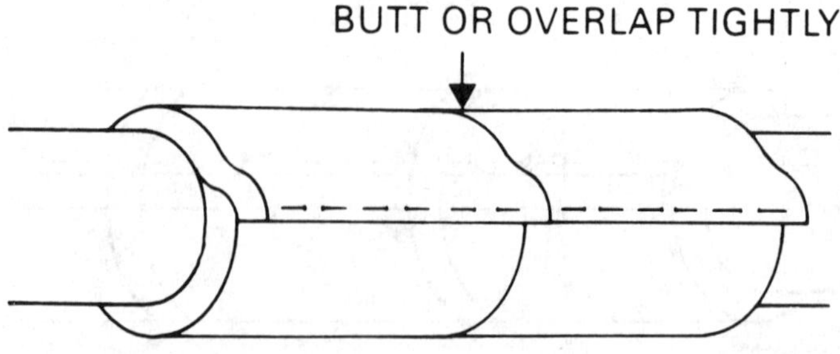

Round ducts are generally preferred because they can carry more air in less space. Also, less material is used, and therefore there is less surface to reduce the amount of friction and heat transfer through the duct wall, and less insulation is required than for other popular shapes.

In practice, rectangular duct is used for the plenums and round duct is used for the branch runs and usually the trunk line.

In some installations, a space above the ceiling or the crawl space beneath the floor may be used as discharge or return air plenums. These spaces must be sealed airtight.

Ducts are made from materials such as sheet metal, aluminum, flexible glass fibers, duct board, cement, and tile.

Fittings for ductwork are available in almost any shape and any size: adjustable and nonadjustable, round or rectangular in cross section, factory made and shop made are some examples.

Changes in air flow through a duct should be kept to a minimum because each change of direction or reduction increases the resistance to airflow. Each fitting used adds a resistance equal to a given number of feet of straight pipe.

All ductwork must be installed permanently, rigid, nonbuckling, and rattle free. All joints should be airtight.

All supply air ducts should be supported and secured by metal hangers, metal straps, lugs, or brackets. Nails should not be driven through the duct walls and no unnecessary holes should be made in them.

All supply air ducts should be equipped with an adjustable volume damper for air control.

Return air systems that have more than one return intake may be equipped with volume dampers.

Noise should be eliminated when possible. When metal ducts are used, they should be connected to the unit by pieces of flexible, fire-resistant fabric (flexible connector).

When choosing and estimating the insulation to be included in the air distribution system, certain considerations must be given to the code requirements for insulations.

Chapter 12

Indoor Air Quality

All air contains almost constant amounts of nitrogen (78% by volume), oxygen (21% by volume) and argon (0.9%), and has varying amounts of carbon dioxide (0.03%) and water vapor (up to 3.5%). These are combined with other trace quantities of inert gases such as: neon, xenon, krypton, helium, and other gases, which are always present in air. When gases other than these are present in the air they are considered to be contaminants or pollution. However, the concentration of these types of gases is usually small, but they do have serious effects on the persons who are exposed to them. It is usually desirable to either reduce or remove these pollutants.

In most instances, indoor pollutants are reduced by the introduction of outdoor air brought in through the ventilation system. The introduction of outdoor air should always be kept to a minimum and a high recirculation rate of the indoor air used to aid in energy conservation. However, it should be realized that the recirculated air cannot be cleaned to near the condition of the outdoor air by the simple removal of particulate contamination. Noxious odors and toxic gaseous contaminants must also be removed by use of the proper equipment. The equipment used for the removal of these contaminants is different from that used for the removal of particulate contaminants. Also, the available indoor air might contain undesirable gaseous contaminants at unacceptable concentrations. If this is possible, the air must be treated with some type of gaseous contaminant removal equipment.

GASEOUS CONTAMINANT CHARACTERISTICS

Before a control system can be designed that will control the present pollutants, the contaminants must be identified, and the minimum following items of information determined:

1. The exact chemical identity of the contaminants.

2. The rates at which the contaminants are generated in the space, and the rates at which they are brought into the space with the outdoor air used for ventilation.

This information is usually very difficult to obtain. System design engineers must usually make do with a chemical family name (such as aldehydes), and a qualitative description of the generation rates (such as from a plating tank), or the perceived concentration (such as at odorous levels). The effectiveness of the system design under these conditions is very uncertain. In cases when the exact identity of a chemical is known, both the chemical and the physical properties which influence its collection by the control devices can usually be obtained from handbooks and technical publications. Some of the factors which are of special importance are:

A. Molecular weight of the contaminant
B. Normal boiling point (i.e., at 1 standard atmosphere)
C. Heat of vaporization
D. Polarity
E. Chemical reactivity and chemisorption velocity

When this information is accurately obtained, the performance of the removal equipment on the contaminants for which no specific tests have been made may be estimated. The following sections on removal methods cover the usefulness of such estimations.

Some gaseous contaminants that have unique properties must be considered in the design of the removal system. For example, ozone will reach an equilibrium concentration in a ventilated space without any type of filtration device. This is possible because a molecule of ozone will combine with other molecules of ozone to form a normal oxygen, while it also reacts with people, plants, and other materials in the space. This oxidation is harmful to all three. Therefore, natural ozone decay is not a satisfactory method of ozone control except when low concentrations are present. Low concentrations of ozone are in the range of less than 0.2 mg/m^3. Activated carbon is a major method of readily controlling ozone contamination because it both reacts with and catalyzes its conversion to oxygen.

Radon is a radioactive gas that decays by alpha-particle emission. Eventually, it gives up individual atoms of polonium, bismuth, and lead. These atoms form extremely fine aerosol particles which are called daughters or radon progeny, which are also radioactive and are especially toxic because they lodge deep in the human lung and emit cancer-producing alpha particles. Radon progeny which are both attached to larger particles of the aerosol and alone (unattached) can be captured by the use of particulate air filters. The radon gas itself must be removed with activated carbon filters. However, this method is very expensive and is not usually cost effective in HVAC systems. The control of the radon emission at its source and by ventilation are the preferred methods of controlling radon contaminants in HVAC applications.

Sulphur trioxide (SO_3) is another gaseous contaminant that is often present in particulate form. It should not be confused with sulphur dioxide (SO_2). It will react with water vapor at ambient temperatures to form a fine mist of sulfuric acid. This fine mist will collect on particulate filters and when no means of removing it are present, it will evaporate and re-enter the protected space. Because many such conversions and desorptions are possible, the designer must understand the problems that they cause.

Some of the major families of gaseous pollutants and examples of specific compounds are given in Table 12-1.

Other sources of such information are: *The Meric Index*, the *Toxic Substances Control Act Chemical Substance Inventory (EPA 1979)*, and *Dangerous Properties of Industrial Materials* all of which are useful in identifying contaminants, including some known by trade names only. It should be noted that the same chemical compound, especially organic compounds, may have several scientific names.

HARMFUL EFFECTS OF GASEOUS CONTAMINANTS

The major reasons for the removal of gaseous contaminants from indoor air is because they have adverse effects on either the occupants, the space, or both. Different concentrations of these contaminants will produce different effects. Usually, the contaminants will become annoying first by their odors then by becoming toxic to the occupants. It should be remembered that this is not always true. As an example, the highly toxic, even deadly, contaminant carbon monoxide has no odor.

The harmful effects of contaminants may be divided into four categories: toxicity, odor, irritation, and material damage. These effects are dependent upon the concentration of the contaminants and are usually expressed as follows:

Table 12-1. Major Chemical Families of Gaseous Air Pollutants (with examples). (Copyright 1991 by the American Society of Heating, Refrigerating and Air Conditioning Engineers, Inc., from HVAC Applications Handbook. Used by permission)

Inorganic Pollutants

1. Single-Element Molecules
   chlorine
   radon
   mercury

2. Oxidants
   ozone
   nitrogen dioxide
   nitrous oxide
   nitric oxide

3. Reducing Agents
   carbon monoxide

4. Acid Gases
   sulfuric acid
   hydrochloric acid
   nitric acid

5. Nitrogen Compounds
   ammonia

6. Sulfur Compounds
   hydrogen sulfide

7. Miscellaneous
   arsine

Organic Pollutants

8. n-Alkanes
   methane
   n-butane
   n-hexane
   n-octane
   n-hexadecane

9. Branched Alkanes
   2-methyl pentane
   2-methyl hexane

10. Alkenes and Cyclohexanes
   1 -octene
   1-decene
   cyclohexane

11. Chlorofluorocarbons
   R-11 (trichlorofluoromethane)
   1, 1, 1 trichloroethane
   R-114 (dichlorotetrafluoroethane)

12. Halide Compounds
   carbon tetrachloride
   chloroform
   methyl bromide
   methyl iodide
   phosgene
   carbonyl sulfide

13. Alcohols
   methanol
   ethanol
   2-propanol
   isopropanol
   phenol
   cresol
   diethylene glycol

14. Ethers
   vinyl ether
   methoxyvinyl ether
   n-butoxyethanol

Table 12-1 (*Continued*)

15. Aldehydes
    formaldehyde
    acetaldehyde

16. Ketones
    2-butanone (MEK)
    2-propanone
    acetone
    methyl isobutyl ketone
    chloroacetophenone

17. Esters
    ethyl acetate
    n-butyl acetate
    di-ethylhexyl phthalate (DOP)
    di-n-butyl phthalate
    butyl formate
    methyl formate

18. Nitrogen Compounds, and Other Than Amines
    nitromethane
    acetonitrile
    acrylonitrile
    pyrrole
    pyridine
    hydrogen cyanide
    peroxyacetal nitrate

19. Aromatic Hydrocarbons
    benzene
    toluene
    ethyl benzene
    naphthalene
    p-xylene
    benz-alpha-pyrene

20. Terpenes
    2-pinene
    limonene

21. Heterocylics
    furan
    tetrahydrofuran
    methyl furfural
    nicotine
    1,4 dioxane
    caffeine

22. Organophosphates
    malathlon
    tabun
    sarin
    soman

23. Amines
    methylamine
    diethylamine
    n-nitroso-dimethyamine

24. Monomers
    vinyl chloride
    methyl formate
    ethylene

25. Mercaptans and Other Sulfur Compounds
    bis-2-chloroethyl sulfide (mustard gas)
    ethyl mercaptan
    methyl mercaptan
    carbon disulfide

26. Miscellaneous
    ethylene oxide

ppm = parts of contaminant by volume per million parts of air by volume

ppb = parts of contaminant by volume per billion parts of air by volume

mg/m^3 = milligrams of contaminant per cubic meter of air

$\mu g/m^3$ = micrograms of contaminant per cubic meter of air

$$ppm = 62.32\,(mg/m^3)(273.15 + t)/(Mp) \quad (1)$$

$$mg/m^3 = 0.01605(ppm)(Mp)/(273.15 + t) \quad (2)$$

where:

M = molecular weight of the contaminant

p = mixture pressure, mm Hg

t = mixture temperature, °C

The concentration data is generally reduced to standard temperature and pressure (i.e., 25°C and 760 mm Hg), in which case:

$$ppm = 24.45(mg/m^3)/M \quad (3)$$

TOXICITY

The harmful effects that a contaminant has on a person depends on two things; First, the short-term peak concentrations, and two, the time-integrated exposure received by the person. The allowable concentration for short exposure is higher than for long term exposure. The Occupational Safety and Health Administration (OSHA) has defined three periods for concentration averaging and has assigned allowable levels that may exist in these categories in workplaces for over 490 compounds, which are mostly gaseous contaminants. The abbreviations for the three concentration averaging periods are:

AMP = acceptable maximum peak for a short term exposure

ACC = acceptable ceiling concentration, not to be exceeded during an 8-hour shift, except for periods where AMP applies

TWA8 = time-weighted average, not to be exceeded in any 8-hour shift of a 40-hour workweek.

In literature published by someone other than OSHA, ACC is sometimes called STEL (short-term exposure limit), and TWA8 is sometimes called TLV (threshold limit value). The medical profession does not agree with OSHA about what values should be assigned to AMP, ACC, and TWA8. OSHA values, which change periodically, are published annually in the *Code of Federal Regulations* (29 CFR 1900, Part 1900.1000 FF) and intermittently in the *Federal Register*. A similar list is also available from the American Conference of Governmental Industrial Hygienists (AGGIH 1989b). The National Institute for Occupational Safety and Health (NIOSH) is charged with researching toxicity problems, and it greatly influences the legally required levels. NIOSH annually publishes the *Registry of Toxic Effects of Chemical Substances* as well as numerous *Criteria for Recommended Standard for Occupational Exposure to (Compound).* Some compounds not listed in the OSHA literature are listed in the NIOSH literature, and their recommended levels are sometimes lower than the legal requirements set by OSHA. The NIOSH *Pocket Guide to Chemical Hazards* (Mackison *et al.* 1978) is a condensation of these references and is convenient for engineering purposes. This publication also lists the values for the following toxic limit:

IDLH = immediately dangerous to life and health

Even though this toxicity level is seldom used by HVAC engineers, they should consider it when deciding the amount of recirculation air that is safe in any given system. The amount of ventilation air must never be so low as to allow the concentration level of a contaminant to rise to the IDLH level. The contamination levels set by OSHA and NIOSH define acceptable occupational exposures and cannot be used alone as acceptable standards for residential or commercial applications. However, they do suggest some upper limits for contaminant concentrations for design purposes.

The OSHA regulations do not address the exposure of personnel to radioactive gases, but by rules promulgated by the Nuclear Regulatory Commission, allowable concentrations, in terms of radioactivity, are listed annually in the *Code of Federal Regulations,* Section 10, Part 50, (10 CFR 50, 1990). Usually, radioactive gases are controlled much the same as nonradioactive gases. However, their toxicity requires a more careful design of the cleaning system.

Another toxic effect that may influence system design is the loss of sensory acuity because of gaseous contaminant exposure. For example, carbon monoxide affects the psychomotor movements and could possibly cause problems in areas such as air traffic control towers. Also, waste anesthetic gases should not be allowed to reach levels in operating rooms where the alertness of any personnel may be affected. NIOSH recommendations are generally based on such subtle effects.

ODORS

The contaminant concentrations set by OSHA are so high that they are seldom encountered in residential and commercial air conditioning applications. The types of gaseous contaminant problems usually encountered in these types of places are mostly in the form of odors and stuffiness and usually result in concentrations far below the TWA8 values. Each individual has different sensitivities to odors, and this sensitivity will usually decrease, even during brief exposure to the pollutant. One type of pollutant may enhance or mask the odor of another in the same place. Odors are usually stronger when the relative humidity is high. This requires that the concentrations of some odorants be substantially reduced before the odor level is believed to be changed. Because of this, determining the concentration of pollutants by their odor is relatively inaccurate in regards to their toxicity. There are fewer data available on odors because they are more of an annoyance than a hazard. As a matter of fact, a toxic explosive material having an odor threshold well below toxic levels is desirable. This is because the odor warns the building occupants that the pollutant is present.

At relatively low concentrations of odors, an individual ceases to be aware that the odor is present. There have been studies made to establish the level at which a percentage of the population, about 50%, is no longer aware of the odor caused by the compound being studied. Table 12-2 includes odor threshold values of a wide range of odorants.

IRRITATION

Gaseous pollutants may not have any noticeable continuing health effects, but the exposure to these pollutants may irritate the occupants of the building. Many times the pollutant will cause coughing; sneezing; eye, throat, and skin irritation; nausea; breathlessness; drowsiness; headaches; and depression. All have been experienced because of air pollutants. It has been said that when approximately 20% of the occupants of a building experience these types of irritations, the building is said to suffer from the "sick building syndrome" or the "tight building syndrome." Case studies of such occurrences, for the most part, have consisted of

Table 12-2. Characteristics of Selected Gaseous Air Pollutants. (Copyright 1991 by the American Society of Heating, Refrigerating and Air Conditioning Engineers, Inc., from HVAC Applications Handbook. Used by permission)

Pollutant	Allowable Concentration, mg/m³				Odor Threshold, mg/m³	Chemical and Physical Properties			
	IDLH	AMP	ACC	TWA8		Family	BP,°C	M	Retentivity, %
Acetaldehyde	18000			360	1.2	15	21	44	8*
Acetone	4800		3200	2400	47	16	56	58	16
Acetonitrile	7000	105		70	>0	18	82	41	1
Acrolein	13		0.75	0.25	0.35	15	52	56	
Acrylonitrile	10			45	50	18	77	53	3
Allyl chloride	810		9	3	1.4	12	44	77	
Ammonia	350		35	38	33	5	33	17	0
Benzene	10000		25	5	15	19	80	78	12
Benzyl chloride	50			5	0.2	12	179	127	0.3
2-Butanone (MEK)	8850			590	30	16	79	72	12
Carbon dioxide	90000		54000	9000	00	4	–78	44	
Carbon monoxide	1650		220	55	00	3	–192	28	0
Carbon disulfide	1500	300	90	60	0.6	6	46	76	4
Carbon tetrachloride	1800	1200	150	60	130	12	77	154	8
Chlorine	75		1.5	3	0.007	1	– 34	71	2*
Chloroform	4800		9.6	240	1.5	12	124	119	11
Chloroprene	1440		3.6	90	12	120	89		
p-Cresol	1100			22	0.056	13	305	108	5
Dichlorodifluoromethane	250000			4950	5400	11	30	121	
Dioxane	720			360	304	21	100	68	13
Ethylenedibromide	3110	271	233	155		12	131	188	11
Ethylenedichloride	4100	818	410	205	25	12	84	99	12
Ethylene oxide	1400		135	90	196	21	10	44	0
Formaldehyde	124	12	6	4	1.2	15	97	30	0.4*
n-Heptane	17000			2000	2.4	8	98	100	7*

(Continued)

Table 12-2. (*Continued*)

Pollutant	Allowable Concentration, mg/m³				Odor Threshold, mg/m³	Chemical and Physical Properties			
	IDLH	AMP	ACC	TWA8		Family	BP,°C	M	Retentivity, %
Hydrogen chloride	140		7	7	12	4	–121	37	0.7*
Hydrogen cyanide	55			11	1	18	26	27	0.4
Hydrogen fluoride	13		5	2	2.7	4	19	20	1*
Hydrogen sulfide	420	70	28	30	0.007	6	–60	34	0.8*
Mercury	28			0.1	00	1	357	201	13
Methane	ASPHY					8	–164	16	0
Methanol	32500			260	130	13	64	32	6
Methyl chloride	59500		1783	1189	595	12	74	133	
Methylene chloride	7500		3480	1740	750	12	40	85	9
Nitric acid	250			5		4	84	6 t	3*
Nitric oxide	120	45		30	>0	2	–152	30	
Nitrogen dioxide	90		1.8	9	51	2	2146	2*	
Ozone	20			2	0.2	2	–112	48	(note)
Phenol	380		60	19	0.18	13	182	94	5
Phosgene	8		0.8	0.4	4	7	8	90	1
Propane	36000				1800	8	42	44	5*
Sulfur dioxide	260			13	1.2	6	– 10	64	1.7*
Sulfuric acid	80			1	1	4	270	98	1.9*
Tetrachloroethane	1050			35	24	12	146	108	
Tetrachloroethylene	3430	2060	1372	686	140	12	121	166	
o-Toluidine	440			22	24	23	199	107	
Toluene	7600	1900	1140	760	8	17	111	92	17*
Toluene diisocyanate	70		0.14	0.14	15	18	251	174	
1,1,1 Trichloroethane	2250			45	1.1	11	113	133	5
Trichloroethylene	5410	1620	1080	541	120	12	87	131	
Vinyl chloride monomer			0.014	0.003	1400	24	14	63	0.13
Xylene	43500		870	435	2	19	137	106	16*

Notes: IDLH, AMP, ACC, and TWA8 are defined in the text.

BP = boiling point at 1 atmosphere pressure

M - molecular mass

Codes for chemical families are as given in Table 12-1.

ASPHY = Simple asphyxiant, causes breathing problems when concentration reaches about 1/3 atmospheric pressure.

Retentivities are for typical commercial-grade activated carbon, either measured at TWA8 levels or corrected to TWA8 inlet concentration using the expression for breakthrough time given in Nelson and Correia (1976):

$$t_{b2} = t_{b1} (C_2/C_1)^{-2/3}$$

Multiplying breakthrough time by inlet concentration gives retentivity, so:

$$R_2 = R_1 (C_2/C_1)^{1/3}$$

Both concentrations must be in the same units, here mg/m^3.

Retentivities marked (*) were calculated from values given in ASTM *Standard* D 1605-60 (1988) and Turk (1954), assuming that the listed retentivities were measured at 1000 ppm.

Ozone life is extremely long; activated carbon assists the essentially complete conversion of ozone to normal oxygen by both chemisorption and catalysis.

References:

ASTM *Standard* D 16005—60 (1988)
Balieu *et al.* (1977)
Dole and Klotz (1946)
Freedman *et al.* (1973)
Gully *et al.* (1969)
Miller and Reist (1977)
Revoir and Jones (1972)
Turk (1954)

questionnaires being given to the building occupants. There have been some attempts to relate irritations to gaseous contaminant concentrations. The correlation of reported complaints to gaseous contaminant concentrations is not very strong; there are many factors that affect these less serious responses to pollution. For the most part, physical irritation does not occur at odor threshold concentrations.

DAMAGE TO MATERIALS

Damage to materials from gaseous contaminants may take many forms, such as corrosion, embrittlement, and discoloration. Because such effects require that water be present for the chemical reaction to occur, material damage from air pollutants will be less in a relatively dry inside environment than when located out-of-doors. However, indoor concentration of the contaminants must be avoided for this to remain possible. Damage to some materials can be substantial from certain contaminants, especially ozone, hydrogen peroxide and other oxidants such as sulphur dioxide and hydrogen sulfide.

The effects of pollution are more serious in places such as museums, because any loss of color or texture changes the nature of the object. Libraries and archives are also affected greatly by pollutants, as well as musical instruments and textiles. Ventilation is generally considered a poor solution for the protection of rare object collections. This is because these types of buildings are located in the center of cities where the pollution in the outdoor air is more concentrated.

Systems which use filtration systems to protect rare objects must be applied with care. A system which modifies, but does not remove, contaminants offers very little or no protection to the fragile objects. Also, delaying the passage of a contaminant, i.e., capturing it and then releasing it at lower concentrations, may reduce its odor and health effects but will have virtually no effect on the damage caused to the stored materials. Usually, a pollutant control uses a chemical reaction to convert a pollutant to another type of compound, which may change the material damage, depending on the reaction of the product. The performance of the control device must be proven by actual tests at the environmental conditions with the concentration of the pollutant involved that is contained in the ventilation air.

Care must be taken when considering electrostatic air cleaners. These control devices can emit ozone into the circulating air stream. When these devices are used in rare-object protection systems, they must be followed by an effective ozone adsorber; the operation and maintenance of the entire system must be such that the ozone levels in the circulating air stream are always kept below the acceptable level.

CONTAMINANT SOURCES AND GENERATION RATES

The following are some of the major sources of indoor contaminants:

Tobacco Smoke

Tobacco smoke is a constant and strong source of indoor air pollution. Cigar and pipe smoke produce compounds that are somewhat different from cigarette smoke; however, almost all tobacco pollution comes from cigarette smoke. Table 12-3 lists some of the more important compounds that are found in the gas phase of cigarette smoke. The average concentrations are for the two types of smoke produced, mainstream and sidestream.

Mainstream smoke is that smoke which goes directly from the cigarette smoke to the respiratory system of the smoker. The smoke that is not trapped in the respiratory tract is returned to the room when the smoker exhales and joins the sidestream smoke that has not yet been inhaled, including the smoke produced by the cigarette between puffs. The type of tobacco and cigarette configurations vary greatly and the research data represent the average of the values from several different sources. The effect that filter tips have on cigarettes is not very impressive.

Cleaning Agents and Other Consumer Products

Generally, the cleaning agents that are used around any building, such as liquid detergents, waxes, polishes, and cosmetics, are made with some type of organic solvent that changes either slowly or quickly. Moth balls and other types of pest control agents give off some type of organic fluids. Studies conducted in the field show that such types of products contribute greatly to pollution of the indoor air; however, there are a large number of these solutions in use. But, there are relatively few studies that allow for the calculation for their emission rates. The pesticides that are used both inside and outside also pollute the indoor air.

Building Materials and Furnishings

The particle board that is so popular in new construction is made by using wood chips which are bonded together using phenol-formaldehyde or some other type of resin. These materials, used along with ceiling tiles, carpet, wall coverings, office partitions, adhesives, and paint finishes,

Table 12-3. Major Gaseous Compounds in Typical Cigarette Smoke. (Copyright 1991 by the American Society of Heating, Refrigerating and Air Conditioning Engineers, Inc., from HVAC Applications Handbook. Used by permission)

Pollutant	Mainstream Smoke Generation Rate[a], mg/cigarette	Sidestream Smoke Generation Rate[b], mg/cigarette	References
Acetaldehyde	0.7	4.4*	2,3
Acetone	0.4	—	1
Acetonitrile	0.1	0.4	1
Acrolein	0.1	0.6 *	2,3
Ammonia	0.1	5.9	1
2-Butanone	0.1	0.3	1
Carbon dioxide	45	360	1
Carbon monoxide	17	43	1,4
Ethane	0.4	1.2*	2
Ethene	—	0.8*	2
Formaldehyde	0.05	2.3	2.3
Hydrogen cyanide	0.4	0.1	1
Isoprene	0.3	3.1	2
Methane	1.0	0 4	1
Methyl chloride	0.4	0.8	1
Nitrogen dioxide	—	0.2	1
Nitric oxide	0.2	1.8	1
Propane	0.2	0.7*	2
Propene	0.2	0.8	1

[a]Averages of values reported in listed references.
[b]Entries without asterisks were obtained by multiplying mainstream values by SS/MS ratios given in Surgeon General (1979), p. 11-6,14-3, and 14-37. Entries with asterisks were obtained by using average chamber concentrations in Loefroth *et al.* (1989), using a chamber volume of 24 m^3.

References:
1. Surgeon General (1979)
2 Loefroth *et al.* (1989)
3. Newsome *et al.* (1965)
4. Cohen *et al.* (1971)

give off formaldehyde and other pollutants. The latex paints that contain mercury give off mercury vapor. However, as these materials age, they give off less of the pollutants. The half-life of these emissions is relatively long (Table 12-4).

Equipment

Commercial and residential buildings all have some form of internal pollution sources. However, the generation rates are considered to be higher in industrial applications. This is due to the fact that equipment sources of pollution are seldom hooded and the contaminants reach the occupants directly. The major source of contaminants in commercial buildings is office equipment, which includes electrostatic copiers that emit ozone; diazo printers that emit ammonia and related product fumes; carbonless copy paper, which emits formaldehyde; correction fluids; inks; and adhesives, which emit various VOCs (volatile organic chemicals). The medical and dental occupations produce pollutants because of the escape of anesthetic gases such as nitrous oxide and halomethanes, and sterilizers, which emit ethylene oxide. The possibility of asphyxiation is always present, even when the exposed gas is nitrogen.

The main sources of equipment-caused pollution are gas ranges, wood stoves, and kerosene heaters. Venting is helpful in controlling these pollutants; however, some of them escape into the occupied area. The pollution caused by gas ranges is somewhat moderate because they operate for a shorter period of time when compared to gas heaters. This same theory holds true for showers, which contribute to the radon and halocarbon pollution in the indoor air.

Emission rates for equipment may vary depending on the way it is used. Generally, typical values of concentrations are difficult to obtain. When a great amount of equipment-generated pollution is suspected, vent hoods should be installed. Filtration is not generally considered to be an economical or safe solution to these types of problems (Table 12-5).

Occupants

The occupants, including both human and animals, give off a great variety of pollutants through their breathing, sweating, and other means. Some of these emissions come from solids or liquids within the body. Many of the gases emitted are caused by pollutants inhaled earlier, with the tracheobronchial system acting like a gas chromatograph or a saturated physical adsorber. For such pollutants, the occupant may be considered to be a filter, and in that sense, a pollutant-control device. Some data on lung efficiency are listed in Table 12-6.

Table 12-4. Generation of Gaseous Pollutants by Building Materials. (Copyright 1991 by the American Society of Heating, Refrigerating and Air Conditioning Engineers, Inc., from HVAC Applications Handbook. Used by permission)

Contaminant	Average Generation Rate, $\mu g/(h \bullet m^2)$						
	Caulk	Adhesive	Linoleum	Carpet	Paint	Varnish	Lacquer
c-10 Alkane	1200						
n-Butanol	7300						760
n-Decane	6800						
Formaldehyde			44	150			
Limonene		190					
Nonane	250						
Toluene	20	750	110	160	150		310
Ethyl benzene	7300						
Trimethyl benzene		120					
Undecane						280	
Xylene	28						310

Contaminant	Average Generation Rate, $\mu g/(h \bullet m^2)$						
	GF Insulation	GF Duct Liner	GF Duct Board	UF Insulation	Particleboard	Underlay	Printed Plywood
Acetone					40		
Benzene					6		
Benzaldehyde					14		
2-butanone					2.5		
Formaldehyde	7	2	4	340	250	600	300
Hexanal					21		
2-propanol					6		

GF = glass fiber UF = urea-formaldehyde foam   Sources: Matthews *et al.* (1983, 1985), Nelms *et al.* (1986), and White *et al.* (1988).

Table 12-5. Generation of Gaseous Pollutants by Indoor Combustion Equipment. (Copyright 1991 by the American Society of Heating, Refrigerating and Air Conditioning Engineers, Inc., from HVAC Applications Handbook. Used by permission)

	Generation Rates, μg/kJ					Typical Heating Rate, 1000 Btu/h	Typical Use, hour/day	Vented or Unvented	Fuel
	CO_2	CO	NO_2	NO	HCHO				
Convective heater	51,000	83	12	17	1.4	31	4	U	Nat. gas
Controlled combustion wood stove		13	0.04	0.07		13	10	V	Oak, pine
Range oven		20	10	22		32	1.0	U	Nat. gas
Range-top burner		65	10	17	1.0	9.5/burner	1.7	U	Nat. gas

Notes: Sterling and Kobayashi (1981) found that gas ranges are used for supplemental heating by about 25% of users in older apartments. This increases the time of use per day to that of unvented convective heaters.

Sources: Wade *et al.* (1975), Sterling and Kobayashi (1981), Cole (1983), Knight *et al.* (1986), Traynor *et al.* (1985b), Leaderer *et al.* (1987), and Moschandreas and Relwani (1989).

Table 12-6. Efficiency of Lung/Tracheobronchial System in Sorption of Some Volatile Organic Pollutants. (Copyright 1991 by the American Society of Heating, Refrigerating and Air Conditioning Engineers, Inc., from HVAC Applications Handbook. Used by permission)

Pollutant	Average Concentrations mg/m3 Ambient	Breath	Lung and Tracheobronchial Efficiency, %	References
Benzene	5	1.5	70	1
Chloroform	4	1.8	80	1
Vinylidene chloride	7	1.5	79	1
1, 1, 1 Trichloroethylene	60	5	92	1
Trichloroethylene	5.5	1	82	1
	5.5	1.7	70	7
n-dichlorobenzene	3	2	33	1
Formaldehyde	< 0.005		98	3
Acetaldehyde	< 0.0004		60	2
Acrolein	< 0.002		82	3
Carbon monoxide			55	4.5
Sulfur dioxide	14.3	4.3	70	6
Mercury	0.8		74	8
Furan	310		80	9

Sources:
1. Wallace *et al.* (1983)
2. Egle (1971)
3. Egle (1972)
4. Cohen *et al.* (1971)
5. Surgeon General (1979)
6. Wolff *et al.* (1975)
7. Stewart *et al.* (1974)
8. Hursh *et al.* (1976)
9. Egle (1979)

There have been studies to measure the pollutants generated by humans (Table 12-7).

Table 12-7. Total-Body Emission of Some Gaseous Pollutants by Humans. (Copyright 1991 by the American Society of Heating, Refrigerating and Air Conditioning Engineers, Inc., from HVAC Applications Handbook. Used by permission)

Contaminant	Typical Emission μg/h	Contaminant	Typical Emission μg/h
Acetaldehyde	35	Methane	1710
Acetone	475	Methanol	6
Ammonia	15,600	Methylene chloride	88
Benzene	16	Propane	1.3
2-Butanone (MEK)	9700	Tetrachloroethane	1.4
Carbon dioxide	32×10^6	Tetrachloroethylene	
Carbon monoxide	10 000	Toluene	23
Chloroform	3	1, 1, 1 Trichloroethane	42
Dioxane	0.4	Vinyl chloride monomer	0.4
Hydrogen sulfide	15	Xylene	0.003

Sources: Anthony and Thibodeau (1980), Brugnone *et al.* (1989); Cohen *et al.* (1971) Conkle *et al.* (1975) Gorban *et al.* (1964) Hunt and Williams (1977), and Nefedov *et al.* (1972).

Outdoor Air

The following ventilation model allows calculation of the effect of outdoor pollution on indoor air quality. Outdoor concentrations of gaseous contaminants must be included as inputs to this model. For a few pollutants, these concentrations are available for cities in the United States from the EPA summaries of the National Air Monitoring Stations network data and in similar reports for other countries (EPA 1989). A condensed summary of this data is available (EPA 1990). In many cases, the concentrations listed in the National Ambient Air Quality Standards, which are levels the EPA considers acceptable for six pollutants (Table 12-8), give satisfactory results. This is the procedure recommended by ASHRAE Standard 62-1989.

Table 12-9 shows outdoor concentrations for gaseous pollutants at some urban sites. These values are typical; however, they may be ex-

Table 12-8. Primary Ambient Air Quality Standards for the United States. (Copyright 1991 by the American Society of Heating, Refrigerating and Air Conditioning Engineers, Inc., from HVAC Applications Handbook. Used by permission)

	Long Term		Short Term	
Contaminant	Concentration, $\mu g/m^3$	Averaging Period	Concentration, $\mu g/m^3$	Averaging Period, h
Sulfur dioxide	80	1 year	365	24
Carbon monoxide			10000	
			40000	8
Nitrogen dioxide	100	1 year		
Ozone[a]			235	
Hydrocarbons				
Total particulate (PM10)	75	1 year	260	24
Lead particulate	1.5	3 months		

[a]Standard is met when the number of days per year with maximum hour-period concentration above 235 $\mu g/m^3$ is less than one.

ceeded if the building under consideration is located near a fossil-fueled power plant, refinery, chemical production plant, sewage treatment plant, municipal refuse dump or incinerator, animal feed lot, or any major source of gaseous contaminants. If sources of this type will affect the ventilation intake air quality, a field survey or dispersion model must be run. Many computer programs have been developed to speed up these types of calculations.

CONTROL TECHNIQUES

The following is a discussion of some of the more popular techniques used for the control of indoor air pollutants:

Elimination of Sources

Control of the gaseous contaminant is one of the most effective and least expensive methods used. The first step in controlling radon gas is to

Table 12-9. Typical Outdoor Concentrations of Selected Gaseous Air Pollutants. (Copyright 1991 by the American Society or Heating, Refrigerating and Air Conditioning Engineers, Inc., from HVAC Applications handbook. Used by permission)

Gaseous Air Pollutants

Pollutant	Typical Concentration, $\mu g/m^3$	Pollutant	Typical Concentration, $\mu g/m^3$
Acetaldehyde	20	Methylene chloride	2.4
Acetone	3	Nitric acid	6
Ammonia	1.2	Nitric oxide	10
Benzene	8	Nitrogen dioxide	51
2-Butanone (MEK)	0.3	Ozone	40
Carbon dioxide	612 000[a]	Phenol	20
Carbon monoxide	3000	Propane	18
Carbon disulfide	310	Sulfur dioxide	240
Carbon tetrachloride	2	Sulfuric acid	6
Chloroform	1	Tetrachloroethylene	2.5
Ethylene dichloride	10	Toluene	20
Formaldehyde	20	1,1,1 Trichloroethane	4
n-Heptane	29	Trichloroethylene	15
Mercury (vapor)	0.005	Vinyl chloride monomer	0.8
Methane	1100	Xylene 10	
Methyl chloride	9		

[a]Normal concentration of carbon dioxide in air. The concentration in occupied spaces needs to be maintained at no greater than three times this level (1000 ppm)
Sources: Braman and Shelley (1980), Casserly *et al.* (1987), Chan *et al.* (1990), Cohen *et al.* (1989), Coy (1987), Fung *et al.* (1987), Hakov *et al.* (1987), Hartwell *et al.* (1985), Hollowell *et al.* (1982), Lonnemann *et al.* (1974), McClenny *et al.* (1987), McGrath and Stele (1987), Nelson *et al.* (1987), Sandalls and Penkett (1977), Shah and Singh (1988), Singh *et al.* (1981), Wallace *et al.* (1983), and Weschler and Shields (1989).

install traps in the sewage drain lines and to seal and vent leaky foundations and crawl spaces to prevent the entry of the gas into the structure. Controlling smoking also greatly reduces the amount of indoor pollution in buildings, even if smoking is not closely monitored. Using waterborne materials rather than those that require organic solvents also reduces volatile organic contaminants in the indoor air. However, sometimes the opposite is true. Substituting compressed carbon dioxide in place of halocarbons in spray-can propellants is an example of using a relatively innocuous pollutant rather than a more noxious one. The growth of mildew and other organisms that give off odorous type pollutants can be slowed by controlling condensation and applying fungicides and bactericides to the affected areas.

Hooding and Local Exhaust

When large amounts of pollutants are generated within a building, they may be reduced through the use of exhaust hoods. Exhaust hoods have proven to be more effective, in most cases, than general ventilation procedures. When these pollutants are toxic, irritating, or give off a strong odor the use of an exhaust hood is almost required. When the pollution is from large particulate matter, the air velocity must be increased to cause them to move with the air to the elimination point. This is the only difference between removing large particulate matter and removing gaseous pollutants.

Generally, exhaust hoods are equipped with exhaust fans and a stack that carry the pollutants to the out of doors. Exhaust hoods require large quantities of air; this requires a lot of fan power to replace the exhausted air and for make up air. There is also a waste in the heating and cooling through the hoods. All air that is exhausted through the hood should be made up by forcing ventilation air into the building. This is required to keep the general ventilation system working as it was designed. When open exhaust hoods are used, back ventilation through the hood into the surrounding work space can be reduced, and perhaps eliminated, by using an isolation enclosure (Figure 12-1).

An enclosure of this type not only isolates the pollutants, but it also helps to keep out unauthorized personnel. The use of glass for the walls tends to reduce the claustrophobic effect which occurs when working in a small place.

Sometimes there are requirements that necessitate the use of filters to clean the air before exhausting it to the out-of-doors to prevent the release of toxic pollutants to the atmosphere. It is generally good practice to include controls on the exhaust hood that allow the quantity of exhausted air to be reduced when the amount of pollutants has been reduced to some level. Some control systems can cause the exhausted air to be discharged back into the building after it has been cleaned of all pollutants, thus saving on the heating and cooling costs. Recirculation of the air back into the building should be applied only to those buildings in which the pollutants are considered to be harmless. This is to prevent the

Figure 12-1. Glass-Walled enclosure for pollutant isolation. (Copyright 1991 by the American Society of Heating, Refrigerating and Air Conditioning Engineers, Inc., from HVAC Applications Handbook. Used by permission)

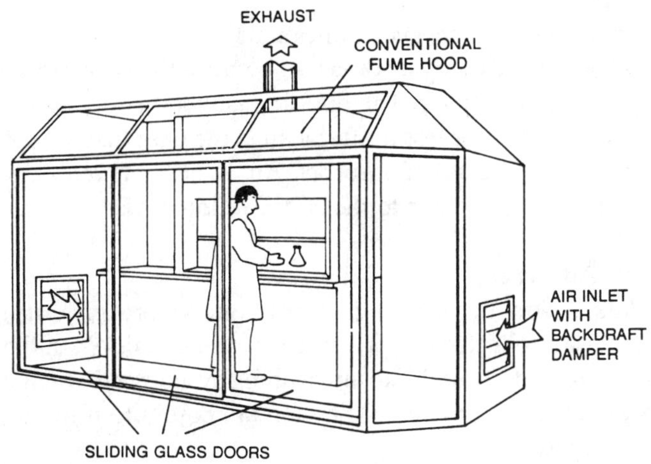

recirculation of polluted air back into the building in case of filtering system failure.

General Ventilation

When problems occur in residential and commercial buildings, the major use of exhaust hoods are in the kitchen, bathrooms, and occasionally around points of greatest pollution emitters such as diazo printers. When there is no localized point of pollutants, a general ventilation system can provide the required pollution control. General ventilation control systems used in such applications must meet both the thermal load requirements and the pollutant control standards. In either application, the complete mixing and the uniform distribution of the air per occupant is desirable. The ASHRAE Standard 62-1989 must be met for ventilation requirements.

When a local exhaust hood is used in combination with general ventilation procedures, a supply of makeup air must be equal to that discharged by the exhaust hood. In some applications, supply air fans may be required so that sufficient supply air is available to keep the internal pressure up inside the building. Systems, such as those used in clean rooms, are sometimes designed so that the static pressure causes the air to flow from the clean room to an area of less pressure. When this type of system is used, caution must be exercised to provide backdraft dampers in the exhaust outlet for when the doors are opened, and changes in wind velocity occur, etc.

Absorbers

There are some special applications of absorbers for cleaning the air. They usually have many draw-backs because they also usually introduce pollutants into the space. The question is whether or not the introduced pollutants are more or less desirable than the ones that are being removed from the space. The use of absorbers for air cleaners requires caution and much consideration given to the effects of the device.

Physical Adsorbers

Adsorption, in contrast to absorption, is a process during which the pollutants are collected on the surface of a material. When the pollutant is a gas, it is forced through the air cleaning system and when its molecules strike the adsorbing material they remain bound to it for an appreciable period of time.

In this procedure the molecules are said to be adsorbed by the surface. This phenomenon takes place on almost all surfaces. However, gaseous contaminant control material has the available surfaces increased in two ways. First, adsorbers are usually available in granular or fibrous forms which increase the amount of surface exposed to the air stream. Second, the surface of the adsorbing material is treated to cause microscopic pores that greatly increase the amount of surface exposed to the gaseous molecules as they flow through the material. Typically, treated activated aluminas have surface areas of 100 m^2 per gram of adsorber; typical activated carbons have areas from 1000 to 1500 m^2 per gram of carbon. The various sizes and shapes of the pores also cause minute traps which can be filled with extra layers of molecules of the pollutant. These extra pores are eventually filled with the condensed vapors of the adsorbed pollutants.

During the physical adsorption process, the forces which bind the molecules of the pollutant to the surfaces of the adsorber material are the same forces that give the pollutant its thermophysical properties which are important to the heat and molecule transfer. There are several steps that must occur before the molecules can be adsorbed (Figure 12-2).

1. The molecule must be moved from the carrier gas stream across the boundary layer surrounding the adsorber granule. This is considered to be a random process, with the movement of the molecule both to

Figure 12-2. Steps in pollutant absorption. (Copyright 1991 by the American Society of Heating, Refrigerating and Air Conditioning Engineers, Inc., from HVAC Applications Handbook. Used by permission)

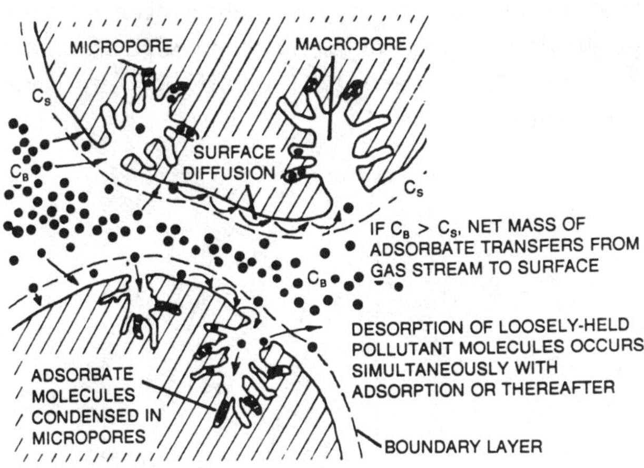

and from the surface, unless the concentration of the pollutant in the gas flow is greater than its concentration at the surface of the granule. Because of this, the adsorption action decreases as the pollutant load on the adsorber surface increases. Also, very low concentrations of the pollutant gas in the air flow result in low adsorption rates.

2. The molecules of the pollutant gas must move into the pores of the adsorbing material if they are to occupy that portion of the surface.

3. The pollutant molecules must bond to the surface of the adsorbing material.

In practice, any one of these three steps can determine the rate at which the adsorption process occurs. Generally, step 3 is the fastest process for physical adsorption; however it is a reversible process that must be considered. What this means is that the adsorbed molecule can be desorbed sometime later. It may be released when relatively cleaner air passes through the adsorber bed, or by the arrival of another pollutant that binds more readily and tightly to the surface of the adsorber. Also, thermal effects must be considered because a physically adsorbed molecule gives up its energy when it bonds to the surface of the adsorber material. This release of energy raises the temperature of the adsorber and of the carrier gas stream.

When a pollutant gas is forced into an adsorber at a steady concentration and a continuous gas flow rate, the resulting concentration of the gas varies with time and the depth of the adsorber bed (Figure 12-3).

Figure 12-3. Dependence of absorbate concentrations on bed depth. (Copyright 1991 by the American Society of Heating, Refrigerating and Air Conditioning Engineers, Inc., from HVAC Applications Handbook. Used by permission)

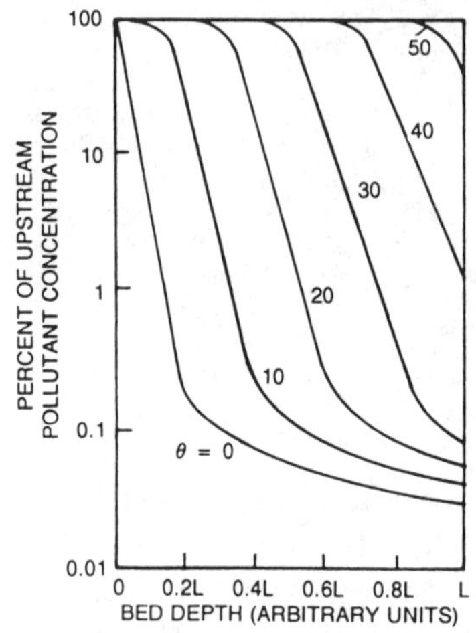

When the bed material is new, the concentration of the pollutant is reduced logarithmically with the depth of the bed material. As the pollutant passes deeper into the bed, the pollutant concentration is reduced, and the slope of concentration-versus-bed depth curve decreases. After the system has operated for a while, the beginning portion of the adsorber material becomes more loaded with the pollutant, allowing higher concentrations of it to reach deeper into the bed.

Usually the downstream air from the adsorber becomes more concentrated with contaminants after the unit has operated for a while and the adsorber becomes clogged with contaminants. The concentration-versus-time pattern downstream can be seen in Figure 12-4.

As the bed reaches saturation, the concentration downstream rises rapidly until the concentration is almost the same as the air before passing through the adsorber. This is called the breakthrough point because it happens suddenly. It should be noted that not all adsorbing pollutant combinations have the same breakthrough point as that shown in Figure 12-4. Therefore, when given a break-through time, the entering pollutant

Figure 12-4. Dependence of adsorbate concentrations on exposure time. (Copyright 1991 by the American Society of Heating, Refrigerating and Air Conditioning Engineers, Inc., from HVAC Applications Handbook. Used by permission)

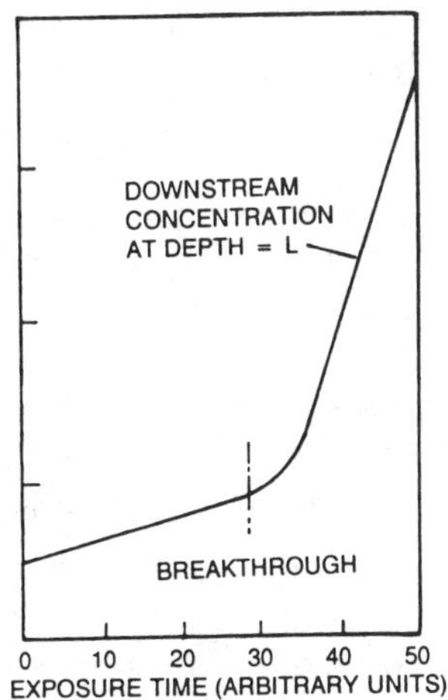

concentration, downstream concentration at break-through, the bed depth, the air velocity, temperature, and the relative humidity of the air stream used in the test also must be specified.

Of the physical adsorbing materials, activated carbon is by far the most popular because it will adsorb a wide range of pollutants. Table 12-10 lists some of the more popular physical adsorbers available for indoor air pollution control.

Chemisorbers

The chemisorption process is much like physical adsorption. The same steps required for the physical adsorption as shown in Figure 12-3 are also necessary for the chemisorption process, with the exception of the third step. In the chemisorption process the binding process is by chemical reaction with an electron exchange between the molecules in the pollutant and the chemisorber. This binding action differs from the physical adsorption process in the following ways:

1. Chemisorption is very specific; that is, only certain pollutant compounds will react with a given type of chemisorber.
2. As the temperature increases, the chemisorption process improves. The physical adsorption process improves when temperature drops.

3. Chemisorption does not generate heat. It may sometimes require heat to be added to aid in the process.

4. Generally, chemisorption is not a reversible process. When the adsorbed pollution has reacted, the original pollutant cannot be desorbed. However, there may be one or more products which are different from the original pollutant formed during the chemisorption process. These by-products may have undesirable effects on the air.

5. In chemisorption, water vapor often aids in the process. Sometimes water vapor is necessary for the process. Water vapor usually has bad effects on physical adsorption.

6. Chemisorption per se is a nonmolecular layer phenomenon. The pore filling process that occurs in physical adsorption is not present in chemisorption. An exception is when adsorbed water vapor condenses in the pores and forms a reactive liquid.

Some of the more common chemisorbers used in indoor air pollution control are shown in Table 12-10.

Plants

Indoor air pollution control is sometimes accomplished by using certain plants (Table 12-11).

The type of pollutant must be known so that the proper plant can be used. Sometimes, air is drawn through the potting soil of several plants to remove the pollutant. The pollutants are removed by activated carbon in the potting soil. It is believed that plant root bacteria are able to regenerate the activated carbon and maintain its effectiveness for a longer than normal period of time.

Occupants

The inhalation and exhalation of a gas by a single occupant is shown in Figure 12-5.

A person who is typically engaged in office work will have an average breath flow rate of about 0.49 cfm. During the breathing process about 70% of this flow rate reaches the lung alveoli, where volatile organics along with other water-soluble pollutant gases are adsorbed in huge amounts. The high efficiency of removal of these subjects indicates a distinct adsorption in the tracheobronchial system and in the alveoli (Table 12-6).

Table 12-10. Low-Temperature Adsorbers, Chemisorbers and Catalysts. (Copyright 1991 by the American Society of Heating, Refrigerating and Air Conditioning Engineers, Inc., from HVAC Applications Handbook. Used by permission)

Material	Impregnant	Typical Vapors or Gases Captured
Physical Adsorbers		
Activated carbon	none	organic vapors, ozone, acid gases
Activated alumina	none	polar organic compounds[a]
Activated bauxite	none	polar organic compounds
Silica gel	none compounds	water, polar organic
Molecular sieves (Zeolites)	none	carbon dioxide, iodine
Porous polymers	none	various organic vapors
Chemisorbers		
Activated alumina	KMnO4	hydrogen sulfide, sulfur dioxide
Activated carbon	I_2 Ag, S	mercury vapor
Activated carbon[b]	I_2 KI_3 amines	radioactive iodine and organic iodides
Activated carbon	$NaHCO_3$	nitrogen dioxide
LiO_3 NaO_3 KO_3	none	carbon dioxide
LiO_2 NaO_2 KO_2, $Ca(O_2)_2$	none	carbon dioxide
Li_2O_2 Na_2, $°_2$	none	carbon dioxide
LiOH	none	carbon dioxide
$NaOH+C^a(OH)_2$	none	acid gases
Activated carbon	KI, I_2	mercury vapor
Catalysts		
Activated carbon	none	ozone
Activated carbon′	$Cu+Cr+Ag+NH_4$	acid gases, chemical warfare agents
Activated alumina	$CuCl_2$ $+PdCl_2$	formaldehyde
Activated alumina	Pt, Rh oxides	carbon monoxide

[a]polar organics = alcohols, phenols, aliphatic and aromatic amines, etc.
[b]Mechanism may be isotopic exchange as well as chemisorption.
[c]"ASC Whetlerite"

Table 12-11. Additional Indoor Air Quality Criteria Specified by ASHRAE Standard 2-89 When Ventilation Air is Recirculated. (Copyright 1991 by the American Society of Heating, Refrigerating and Air Conditioning, Inc., from HVAC Applications Handbook. Used by permission)

Pollutant	Maximum Allowed Concentration	Sampling Period
Carbon dioxide	1.8 g/m^3	Continuous
Chlordane	5 μg/m^3	Continuous
Ozone	100 μg/m^3	Continuous
Radon	0.027 WL	Annual average

WL = Working Level, a unit of measure of human exposure to radon gas and radon progeny.

Figure 12-5 Recirculatory Air-Handling and Gaseous Pollutant System. (Copyright 1991 by the American Society of Heating, Refrigerating and Air Conditioning, Inc., from HVAC Applications Handbook. Used by permission)

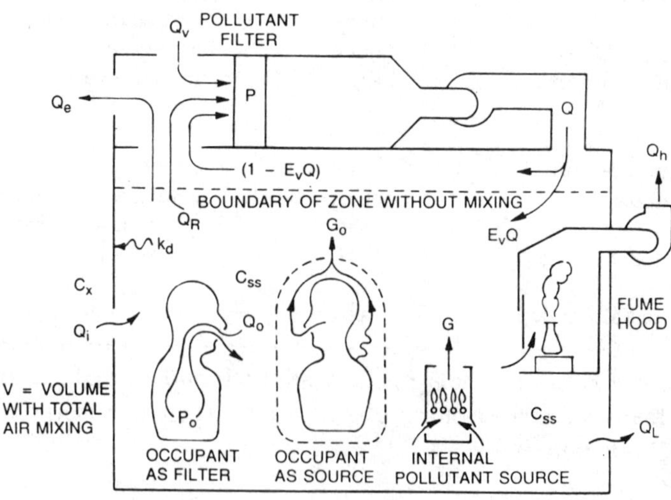

The human body acts much like a physical adsorber. It is not a perfect trap for organic pollutants in that it also desorbes some of the nonmetabolized pollutants when cleaner air is breathed after their adsorption. This desorption can occur up to several days after the adsorption. Inorganic volatiles, such as acid gases are adsorbed in the body more efficiently than volatile organic pollutants by the breathing system and are more readily taken up by the body.

COUNTERACTING ODOR AND ODOR MASKING

An odor counteractant is a vapor that is introduced into a space to reduce the sensitivity of the nose to a known odor. An odor mask is a vapor that is introduced into a space to cover up an existing odor. Both of these processes are effective only on specific odors and are usually effective for only a percentage of the occupants. They do not change or improve the toxic effects or material damage. They are themselves pollutants.

TYPES OF CONTROL SYSTEMS

The media used in adsorbers and chemisorbers are available in either granular or pellet type material. These materials must be contained in some type of frame that will permit air to pass through the media without a huge air flow pressure drop. Some of the types and configurations used with granular or pellet type media are shown in Figure 12-6.

The media is held between perforated metal sheets or wire screens. There must be an area left around the edge of the perforated metal or retaining screens to reduce the amount of air that will bypass the edges of the media. To be effective, the media must be firmly packed in the frame to prevent open spaces in the bed. These retainers may be made from aluminum, stainless steel, painted or conversion-coated steel, and plastics. Sometimes the media used for adsorptive units are placed in fibrous filter media which is then pleated into large filter surfaces (Figure 12-6B).

Figure 12-6A is the type of filter that can be used in applications such as the collection of noble gases in a cryogenic atmosphere in nuclear installations. In these applications, the air velocity is low allowing for a longer contact time between the media and the contaminants. Figures 12-6B, 12-6C, and 12-6D are typical fixed-bed units that are used in nuclear safety systems and highly toxic applications. The contact time in these

Figure 12-6. Configurations of Gaseous Pollutant Filters Using Dry Granular Adsorbers, Chemisorbers, and Catalysts. (Copyright 1991 by the American Society of Heating, Refrigerating and Air Conditioning, Inc., from HVAC Applications Handbook. Used by permission)

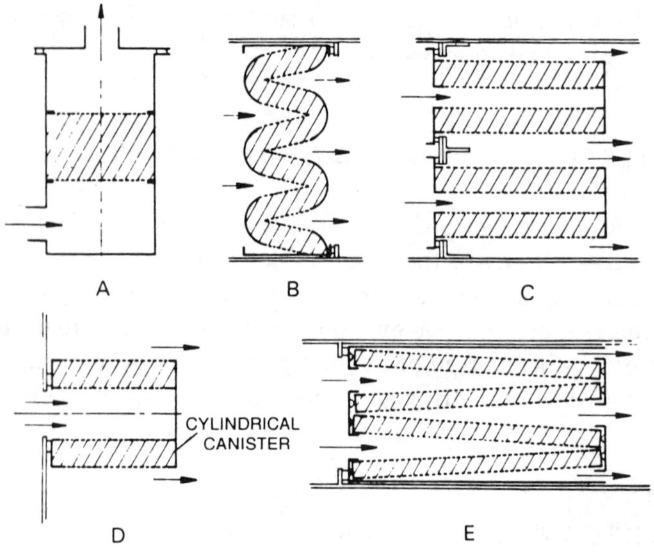

applications is also long, and the size of the media particle is much smaller than those used in commercial HVAC applications.

Figure 12-6E is a typical commercial HVAC filter. The retaining elements are flat panels or trays. These trays are placed into a holding track that has gaskets to seal all joints to prevent the bypass of air. This type of construction allows for easy removal of the trays for replacement or regeneration of the active media.

ENERGY CONSUMPTION

The choice to use outside air only or outside air plus filtration is generally made on the basis of technical or maintenance factors, convenience, economics or a combination of all these factors. The total cost of operation must be made before an economic analysis can be made. In most cases, a building heat load estimation is available. The heat load estimation will include all the various air flows shown in Figure 12-5, and the total heating and cooling energy requirements for the different seasons of the year.

To satisfy the metabolic requirements of the building occupants a sufficient quantity of ventilation air must be supplied to every part of the building. To provide the proper dilution of carbon dioxide to nontoxic levels requires 0.11 cfm for every person. This is automatically supplied when adhering to ASHRAE Standard 62—1989 which requires 15 cfm per person. When smoking is permitted, or expected, more ventilation air is required. The minimum amount of air required by the Standard is listed in Table 12-11.

When a lesser amount is used a different set of criteria must be met. However, the quality of the air must never be less than that listed in Table 12-8. Also, it must have contaminant concentrations less than those listed for four additional pollutants listed in Table 12-8. The control system employed may use either outdoor air or filtered, recirculated air in any ratio as long as the air quality is maintained.

In applications where the indoor air quality can be maintained with ventilation alone an economizer cycle is recommended when appropriate outdoor conditions exist. During operation of the economizer system some energy is used to move the air through the system. This energy use is almost constant when fixed-flow systems are used. When VAV systems are used the power use rate will be determined by the air flow-rate and the design of the control, fan, and motor systems.

The amount of resistance offered by the filtering system along with the heating and cooling coils is related to the air flow through it. This type of information can be obtained from the various manufacturers.

There are more calculations involved when a VAV system is used when compared to a fixed-flow system. There are computer programs designed to make these calculations easier. The VAV system calculation should involve operation in the heating, cooling, and economizer modes to indicate which system is best economically.

Chapter 13

System Cost Estimating

The proper cost estimation of the overall installation for an air conditioning system cannot be overemphasized. This is where the money is either made or lost. Many people claim that there is no break-even point. When no profit is made, the contractor loses money because he/she could have been working on another job where he could possibly have made some money.

INTRODUCTION

Care should be taken when estimating the price of an air conditioning unit installation. When the bid is too high, the job will probably be given to another contractor. If the bid is too low, the contractor will probably lose money. Thus, a markup of a required percentage of the materials, equipment, and labor costs plus a profit should be included in the price to the customer. The price of the job together with a list of what the contractor proposes to do and a list of what is required of others should be presented to the customer. This is sometimes entered on proposal forms, whereas other contractors prefer a typewritten letter for each job. This is a matter of personal preference.

ESTIMATION FORM

A checklist type of form should be used which lists every possible piece of equipment, material, subcontracts, and labor that may be required for the installation (Exhibit 13-1).

Often there are things on the list that will not be used on every installation; however, they do not need to be given consideration in this instance.

The estimate form should include the item, size, type, quantity, cost each, combined cost, percent markup, selling price, and a total. This form need not be given to the customer.

Estimation Example

We will now estimate the cost of the unit that we designed for our model house in previous chapters. (Note: It should be remembered that these prices are for illustration purposes only. The individual making an estimate should use the current prices when the estimate is being made.) When obtaining prices, it is usually a good idea to see if the supplier will make the prices good for a certain period of time such as 30 days, 60 days, 90 days, etc. depending on how long it will take to complete the job. This is desirable because the great fluctuation in prices can sometimes be sufficient to cause the contractor to lose money when the installation takes longer than thought.

It is usually easier to start at the top of the price estimate form and calculate each item, as needed, on an individual basis.

Step 1. Enter the customer's name, address, date, and estimator as required on the top of the form. This identifies the job, the date, and the estimator for future reference should there be any question about the estimation (Exhibit 13-2).

Step 2. Enter the size of the condensing unit, the model number, cost, quantity, combined cost, markup, and selling price. When we check the supplier for the cost of the equipment, the price is found to be $431.46.

We will assume that a 30% markup will provide us with sufficient money for all our office expenses, insurance, truck costs, labor, and so on, and for a fair profit margin. However, the markup should be determined for each individual company by previous experience or other suitable method.

Thus, when the condensing unit cost is multiplied by 30%, the selling price will be $560.90. Enter this figure in the proper column.

Step 3. The cost of the furnace, from the supplier, is $138.51. With a markup of 30%, the selling price is $180.06. Enter these figures in the proper column on the form.

Step 4. The evaporator costs $113.72. Thus, a selling price of $147.84 is entered on the form in the proper column.

Exhibit 13-1. Air conditioning price estimating form.

Customer Name				Address			
Date			Estimator				
			EQUIPMENT				
Item	Size	Model Number	Cost Each	Quantity	Combined Cost	Percent Markup	Selling Price
Condensing Unit							
Furnace							
Evaporator							
Humidifier							
Thermostat							
Electronic Filter							
Subtotal							
			MATERIAL				
Item	Size	Model Number	Cost Each	Quantity	Combined Cost	Percent Markup	Selling Price
Plenum, Supply							
Plenum, Return							
Round Duct							
Rectangular Duct							
Starting Collars							
Flexible Duct							

Exhibit 13-1 (*Continued*)

Material (*Continued*)							
Item	Size	Model Number	Cost Each	Quantity	Combined Cost	Percent Markup	Selling Price
Elbows							
Stack Boots							
Register Boots							
Stack Heads							
Wyes							
Tees							
Reducers							

Exhibit 13-1 (*Continued*)

Material (*Continued*)							
Item	Size	Model Number	Cost Each	Quantity	Combined Cost	Percent Markup	Selling Price
Supply Grills							
Ceiling Diffusers							
Return Air Grill							
Combustion Air Box							
Refrigerant Lines							
Copper Fittings							
Suction Line Insulation							
Refrigerant							
Vent, Double Wall							
Vent Elbows, Double Wall							
Flashing							
Collar							
Vent Cap							
Drain Line							
Duct Insulation							
Duct Board							

Exhibit 13-1 (*Continued*)

Material (*Continued*)							
Item	Size	Model Number	Cost Each	Quantity	Combined Cost	Percent Markup	Selling Price
Duct Tape							
Duct Screws							
Duct Hanger							
Control Wire							
Concrete Slab							
Condensate Pump							
Misc. Materials							
Subtotal							
SUBCONTRACT AND LABOR							
Freight							
Trucking and Rigging							
Foundations							
Ductwork							
Insulation							
Cutting and Patching							
Painting							
Electric Wiring							
Plumbing							
Oil Tank							
Installation							
Gas Connection							
Labor							
Startup and Test							
Permits							
Cleaning of Existing Ducts							
Service Reserve							
Taxes							
Misc.							
Subtotal							

Exhibit 13-1 (*Concluded*)

BID PRICE	
Equipment Subtotal	
Materials Subtotal	
Subcontract and Labor	
Miscellaneous Charges	
TOTAL	

Exhibit 13-2. Air conditioning price estimating form.

Customer Name **Model House** Address **Dallas, Texas**

Date **10-5-82** Estimator **Langley**

EQUIPMENT

Item	Size	Model Number	Cost Each	Quantity	Combined Cost	Percent Markup	Selling Price
Condensing Unit	3.5 tons	CTD-351-QC	431.46	1	431.46	30	560.90
Furnace	100,000 Btu	CF 100-BD-33	138.51	1	138.51	30	180.06
Evaporator	—	UCC-41-23	113.72	1	113.72	30	147.84
Humidifier							
Thermostat		T-87F Honeywell	13.52/ 10.46	1	23.98	30	31.17
Electronic Filter							
Subtotal					707.67		919.97

MATERIAL

Item	Size	Model Number	Cost Each	Quantity	Combined Cost	Percent Markup	Selling Price
Plenum, Supply	19-1/8 × 22 × 48	—	39.14	1	39.14	30	50.88
Plenum Return	Platform		20.00 Estim.	1	20.00	30	26.00
Round Duct	10"		1.47	15 ft.	22.00	30	28.60
	7"		1.00	25 ft.	24.99	30	32.49
	8"		1.16	25 ft.	28.90	30	37.57
	9"		1.32	15 ft.	19.85	30	25.81
	6"		0.84	15 ft.	12.57	30	16.34

Exhibit 13-2 (*Continued*)

MATERIAL (Continued)							
Item	Size	Model Number	Cost Each	Quan- tity	Combined Cost	Percent Markup	Selling Price
Rectangular Duct							
Starting Collars	0"		1.41	2	2.82	30	3.67
	9"		1.33	1	1.33	30	1.73
	7"		1.07	1	1.07	30	1.39
Flexible Duct							
Elbows	7"	Adj.	1.64	4	6.56	30	8.53
	10"	Adj.	2.74	2	5.48	30	7.12
	8"	Adj.	1.89	1	1.89	30	2.46
	9"	Adj.	253	2	5.06	30	6.50
	6"	Adj.	1.49	2	3.98	30	5.17
Stack Boots							
Register Boots	12 × 3-1/4 × 70	201	3.95	3	11.85	30	15.41
	8 × 4 × 60	601	4.74	2	9.48	30	12.32
	14 × 6 × 80	601	3.96	1	3.96	30	5.15
	12 × 6 × 80	601	5.24	1	5.24	30	6.81
Stack Heads							
Wyes	10-7-8	#800	5.07	2	10.14	30	13.18
	8-6-6	#800	4.39	1	4.39	30	5.70

Exhibit 13-2 (*Continued*)

MATERIAL (Continued)							
Item	Size	Model Number	Cost Each	Quan-tity	Combined Cost	Percent Markup	Selling Price
Tees							
Reducers	9¥8	#400	2.63	1	2.63	30	3.42
Supply Grills	8 × 4	#351	2.80	2	5.60	30	7.28
	12 × 4	#351	3.00	3	9.00	30	11.70
	12 × 6	#351	3.30	2	6.60	30	8.58
Ceiling Diffusers							
Return Air Grill	24 × 9		4.77	2	9.54	30	12.40
Combustion	200 sq. in.						
Air Box	(5" × 40")		18.81	1	18.81	30	24.45
Refrigerant	3/8"		91.50 set	1	91.50 set	30	118.95
	3/4"			1			
Copper Fittings							
Suction Line	Included						
Insulation	in line set						
Refrigerant	Included in	line set					
Vent	5"		12.67	2(10')	25.34	30	32.94
Double Wall							

Exhibit 13-2 (*Continued*)

MATERIAL (Continued)							
Item	Size	Model Number	Cost Each	Quantity	Combined Cost	Percent Markup	Selling Price
Elbows, Double Wall							
Vent Connector	5"		1.87	1	1.87	30	2.43
Flashing	5"		5.95	1	5.95	30	7.74
Collar	5"		1.53	1	1.53	30	1.99
Vent Cap	5"		6.63	1	6.63	30	8.62
Drain Line	3/4"	copper pipe	0.72	8'	5.76	30	7.45
	3/4"×90°Ell	copper	0.65	1	0.65	30	0.85
Duct Insulation	2"		77.95	1 roll	77.95	30	101.34
Duct Board							
Duct Tape	2"		5.32	1 roll	5.32	30	6.92
Duct Screws	3/8"×1/2"		1.86	1 lb.	1.86	30	2.42
Duct Hangar	1 roll		7.34	1 roll	7.34	30	9.54
Control Wire	18-2/18-4		0.03/0.06	25'/15'	1.62	30	2.11
Concrete Slab	26×32×4		32.50	1	32.50	30	42.25
Condensate Pump							
Misc. Materials	coil housing	HUH-	37.82	1	37.82	30	50.34
	Dampers	36-48-23	3.39/3.12/	3/2/1/1	24.55	30	31.92
	7"/6"/9"/8"		4.46/3.68				
Subtotal					394.90		514.52
SUBCONTRACT AND LABOR							
Freight							
Trucking And Rigging							
Foundations							
Ductwork							
Insulation							
Cutting and Patching							
Electric Wiring	Line voltage to both units		275.00	1	275.00		275.00

Exhibit 13-2 (*Concluded*)

SUBCONTRACT AND LABOR (Continued)							
Item	Size	Model Number	Cost Each	Quantity	Combined Cost	Percent Markup	Selling Price
Plumbing							
Oil Tank Installation							
Gas Connection	Gas line to furnace		75.00 75.00	1 1	75.00 75.00		75.00 75.00
Labor			450.00	15.00 × 30 hrs.	450.00		450.00
Start Up And Test			45.00	1 hr.	45.00		45.00
Permits			25.00		25.00		25.00
Cleaning of Existing Ducts							
Service Reserve			175.00	50.00 per ton	175.00		175.00
Taxes					1434.49	0.05	71.72
Misc.							
Subtotal							1116.72

BID PRICE	
Equipment Subtotal	919.97
Materials Subtotal	514.52
Subcontract and Labor	1020.00
Miscellaneous Charges	96.72
TOTAL	2551.21

Step 5. The thermostat is a T-87F Honeywell, which requires a subbase. The cost of the thermostat is $13.52. The subbase costs $10.46. The combined cost is $23.98. The selling price is found to be $31.17. Enter these figures on the form in the proper column.

Step 6. The subtotals of the equipment can now be determined by adding the columns "Combined Cost" and "Selling Price" and entering the sum of each column in the proper place. The combined cost is $707.67 and the selling price is $919.97.

Step 7. The supply plenum presents another problem. The supplier must be consulted to get the current, proper price. We will use the enclosed price list for all the metal duct and fittings (Figure 13-1, pg. 380).

Refer to the section of the price list for plenums. There is an example on estimating the cost. Our plenum size is 19 3/8 + 22 + 22 = 82.75. We will select 26-gauge metal, and a height of 48 in. shows a multiplier of 0.473. Thus 0.473 × 82.75 = $39.14. The selling price is $50.88.

Step 8. The furnace is to be set on a platform in a closet. Thus, the return plenum would consist of the material to make the platform. This is an educated guess in most cases.

Steps 9-15. Refer to the price list in Figure 13-1 to calculate the cost and selling price for these steps.

Step 16. The supply grills are found in tables like the one presented in Table 13-1.

Table 13-1. Grill Price List

Duct Size W × H	List Price	Free Area Sq. Inch	Duct Size W × H	List Price	Free Area Sq. Inch
8 × 6	$2.80	33	16 × 6	$4.30	70
			16 × 8	4.85	96
10 × 4	2.90	27			
10 × 6	3.00	42	18 × 6	4.50	77
10 × 8	3.55	57	18 × 8	4.90	105
12 × 4	3.00	33	20 × 6	5.60	84
12 × 6	3.30	51	20 × 8	6.35	115
12 × 8	3.85	70			
			24 × 6	6.00	103
14 × 4	3.10	39	24 × 8	7.30	140
14 × 6	3.50	61			
14 × 8	4.20	83	30 × 6	7.15	129
			30 × 8	8.85	176

(Courtesy of Standard Perforating & Mfg., Inc.)

Locate the cost in the table and enter it in the proper column and calculate the selling price and enter it in the proper place.

Step 17. The prices for return air grills are also placed in tables. In some cases, however, it is better to get the price from your local supplier. In our case they cost $4.77 each, with a selling price of $12.40 each. Notice that the required opening is 24 in. × 18 in. Since our platform is only 12 in. high, we can divide the return grill into two grills, one measuring 24 × 12 and the other measuring 12 × 12. The two grills may be placed on different walls that lead into the return air plenum.

The remainder of the material section is completed in the same way. The combined cost and selling price are calculated and entered in the subtotal line in their respective places.

The "Subcontract and Labor" section is approached in the same manner. When subcontractors give an estimate, or when the estimator makes an estimate, the figure is entered in the proper column. Some companies make a markup on subcontract labor and others do not. Again, this is a matter of preference and depends on the method used for calculating profit. In our example we did not place a markup on any item in this section.

The labor is probably the most difficult item to estimate. Usually, it is much easier when the estimator has worked with the same installation crew for some time. Thus he can more accurately judge the amount of time required for each job. We estimated that a journeyman installer and a trainee would take 30 hours to install this unit in an existing home. We used an hourly labor cost of $10.00 for the journeyman and $5.00 for the trainee. The installation labor would be $15.00 × 30 = $450.00.

We estimated 1 hour for a service technician to start up and test the unit, for a cost of $45.00. The permits were estimated at $25.00. The local cost will probably be something different.

We used the price of $50.00 per ton of system capacity for warranty reserve, for a total of $175.00. This warranty reserve is to pay for any labor or materials used to maintain the unit through the warranty period. The equipment manufacturer will usually furnish compressors and other electrical equipment for the first year after installation. Few of them, however, will pay any labor charges for replacing these components.

The total tax rate should be used to calculate any taxes due. In our example the rate is 5% on equipment and materials only. Labor, permits, and so on, are not taxable. However, they may be in another area. Be sure to check with the local authorities.

The bid price can now be determined by adding the equipment subtotal, materials subtotal, subcontract and labor, and any miscellaneous charges related to the job.

Each of these items is found under the respective heading. The miscellaneous charges are the permits and taxes, for a total of $96.72. The subcontract and labor subtotal is $1020.00. The materials subtotal is $514.52 and the equipment subtotal is $919.97. The sum of the subtotals is $2551.21, the bid price of the job. This is the price that the customer pays.

Figure 13-1. Sample duct work price list.

DUCT PIPE		3"	4"	5"	6"	7"	8"	9"	10"	12"	14"	16"	18"	20"	22"	24"
Snaplock #100 Duct Pipe	30 ga	73.66	74.86	77.89	83.81	99.98	115.58	132.34	146.69	198.65	—	—	—	—	—	—
	28 ga	—	—	—	—	—	—	—	—	—	242.71	271.70	351.38	458.83	—	—
	26 ga	—	—	131.98	158.88	180.86	202.88	229.75	259.09	303.07	383.75	481.52	559.72	620.82	689.26	743.03
	24 ga	—	—	175.98	210.21	241.98	273.75	305.54	337.30	400.84	510.84	—	—	—	—	—
ELBOWS #125 90 Elbows	30 ga	1.27	1.27	1.33	1.49	1.64	1.89	—	—	—	—	—	—	—	—	—
	28 ga	—	—	—	—	—	—	2.53	2.74	3.96	6.82	—	—	—	—	—
	26 ga	—	2.78	2.78	3.05	3.39	3.75	4.57	4.90	7.16	9.18	9.52	13.81	21.28	—	—
	24 ga	—	—	—	4.02	4.34	5.07	6.53	7.97	10.84	13.30	14.17	20.21	25.98	37.56	47.54
#150 45 Elbows	30 ga	—	1.20	1.24	1.35	1.47	1.56	—	—	—	—	—	—	—	—	—
	28 ga	—	—	—	—	—	—	2.19	2.46	3.22	5.73	—	—	—	—	—
	26 ga	—	2.08	2.08	2.29	2.55	2.82	3.41	3.70	5.39	6.88	8.15	11.39	—	—	—
	24 ga	—	—	—	3.22	3.45	4.02	5.20	6.36	8.65	10.63	11.33	16.19	—	—	—
DAMPERS (No. 100D Inline Damper; No. 300Q Damper with Quadrants)	In Line Damper 100D	—	2.74	2.82	3.12	3.39	3.68	4.46	4.63	5.77	7.98	11.75	14.72	—	—	—
	Damper Blank 300	—	.32	.32	.40	.50	.57	.67	.76	.80	1.37	2.34	2.97	3.83	4.72	6.76
	Damper Blank w/Quadrant 300Q	—	1.30	1.30	1.39	1.64	1.81	2.04	2.29	2.69	3.37	3.96	4.72	6.02	7.94	8.50
	Butterfly Damper	—	—	—	—	5.07	—	—	8.74	8.90	—	—	—	—	—	—
	Blastgate Damper	—	12.88	—	15.05	16.13	17.18	—	21.49	—	—	—	—	—	—	—
REDUCERS	#400 Price based on big end	—	1.94	2.44	2.21	2.32	2.55	2.63	2.80	3.49	4.65	6.27	9.56	12.84	16.97	18.29

STARTING COLLARS

#500 #500D #500TD #500DB #500DBD #500DBS 510 #500DBSD 510D

Item	Code															
Metal Starting Collar	500	.93	.93	.97	1.07	1.20	1.33	1.41	1.77	2.17	2.55	3.43	4.10	6.15	7.03	
Metal Starting Collar with Damper	500D	2.34	2.36	2.59	2.86	3.16	3.41	3.68	4.23	5.20	6.57	10.27	13.66	15.01	16.38	
Tapered Starting Collar with Damper	500TD	—	4.55	4.72	4.97	5.33	6.29	6.63	7.52	9.94	—	—	—	—	—	
Duct Board Starting Collar	500DB	2.21	2.21	2.21	2.50	2.80	3.12	3.39	3.66	4.25	5.16	6.29	7.77	—	—	
Duct Board Starting Collar	500DBD	3.73	3.83	3.83	4.38	4.82	5.43	5.81	6.61	7.70	8.76	10.44	12.44	—	—	
Spin-In Duct Board Starting Collar	500DBS	2.53	2.53	2.53	2.86	3.20	3.54	3.89	4.23	4.95	5.98	7.22	8.88	—	—	
Spin-In Duct Board Starting Collar With Damper	500DBSD	4.02	4.17	4.23	4.80	5.24	5.91	6.29	7.14	8.36	9.52	11.37	13.62	—	—	
45 Side Take-off	510	—	5.18	5.41	5.75	6.00	6.86	8.88	10.55	13.56	—	—	—	—	—	
45 Side Take-off (with Damper)	510D	—	7.28	7.52	8.29	9.01	10.55	12.06	13.54	15.85	—	—	—	—	—	

STACK BOOTS

201 Universal Stack Boot

202 Center-End Stack Boot

203 90° Angle Stack Boot

	201	202	203
8 x 3¼—5" ϕ	3.77	3.96	3.96
10 x 3¼—5" ϕ	4.90	5.66	5.66
10 x 3¼—6" ϕ	3.87	4.46	4.46
10 x 3¼—7" ϕ	4.90	5.66	5.66
12 x 3¼—7" ϕ	3.95	4.63	4.63
14 x 3¼—8" ϕ	4.63	5.09	5.09
16 x 3¼—9" ϕ	5.20	6.15	6.15

REGISTER BOXES

651 651ET 661 661R 601 626 671

Most Popular Sizes	651	651ET	661	661R	601	626	671	
8 x 4—5" ϕ	2.95	3.22	3.77	4.36	3.49	4.10	4.10	
10 x 6—6" ϕ	3.20	3.43	4.10	4.67	3.66	4.23	4.23	
12 x 6—7" ϕ	3.30	3.54	4.23	4.80	3.87	4.44	4.44	
14 x 6—8" ϕ	3.43	3.68	4.34	4.90	3.96	4.55	4.55	
14 x 8—9" ϕ	4.32	4.61	4.97	5.62	4.97	5.62	5.62	
16 x 6—9" ϕ	4.32	4.61	4.97	5.62	4.97	5.62	5.62	

Figure 13-1 (*Continued*)

REGISTER BOOTS

201-R

202-R

203-R

	201-R	202-R	203-R
8 x 4—5" ϕ	x	5.73	4.63
10 x 2¼—(5, 6)" ϕ	3.87	5.75	4.63
10 x 4—(5, 6)" ϕ	3.87	5.75	4.63
10 x 6—6" ϕ	x	5.75	4.63
12 x 2¼—(6, 7)" ϕ	4.10	6.10	5.09
12 x 4—(6, 7)" ϕ	4.10	6.10	5.09
12 x 6—7" ϕ	x	6.10	5.09
14 x 2¼—(6, 7, 8)" ϕ	4.63	6.48	5.75
14 x 4—(6, 7, 8)" ϕ	4.63	6.48	5.75
14 x 6—8" ϕ	x	6.48	5.75
14 x 8—9" ϕ	x	10.46	9.49
16 x 6—9" ϕ	x	10.46	9.49

Other Sizes Available

6 x 6—(5, 6)" ϕ	3.64	3.94	4.76	5.33	4.48	5.07	5.07	
8 x 4—(4, 6)" ϕ	3.92	4.21	5.03	5.60	4.74	5.33	5.33	
8 x 6—(5, 6)" ϕ	3.92	4.21	5.03	5.60	4.74	5.33	5.33	
8 x 8—(6, 7, 8)" ϕ	3.92	4.21	5.03	5.60	4.74	5.33	5.33	
10 x 6—(5, 7)" ϕ	3.92	4.21	5.03	5.60	4.74	5.33	5.33	
10 x 8—8" ϕ	4.29	4.63	5.52	6.08	5.24	5.81	5.81	
10 x 10—(6, 7, 8, 9, 10)" ϕ	4.29	4.63	5.52	6.08	5.24	5.81	5.81	
12 x 6—(6, 8") ϕ	4.29	4.63	5.52	6.08	5.24	5.81	5.81	
12 x 8—8" ϕ	5.54	5.87	6.82	7.41	6.57	7.14	7.14	
12 x 12—(8, 9, 10, 12)" ϕ	5.73	6.06	6.82	7.41	6.57	7.14	7.14	
14 x 6—(6,7)" ϕ	4.55	5.16	5.85	6.46	5.85	6.46	6.46	
14 x 8—(8, 10)" ϕ	7.12	7.45	8.27	8.86	8.27	8.86	8.86	
14 x 14—(8, 9, 10, 12)" ϕ	8.61	9.01	9.56	10.15	10.15	10.72	10.72	
16 x 6—(8, 10)" ϕ	7.12	7.45	8.15	8.72	8.72	9.28	9.28	
16 x 8—(8, 9, 10)" ϕ	7.12	7.45	8.15	8.72	8.72	9.28	9.28	
18 x 6—10" ϕ	7.94	8.23	9.45	10.02	10.02	10.59	10.59	
18 x 8—10" ϕ	8.84	9.14	10.80	11.37	11.37	11.96	11.96	
20 x 6—10" ϕ	8.84	9.14	10.80	11.36	11.37	11.96	11.96	
20 x 8—12" ϕ	9.18	9.87	11.16	11.75	11.75	12.32	12.32	
24 x 6—10" ϕ	11.16	11.83	13.81	x	14.38	x	x	
24 x 8—12" ϕ	14.10	12.48	14.44	x	15.01	x	x	
SNAP-ON RAILS .57/pr.								

STACKHEADS

#704 Stackhead

#751 45° Perimeter Boot

#770 Stackhead

#771 Straight Perimeter Boot

	704	751	770	771
8 x 4	4.50	6.76	6.76	6.76
10 x 6	4.67	7.37	7.37	7.37
12 x 6	5.09	8.08	8.08	8.08
14 x 6	6.00	9.66	9.66	9.66
14 x 8	x	10.23	10.23	10.23
16 x 6	7.49	10.23	10.23	10.23
18 x 6	9.37	11.28	11.28	11.28
20 x 6	x	14.17	14.17	14.17
24 x 6	x	16.65	16.65	16.65
30 x 6	x	19.62	19.62	19.62

WYES & TEES

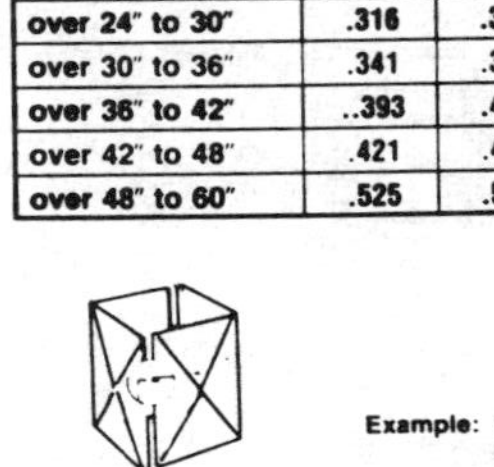

#800 WYE

#900 TEE

PRICE FROM LARGE END

	WYE #800	WYE #800S	TEE #900
3"	—	—	4.77
4"	—	5.87	4.77
5"	4.39	4.82	5.27
6"	4.39	4.82	5.27
7"	4.21	4.98	5.57
8"	4.39	5.07	6.14
9"	4.66	5.32	7.00
10"	5.07	6.27	7.71
12"	6.46	7.37	9.23
14"	8.89	9.75	10.23
16"	11.60	11.60	—
18"	15.28	15.28	—
20"	19.12	19.12	—
22"	24.60	24.60	—
24"	27.31	27.31	—

CEILING DROPS

680 690 680D 690D

Round Ceiling Drop No Damper

Round Ceiling Drop with Pull Chain Damper

	680	690	680D	690D
6	4.55	4.55	5.41	5.41
8	4.55	4.55	5.59	5.59
10	5.03	5.03	6.32	6.32
12	6.00	6.00	7.53	7.53
14	6.78	6.78	8.87	8.87

Price based on Large End

PLENUMS

PRICE PER INCH OF GIRTH

PLENUM HEIGHT	28GA	26GA	24GA	Add for Duct Liner ½"	Add for Duct Liner 1"
to 18"	.209	.236	.289	.209	.236
over 18" to 24"	.264	.289	.341	.264	.289
over 24" to 30"	.316	.341	.421	.316	.341
over 30" to 36"	.341	.368	.446	.341	.368
over 36" to 42"	.393	.421	.498	.393	.421
over 42" to 48"	.421	.473	.578	.421	.473
over 48" to 60"	.525	.578	.682	.525	.578

Quantity Discount
6 LESS 5%
24 LESS 10%

Example: Plenum 16 x 20 x 36 26GA
Girth = 16 + 20 + 16 + 20 = 72
Price = 72 x .368 = 26.50
For ½" D. L. Add 72 x .341 = 24.55

Figure 13-1 (*Continued*)

INSULATED BOXES

IB

	COLLAR SIZE					
	4" φ	5" φ	6" φ	7" φ	8" φ	9" φ
6 x 6	5.73	5.73	x	x	x	x
8 x 4	5.73	5.73	x	x	x	x
8 x 6	x	5.73	5.73	x	x	x
8 x 8	x	x	6.18	6.18	x	x
10 x 6	x	x	6.18	x	x	x
10 x 10	x	x	7.05	7.05	7.05	x
12 x 6	x	x	6.25	6.25	x	x
12 x 8	x	x	x	x	7.05	x
12 x 12	x	x	x	x	8.41	8.41
14 x 6	x	x	6.82	6.82	6.82	x
14 x 8	x	x	x	x	8.19	8.19
14 x 14	x	x	x	x	x	9.09
16 x 6	x	x	x	x	8.19	8.19
16 x 8	x	x	x	x	10.50	10.50

Turbine Flashing

XL Ventilator Cap

BX Adjustable Pitch Turbine Flashing

CF Conduit Flashing

	TF	XL	BX	CF	
3"	x	x	x	4.89	
4"	x	x	x	5.43	
5"	x	x	x	7.14	
6"	11.30	23.51	11.30	9.87	
7"	x	24.31	x	x	
8"	12.48	24.94	12.48	x	
10"	13.80	29.38	13.80	x	
12"	11.82	41.11	11.82	x	
14"	20.62	70.53	20.62	x	
16"	26.53	88.13	x	x	
18"	32.31	110.50	x	x	
20"	36.76	x	x	x	
24"	52.52	x	x	x	
30"	78.78	x	x	x	

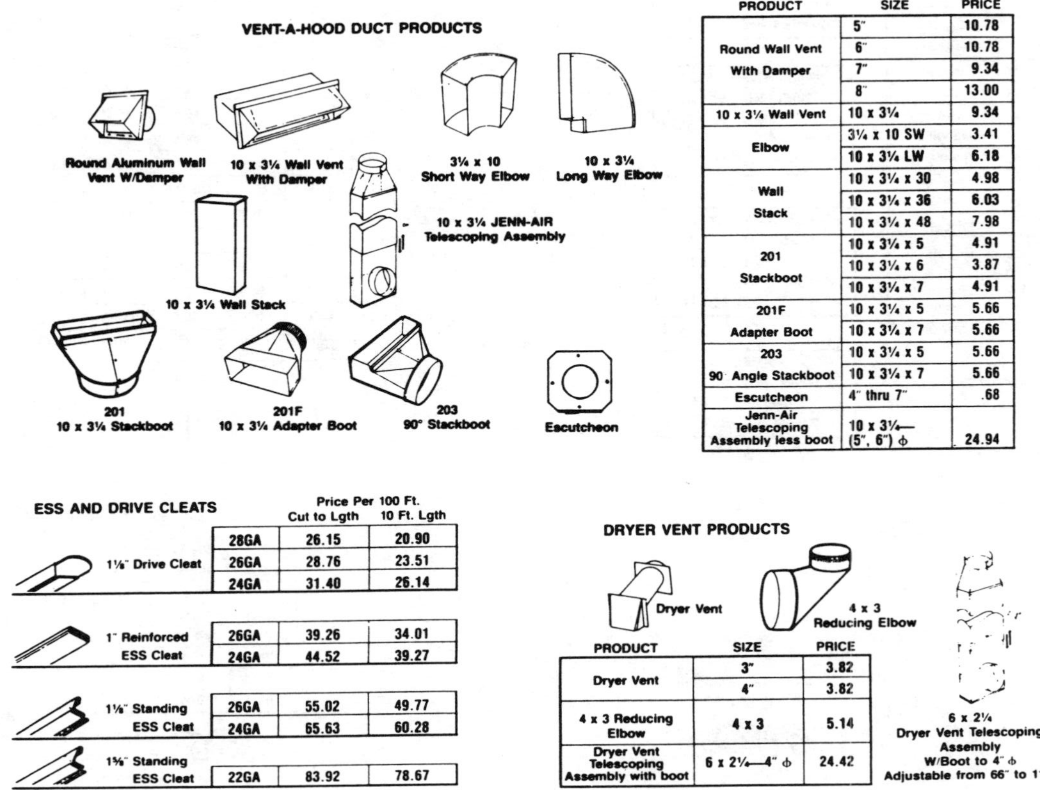

PRODUCT	SIZE	PRICE
Round Wall Vent With Damper	5"	10.78
	6"	10.78
	7"	9.34
	8"	13.00
10 x 3¼ Wall Vent	10 x 3¼	9.34
Elbow	3¼ x 10 SW	3.41
	10 x 3¼ LW	6.18
Wall Stack	10 x 3¼ x 30	4.98
	10 x 3¼ x 36	6.03
	10 x 3¼ x 48	7.98
201 Stackboot	10 x 3¼ x 5	4.91
	10 x 3¼ x 6	3.87
	10 x 3¼ x 7	4.91
201F Adapter Boot	10 x 3¼ x 5	5.66
	10 x 3¼ x 7	5.66
203 90° Angle Stackboot	10 x 3¼ x 5	5.66
	10 x 3¼ x 7	5.66
Escutcheon	4" thru 7"	.68
Jenn-Air Telescoping Assembly less boot	10 x 3¼— (5", 6") ⌀	24.94

ESS AND DRIVE CLEATS

		Price Per 100 Ft. Cut to Lgth	Price Per 100 Ft. 10 Ft. Lgth
1⅛" Drive Cleat	28GA	26.15	20.90
	26GA	28.76	23.51
	24GA	31.40	26.14
1" Reinforced ESS Cleat	26GA	39.26	34.01
	24GA	44.52	39.27
1⅛" Standing ESS Cleat	26GA	55.02	49.77
	24GA	65.63	60.28
1⅝" Standing ESS Cleat	22GA	83.92	78.67

DRYER VENT PRODUCTS

PRODUCT	SIZE	PRICE
Dryer Vent	3"	3.82
	4"	3.82
4 x 3 Reducing Elbow	4 x 3	5.14
Dryer Vent Telescoping Assembly with boot	6 x 2¼—4" ⌀	24.42

Figure 13-1 (*Continued*)

PITCH PANS AND DRAIN PANS

Pitch Pan

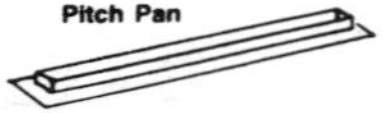

1 Ft.	7.35
2 Ft.	8.03
3 Ft.	9.21
4 Ft.	13.16
5 Ft.	16.48
6 Ft.	19.78
7 Ft.	23.10
8 Ft.	26.44
10 Ft.	33.10

Drain Pan

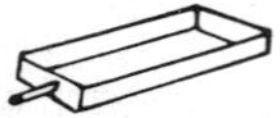

12 x 30	26 ga.	16.03
12 x 36	26 ga.	17.08
12 x 48	26 ga.	18.39
14 x 30	26 ga.	18.39
14 x 36	26 ga.	19.69
24 x 24	26 ga.	19.69
30 x 30	26 ga.	26.26
24 x 24	18 ga.	31.51

ROOF JACKS AND FLASHINGS

Roof Jack/Coolie Cap RJCC | Roof Jack Banded Cap RJBC | Stack W/CC | Stack W/BC

Model No.	Diameter	RJCC	RJBC	BC	Stack W/CC	Stack W/BC
# 3	4½"	7.50	11.55	5.55	5.62	9.23
# 4	5½"	8.55	12.48	6.00	6.39	9.98
# 5	6½"	9.59	14.73	6.53	7.21	11.82
# 6	7½"	11.16	16.51	7.28	8.39	13.21
# 7	8½"	14.19	19.39	7.94	9.96	15.48
# 8	9½"	16.42	24.56	11.00	12.32	19.69
#10	10"	19.69	27.19	13.23	—	—
#12	12"	23.50	34.81	19.69	—	—
#14	14"	26.78	47.68	24.94	—	—
#16	16"	35.45	64.60	32.31	—	—
#18	18"	44.00	82.19	41.11	—	—
#20	20"	52.93	99.79	51.36	—	—
#24	24"	81.42	146.79	80.76	—	—
#30	30"	124.73	196.97	118.19	—	—

Less 10% for 12 qty. — one size one pitch
Prices above are for standard pitches (0/12, 3/12, 4/12, 5/12, 6/12 and 7/12).
Request quotation for other pitch.

MOBILE HOME DUCT KIT PRODUCTS

Floor Return Air Grille 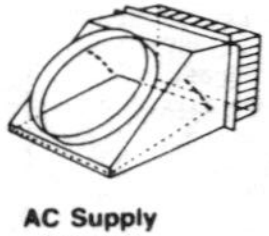 | AC Supply Damper 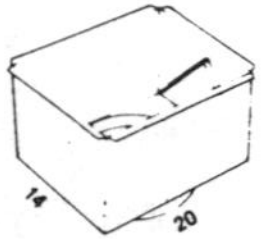 | Double Filter Return Air Box 14 x 20  | 12" ϕ Starter  | Butterfly Damper 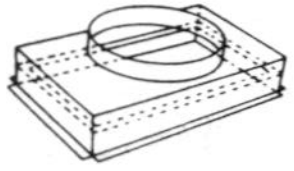 | Single Filter Return Air Box 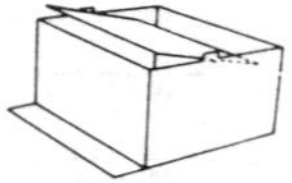 | Backdraft Damper

#8451 DUCT KIT

1	14 x 20	DBL Filter Box
2	10 x 20	Filters
1	14 x 20	Floor Grille
1	12" ϕ	Starter

12" AC KIT

1	12 x 20	Single Filter Box w/12" ϕ Collar
1	12 x 20	Filter
1	12 x 20	Floor Grille
1	12" AC	Supply Damper

14" AC KIT

1	12 x 20	Single Filter Box w/14" ϕ Collar
1	12 x 20	Filter
1	12 x 20	Floor Grille
1	14" AC	Supply Damper

MSD KIT

1	14 x 20	Single Filter Box w/14" ϕ Collar
1	14 x 20	Filter
1	12" ϕ	Starter
1	14 x 20	Floor Grille

PRODUCT	SIZE	PRICE
Double Filter Box	14 x 20 w/14" ϕ collar	18.42
Single Filter Box	12 x 20—12" ϕ	15.12
	12 x 20—14" ϕ	17.05
	14 x 20—14" ϕ	17.05
Floor Grille	12 x 20	17.69
	14 x 20	20.55
12" ϕ Starter		2.55
Butterfly Damper	10" ϕ	9.44
	12" ϕ	9.62
AC Damper	12" ϕ	21.83
	14" ϕ	22.51
Backdraft Damper		23.42
#8451 Duct Kit		45.47
12" AC Kit		59.12
14" AC Kit		61.39
MSD Kit		42.29

Index

A

Absolute Humidity 7
Air Conditioning, definition 1
  Purpose of 37
  Terminology 7-12
Air, Dry 7
Air Movement, 6
Air Supply 276-79
Annual Cooling Cost Estimation 183-88
Apparatus Dew Point and Air Quantity 31-33

C

Calculation of U Factor 43-59
Commercial Air Distribution Systems 226
  Variable Air Volume 228
    Fan-Powered Systems 229
    Induction System 230
    Parallel Flow 229
    Reheat 228
    Series Flow 229
    Variable Diffuser 231
  Zoning 227
Commercial Equipment Location 213-32
  Accessibility 216
  Gas Pipe Connections 213
  Indoor Unit Location 218
  Length of Duct Runs 218
  Minimum Clearances 216
  Noise 216
  Outdoor Unit (Condensing Unit) 223
  Return Air Location 214
  Self-Contained Indoor Unit (Horizontal and Upflow) 220-23
  Self Contained Outdoor Unit 224-26
  Supply Air Location 216
  Thermostat Location 218
  Venting Requirements 214
Commercial, Equipment Sizing 189-212
  Cooling Equipment Sizing 195
  Data Required for 195
  Data Required for Heating Unit Selection 190
  Heating, Selection 192-95
  Inside and Outside Design Temperatures 191
  Split-System Equipment Selection Procedure 200-02
  The cfm Air Delivery Required 197
  The Condensing Medium (air or water) and the Temperature of the Medium Entering the Condenser 197
  The Conditioned Space Design DB and WB Tempera tures 196
  The Design Outdoor Air Temperatures 196
  The Duct External Static Pressure 200
  The Required cfm 191
  The Required Ventilation Air in cfm 197
  The Total Heat Loss in Btuh 191
  Types of Heating Equipment 191

Vertical Packaged (PTAC)
Unit Selection Procedure 202-08
Commercial Heat Load, Calculation 165-88
Form 166-69
Commercial Operating Cost Estimate 183-88
Comfort Zone 10
Conduction Load 43
Cubic Foot of Dry Air 10

D

Dehumidification 8
Design Temperatures 59-62
Dew Point, Apparatus and Air Quantity 31-33
Lines 14
Temperature 9
Dry Bulb Temperature 8
Lines 12
Duct Construction 279-95
Duct Heat Load 78
Duct Liner Board 322
Duct Pipe, Fittings and Insulation 313-31
Insulation 318-31
Installing Duct Wrap Insulation 323-30
Types and Uses 230
Materials 316
Shapes 314
Duct Systems and Design 261-311
Air Distribution 295-300
Air Supply 276
Design Methods 284
Equal-Friction Method 285-93
Static-Regain Method 294
Velocity-Reduction Method 293
Designing A Duct System 301
Duct Construction 279-95
Duct Design 274
Procedure for 282
Selecting the Proper Supply Air Grills 301-07
Selecting the Return Air Grill 307
Sizing A Duct System 300
Selecting Air Distribution Systems 263
Types of Air Distribution Systems 264
Ceiling Diffuser Systems 264
High-Inside-Wall Furred Ceiling Systems 265
High-Inside-Wall Systems 265
Overhead High-Inside-Wall Systems 265
Parallel-Flow Systems 267
Perimeter Systems 264
Perimeter System Return Air Intakes 267
Types of Structures 272
Apartment 274
Basement 273
Crawl Space 273
Slab Floors 273
Split-Level and Multilevel 273
Velocities 283
Ductwork Fittings 316-18

E

Effective Temperature 10
Electric Lights and Appliances 63

F

Fan Heat Load 78
Flexible Duct Liner 321

G

Gas Furnace (Downflow) 153
Gas Furnace (Horizontal) 153
Gas Furnace (Upflow) 151-52
Gas and Steam Appliances 64
Gaseous Contaminant Characteristics 333
  Harmful Effects of 335
Glass Fiber Duct Board 320
Glass Fiber Duct Wrap 320

H

Heat Rejection by the Human Body 2, 4
Heat
  Latent 24-27
  Sensible 24
    Ratio 27
  Sources 42
Heat Load, Infiltration 75-77
  Duct 78
  Fan, Load 78
  Ventilation 77
Heat Transmission, Overall Coefficient of 11
Horizontal Packaged (PTAC) Unit Selection Procedure 209
Human Comfort 2
Humidification 8
Humidity, Absolute 7
  Absolute lines 14
  Effects of Temperature on 6
  Relative 4, 7
  Relative, Lines 14

I

Indoor Air Quality 333-65
  Building Materials and Furnishings 345
  Cleaning Agents and Other Consumer Products 345
  Contaminant Sources and Generation Rates 345
  Control Techniques 352
    Absorbers 356
    Chemisorbers 359
    General Ventilation 355
    Hooding and Local Exhaust 254
    Occupants 360
    Physical Adsorbers 356-59
    Plants 360
  Counteracting Odor and Odor Masking 363
  Damage to Materials 344
  Energy Consumption 364
  Equipment 347
  Gaseous Contaminant Characteristics 333
    Harmful Effects of 335
  Irritation 340
  Occupants 347
  Odors 340
  Outdoor Air 351
  Tobacco Smoke 345
  Toxicity 338
  Types of Control Systems 363
Infiltration 65-77
  Calculations 72-77
  Heat Load 75-77

L

Latent Heat 24
Load, Conduction 43

M

Mixing Two Quantities of Air at Different Conditions 29-31

O

Occupancy 62
Odors 340
Oil return 234

Overall Coefficient of Heat Transmission 11

P

Plenum Chamber 12
Pound of Dry Air 10
Pressure Static 11
  Total 11
  Velocity 11
Psychrometric Chart 12, 17
  Identification of lines and Scales on 12
  Use of 16-33
Psychrometrics 12

R

Refrigerant Lines 233-59
  Double Risers 250-54
  Equivalent Length of Pipe 235
  Oil Return 234
  Piping for Horizontal and Vertical 254
  Pressure Drop 233
    Tables 237
  Sizing of Hot Discharge 237-43
  Sizing of Liquid 243-48
  Sizing of Suction 148-50
  Suction Line Piping At the Evaporator 256
Relative Humidity 4
  Determining Percentage of 4-6
  Lines 14
Residential, Condensing Gas Furnaces 154
  Cooling Equipment 156
    Split-System Units 156
  Cooling Equipment Sizing 125-30
  Data Required for Heating Unit Selection 112-20
  Electric Furnace Selection 124
  Electric Furnaces 155
  Equipment Sizing 111-42
  Furnace Selection Procedure 120-24
  Gas Furnace (Downflow) 153
  Gas Furnace (Horizontal) 153
  Gas Furnace (Upflow) 151-52
  Heating Equipment Sizing 111
  Heat Pump Selection 125
  Horizontal Packaged (Self-Contained) Unit Selection Procedure 139-40
  Required cfm 119
  Split-System Equipment Selection Procedure 130-33
  Total Heat Loss in Btuh 116
  Type of Air Flow 116-19
  Type of Energy Used 116
  Type of Heating Equipment 116
  Vertical Packaged (Self-Contained) equipment Selection Procedure 138-39
  Vertical Packaged (Self-Contained) Unit Selection Procedure 133-38
Residential Equipment Location 143-62
  Accessibility 151, 158
  Air Supply 158
  Comfort Considerations 158
  Electric Lines, Length of 159
  Furnaces 143-56
  Gas Pipe Connections 145
  Length of Duct Runs 151
  Minimum Clearances 150
  Noise 150
  Refrigerant Lines, Length of 159
  Return Air Location 146
  Support 157
  Supply Air Location 148
  Thermostat Location 151

Venting Requirements 145
Residential Heat load, Calculation Procedure 83-109
Estimate Form 84-88
Entering Information on 88-109

S
Safety Factor 79
Saturated Air 8
Saturated Gas (Vapor) 10
Saturation Temperature 10
Self-Contained (Packaged) Units 159-62
Horizontal Self-Contained 160
Upflow Self-Contained 160
Sensible Heat 24
Ratio 27-29
Specific Volume Lines 14
Stack Effect 11
Static Pressure 11
Survey and Check List 37-41
System Cost Estimating 367-86
Estimation Form 367-73
Example 368-86

T
Temperature, Effective 10
Total Heat (Enthalpy) 10
Total Pressure 11
Toxicity 338

U
U Factor 43-59
Calculation of 43-59

V
Velocity Pressure 11
Ventilation 10
Heat Load 77
Vertical Packaged (PTAC) Equipment Selection Procedure 208-09
Vertical Packaged (PTAC) Unit Selection Procedure 202-08
Volume, Specific lines 14

W
Wet Bulb Depression 9
Wet Bulb Temperature 8
Lines 12

Z
Zone, Comfort 10